www.wadsworth.com

wadsworth.com is the World Wide Web site for Wadsworth Publishing Company and is your direct source to dozens of on-line resources.

At *wadsworth.com* you can find out about supplements, demonstration software, and student resources. You can also send e-mail to many of our authors and preview new publicatio and exciting new technologies.

wadsworth.com
Changing the way the world learns®

About the Authors

David Knox, Ph.D., is Professor of Sociology at East Carolina University, where he teaches courses in Courtship and Marriage, Marriage and the Family, Marriage Problems, and Human Sexuality. He is a marriage and family therapist and the author or co-author of 10 books and 30 professional articles. He and Caroline Schacht are married.

Caroline Schacht, M.A. in Sociology and M.A. in Family Relations, is Instructor of Sociology at East Carolina University and teaches Courtship and Marriage and Introduction to Sociology. Her clinical work includes marriage and family relationships. She is also a divorce mediator and the co-author of several books.

Marriage and the Family

A BRIEF INTRODUCTION

David Knox
EAST CAROLINA UNIVERSITY

Caroline Schacht
EAST CAROLINA UNIVERSITY

Wadsworth Publishing Company

I(T)P® An International Thomson Publishing Company

Belmont, CA • Albany, NY • Boston • Cincinnati • Johannesburg • London • Madrid • Melbourne
Mexico City • New York • Pacific Grove, CA • Scottsdale, AZ • Singapore • Tokyo • Toronto

Publisher: Eve Howard
Acquisitions Editor: Denise Simon
Assistant Editor: Barbara Yien
Editorial Assistant: Angela Nava
Marketing Manager: Christine Henry
Project Editor: Tanya Nigh
Print Buyer: Karen Hunt
Permissions Editor: Yanna Walters
Production: Greg Hubit Bookworks
Designer: Lisa Devenish
Copy Editor: Jamie Fuller
Cover Design: Lisa Devenish
Compositor: Carlisle Communications
Printer: Courier Companies, Inc.

Printed in the United States of America
1 2 3 4 5 6 7 8 9 10

For more information, contact Wadsworth Publishing Company, 10 Davis Drive, Belmont, CA
94002, or electronically at http://www.wadsworth.com

International Thomson Publishing Europe
Berkshire House
168-173 High Holborn
London, WC1V 7AA, United Kingdom

International Thomson Editores
Seneca, 53
Colonia Polanco
11560 México D.F. México

Nelson ITP, Australia
102 Dodds Street
South Melbourne
Victoria 3205 Australia

International Thomson Publishing Asia
60 Albert Street
#15-01 Albert Complex
Singapore 189969

Nelson Canada
1120 Birchmount Road
Scarborough, Ontario
Canada M1K 5G4

International Thomson Publishing Japan
Hirakawa-cho Kyowa Building, 3F
2-2-1 Hirakawa-cho, Chiyoda-ku
Tokyo 102 Japan

International Thomson Publishing Southern Africa
Building 18, Constantia Square
138 Sixteenth Road, P.O. Box 2459
Halfway House, 1685 South Africa

Library of Congress Cataloging-in-Publication Data
Knox, David
    Marriage and the family : a brief introduction / David Knox,
Caroline Schacht.
        p. cm.
    Includes bibliographical references and index.
    ISBN 0-534-55287-0
    1. Marriage. 2. Family. I. Schacht, Caroline. II. Title.
HQ734.K668 1999
306.8—dc21   98-8095

 This book is printed on acid-free recycled paper.

To Lisa, Dave, and Isabelle

Marriage and the family involve intimate relationships that affect both individuals and society. The goal of *Marriage and the Family: A Brief Introduction* is to provide basic information on these relationships and fundamental institutions in our society. The text includes the following features:

Brevity: To counteract the trend toward bigger and more expensive college textbooks, and to fulfill students' need for material that is thorough but concise and affordable, we provide twelve chapters of core information on marriage and the family rather than the standard eighteen to twenty-two chapters. This format gives students the basic content of a marriage and family course and allows professors to supplement as desired.

National Data: The text provides data from national samples to give state-of-the-art information about marriage and the family today. For example, responses from 252,080 undergraduate first-year students at 464 colleges and universities (a national random sample) throughout the United States provide a fairly accurate picture of marriage and family values among today's college youth. Data from the Census Bureau and other large databases are offset in the text.

Social Policy: The book identifies social policy issues that affect individuals, spouses, marriages, and families. Examples include "Legalization of Same-Sex Marriage?" (Chapter 1), "One's Sexual Health Status: To Tell or Not to Tell?" (Chapter 6), and "Children After Fifty: How Old Is Too Old?" (Chapter 8).

Cross-Cultural Perspective: We include information about marriage and the family in other societies and cultures. For example, whereas societal attitudes about cohabitation in the United States is ambivalent, in Iceland, cohabitation is an expected and approved stage in a couple's relationship. These cross-cultural references are identified by the logo Other Cultures in the text.

Insight: This feature provides special insights relevant to the topic being discussed. For example, an Insight feature in Chapter 3, on love, points out that a "Ludic" (game-playing) love style is associated with the lowest relationship satisfaction and the highest likelihood of breaking up. However, "Eros" and "Agape" love styles, which reflect powerful attraction and selfless love, respectively, are associated with the highest relationship satisfaction and durability.

Personal Application: At the end of each chapter we provide a scale that allows for self-discovery, as an individual or a partner in a relationship, of one's values or attitudes regarding topics discussed in the chapter. Each scale has been completed by other adults for score comparison. Examples include Family Functioning Scale (Chapter 1), Attitudes toward Feminism Scale (Chapter 2), and Love Attitudes Scale (Chapter 3).

Other self-assessments allow the student to measure

- the level of supportive communication in their relationship (Chapter 6)
- the strength of their relationship (Chapter 4)
- the degree to which their companionship needs are being met in their relationship (Chapter 5)
- the level of risk of exposure to HIV or other STDs being taken in one's sexual relationships (Chapter 7)
- one's motivation for having children (Chapter 8)
- one's attitudes toward the effect of maternal employment on children (Chapter 9)
- the level of emotional and physical abuse in one's relationship (Chapter 10)
- the degree to which an individual is in control of his or her own life rather than being controlled by other factors such as fate (Chapter 11)
- what children believe would be the effects if their parents were to divorce (Chapter 12)

Epilogue: We propose that marriage and family relationships constitute our most important "natural resource." Just as our environment is being threatened by pollution, so are marriages and families being threatened by the toxicity of domestic violence, substance abuse, intolerance of diversity, and economic hardship. Governmental support for our greatest resource is needed.

Format: The format of each chapter facilitates learning by providing questions in the chapter outline, defining key terms in the glossary, and summarizing the main points at chapter's end.

Acknowledgments

Denise Simon, Executive Senior Editor, provided the vision for this book. Her support for the project has brought it to fruition. The authors would also like to thank Tanya Nigh and Barbara Yien for facilitating chapter reviews and moving the text through production, Esther Devall for writing the section on family systems theory and for developing a chart comparing family theories in Chapter 1, Karen Crowell for researching various topics for the text, Stacey Hatfield for providing Proquest resources, Wandy Nieves for her diligence in checking the references for the manuscript, and Lisa Shriver for her assistance in preparation of the Instructor's Manual and Test Bank.

We would also like to thank the following reviewers who provided superb feedback on changes to make for this edition. Their suggestions are evident throughout the text:

Scot Allgood, Utah State University
Deborah B. Ascione, Utah State University
Janette C. Borst, Emporia State University
Janet Cosbey, Eastern Illinois University
John H. Curtis, Valdosta State University
Esther Devall, New Mexico State University

David W. Miller, American River College
Owen Morgan, Arizona State University
Kenrick S. Thompson, Northern Michigan University
Harold A. Widdison, Northern Arizona University

We are always interested in ways to improve the text and invite your feedback and suggestions for new ideas and material to include in subsequent editions. We are also interested in dialogue with professors and students and invite you to write or E-mail us.

David Knox
Caroline Schacht

Department of Sociology
East Carolina University
Greenville, NC 27858

E-mail addresses:
For David Knox: DAVIDKNOX@prodigy.net
For Caroline Schacht: UHXP97B@prodigy.com

Contents in Brief

Contents

CHAPTER 4 **Dating and Mate Selection / 92**

A successful 21st century will require major changes in how we re-establish the family as the core of our national life.

Gerald Jensen

Larry Jensen

Social Policies to Strengthen the Family

In the fall of 1997, researchers at the University of California at Los Angeles Higher Education Research Institute asked 252,080 students at 464 colleges and universities throughout the United States what is important to them. Seventy-two percent reported that "raising a family" was their most important goal (competing with the importance of financial security) (Gose, 1998). Similarly, 97 percent of 620 undergraduates at a large southeastern university strongly agreed or agreed that "I want to marry at some time in my life," and 92 percent reported strong agreement or agreement with "I want to have children at some time in my life" (Knox & Zusman, 1998). These data reflect the importance of marriage and the family among young adults.

Family is central not only to our lives as adults but to those of our children. The positive effect of family connectedness on children has been confirmed. On the basis of a sample from over 90,000 adolescents, Resnick et al. (1997) found that "feelings of warmth, love, and caring from parents" were associated with less emotional distress, lower substance abuse, and later age for first intercourse (p. 830). Politicians often point to the "breakdown" of the family as one of the primary social problems in the world today—a problem that underlies other social problems such as crime, drug abuse, and poverty. Textbooks on marriage and the family are sometimes criticized for not emphasizing the value of such relationships (Glenn, 1997), but in this book we want to make clear that nurturing families are beneficial to spouses, children, and society.

In this chapter we review core concepts, emerging views, and lifestyle alternatives to marriage and the family in the United States. We also present various theoretical frameworks to help understand different aspects of marriage and family. Finally, since the text is based on research, we discuss cautions in research the reader should keep in mind that may affect interpretation of and confidence in research findings.

The most ancient of all societies, and the only natural one, is that of the family.
Jean-Jacques Rousseau
French philosopher

RECENT RESEARCH

Undergraduates who have taken a course in marriage and the family believe significantly fewer myths about marriage and family relationships than students who have not taken such a course (Carter & Morris, 1997).

Marriage

Though the nature of marriage may vary by the individuals involved and the society in which it exists, marriage is recognized as one of the most valued of all relationships.

NATIONAL DATA

Over 95 percent of women and men decide to marry at some time in their adult life (*Statistical Abstract of the United States: 1997*, Table 59).

The evidence indicates that marriage has positive effects for both men and women.
Norval Glenn
Sociologist

Importance of Marriage to Individuals

The overwhelming majority of both women and men in our society decide to marry. Although individuals today are waiting until they are older to marry, marriage remains a goal for most. Indeed, among youth today "commitment to the idea of life-long marriage appears to be stronger" (Rogers & Amato, 1997, 1099). The benefits that await the married are enormous (Waite, 1995; Rogers, 1995; Ross et al., 1990). When married persons are compared with singles who are never-married, separated, or divorced, the differences are striking (see The Case for Marriage in Table 1.1).

Table 1.1

The Case for Marriage

Benefits of Marriage	Liabilities of Singlehood
Health: Spouses have fewer hospital admissions, see a physician more regularly, are "sick" less often.	Singles are hospitalized more often, have fewer medical checkups, and are "sick" more often.
Longevity: Spouses live longer than singles.	Singles die sooner than marrieds.
Happiness: Spouses report being happier than singles.	Singles report less happiness than marrieds.
Sexual Satisfaction: Spouses report being more satisfied with their sex lives, both physically and emotionally.	Singles report being less satisfied with their sex lives, both physically and emotionally.
Money: Spouses have more economic resources than singles.	Singles have fewer economic resources than marrieds.
Lower Expenses: Two can live more cheaply together than separately.	Cost is greater for two singles than one couple.
Drug Use: Spouses have lower rates of drug use and abuse.	Singles have higher rates of drug use and abuse.
Connectedness: Spouses are connected to more individuals who provide a support system—partner, in-laws, etc.	Singles have fewer individuals upon whom they can rely for help.
Children: Rates of high school dropouts, teen pregnancies, and poverty are lower among children reared in two-parent homes.	Rates of high school dropouts, teen pregnancies, and poverty are higher among children reared by single parents.
History: Spouses develop a shared history across time with significant others.	Singles may lack continuity and commitment across time with significant others.

Society and Marriage

Every society recognizes at least one form of marriage. Within these various forms "there is considerable variation in who can be married to whom, and the nature of their relationship" (Hunt, 1996, 3). In all societies, getting married involves attaining a new status. Along with this new status comes a set of role expectations—the rights, responsibilities, and obligations associated with being married. After defining marriage in the United States, we discuss marriage forms in other cultures.

NATIONAL DATA

About 2.3 million marriage licenses are issued each year (National Center for Health Statistics, 1998).

While this bride and groom are pledging their love and commitment to each other, they are also entering into a legal agreement with each other and with the state in which they reside.

Marriage in the United States

Marriage in the United States is a legal contract entered into voluntarily between two unmarried adults who meet the legal requirements (age, mental competence, opposite sex). The marriage license, which is obtained at a courthouse for a small fee (about $40.00), certifies that the individuals were married by a legally empowered representative of the state, often with two witnesses present. Marriage laws differ among the states in such aspects as minimum age requirement (usually 18) and mandatory blood testing of prospective spouses. Although many marriage ceremonies are performed in a religious context, religious marriage ceremonies are not recognized as legal marriages until the couple, witnesses, and person who conducts the ceremony sign the marriage license issued by the state.

A marriage license confers many legal rights and responsibilities. Being married means that all future assets and property will be jointly owned and that both spouses will be responsible for each other's debts. In the event of divorce, the state may dictate the economic responsibilities of each spouse to the other and to their children. Marriage also conveys the right of one partner to make crucial medical decisions for the other in the event of the latter's critical illness, immunity from having to testify against a spouse in a criminal proceeding, the right to visit one's partner in prison or in the hospital, the right to Social Security survivor benefits, the option to reduce the couple's tax liability by filing joint returns, the right of an employee to include a partner in his or her health insurance coverage, and the right to be regarded as "family" in reference to family leave policies. In most states, assets and liabilities of a deceased spouse are automatically legally transferred to the surviving spouse at the time of death. In the event of divorce, assets and liabilities are usually equitably divided regardless of the contribution of each partner.

Common-law marriage, which is recognized by thirteen states and the District of Columbia, is a marriage by mutual agreement between a cohabiting man and woman without a marriage license or ceremony. In common-law states, a woman and man who live together and consider themselves "married" by common law may inherit from each other or receive alimony and property in the case of "divorce." They may also receive health and Social Security benefits as would other spouses who have a marriage license. Common-law marriage may require a legal divorce, especially if the couple have children together.

Marriage in the United States also involves the cultural and legal expectation of sexual monogamy, and most spouses are faithful most of the time (Michael et al., 1994). Infidelity is illegal throughout the United States. Another cultural expectation associated with marriage is that it is based on love. Less than 4 percent of the U.S. respondents in one study said that they would marry someone with whom they were not in love (Levine et al., 1995). And, despite the high rate of divorce in the United States, most people enter marriage with the expectation that it will last "till death do us part."

Marriage in Other Cultures

While we think of marriage in the United States as involving one man and one woman, other societies view marriage differently. *Polygamy* is a form of mar-

Most people in the United States assume that polygyny exists to satisfy the sexual desires of the man, that the women are treated like slaves, and that jealousy among the wives is common. In most polygynous societies, however, polygyny has a political and economic rather than a sexual function. Polygyny is a means of providing many male heirs to continue the family line. In addition, by having many wives, a man can produce a greater number of children for domestic/farm labor. Wives are not treated like slaves (although women have less status than men in general), as all household work is evenly distributed among the wives and each wife is given her own house or own sleeping quarters.

Bigamy is having one husband too many. Monogamy is the same.
Elizabeth Taylor
Actress married and divorced eight times

riage in which there are more than two spouses. Polygamy occurs in societies or subcultures whose norms sanction multiple partners. One form of polygamy is *polygyny,* in which one husband has two or more wives.

Polygyny is illegally practiced in the United States by some religious fundamentalist groups in Arizona, New Mexico, and Utah that have splintered off from the Church of Jesus Christ of Latter-day Saints (commonly known as the Mormon Church). Among these small sects, polygyny serves a religious function in that large earthly families are believed to result in large heavenly families. Notice that polygynous sex is only a means to accomplish another goal—large families.

The Buddhist Tibetans foster yet another brand of polygamy, referred to as *polyandry,* in which one wife has two or more (up to five) husbands. These husbands, who may be brothers, pool their resources to support one wife. Polyandry is a much rarer form of polygamy than polygyny and is sometimes motivated by the need to keep the population rate down (Crook & Crook, 1988). Several men marrying one wife will have fewer children than one man marrying several women, since one woman can have only a limited number of children but several women can, collectively, have numerous children.

The major reason for polyandry, however, is economic. A family that cannot afford wives or marriages for each of its sons may find a wife for the eldest son only. Polyandry allows the younger brothers to also have sexual access to the one wife or marriage that the family is able to afford. When a Tibetan woman marries a man, it is understood that she becomes the wife of his brothers as well.

Sexual access by Tibetan brothers who are polyandrously linked to one wife is handled by age and maturity, with the eldest brother having the greatest access. This necessarily implies that younger brothers will have less access and are less likely to be the biological fathers of the children born to the wife (Crook & Crook, 1988).

Finally, just as some societies permit more than one spouse to a marriage, others permit same-sex marriages. Norway, Sweden, Iceland, and Denmark recognize same-sex marriages. The legalization of same-sex marriage in the United States will be discussed later in this chapter.

OTHER CULTURES The Cheyenne Indians permitted married men to take on *berdaches,* or male transvestites, as second wives. In the African Sudan, Azande warriors who could not afford wives were allowed to marry "boy-wives" to satisfy their sexual needs and perform household chores. ●

INTERNATIONAL DATA
Though African countries are changing and polygyny is becoming less frequent, a national fertility study in Nigeria revealed that 44 percent of marriages are polygynous (Gage-Brandon, 1992).

Family

A common assignment for elementary school children involves asking them to draw a picture of their family. Though this assignment seems clear and straightforward, many children are probably confused about whom to include in their drawing. Should they include their stepsibling and stepparent? What about their mother's live-in boyfriend? How about their grandfather who died last year? Or the nanny who takes care of them after school or pet dog who sleeps with them every night?

The idealized image of the family is an ahistorical amalgam of structures, values, and behaviors that never coexisted in the same time and place.

Stephanie Coontz
Historian

In the postmodern family, no single pattern is dominant.

Judith Stacey
Sociologist

Children are not the only ones who are puzzled by what the term *family* means. Politicians and policymakers, as well as family experts from diverse disciplines, disagree on how narrow or broad the criteria should be for determining what constitutes a family. In an effort to recognize social diversity, some suggest we should talk about "families" instead of "family." Others suggest that family is not an entity at all but an ideology. Hunt (1996) explains that, like other institutions, family is a social construction. Although we can specify the citizens of a country, the members of a church, the residents of a neighborhood, and the stockholders of a corporation, we cannot identify with broad consensus the members of families because families are not discrete units. This, explains Hunt, is because the boundaries of families "blur with distance of relationship, and each person's family network is a unique set of relationships determined by birth, marriage, adoption, and a variety of kin-like ties. The families of sisters, for example, overlap but have distinctive elements—such as different siblings, spouses, children, nieces and nephews, and the like" (Hunt, 1996, 1).

Despite its elusive and varied definitions, "family" is a central aspect of our culture and our lives. We talk of "family values," trace our "family tree," plan "family vacations," go to "family reunions," keep "family secrets," and take "family photographs." For the purposes of research and social policy, social scientists measure things like "family income" and "family size" and compare "intact families" with "single-parent families" and "blended families." Psychologists explore their clients' "family background" and help clients cope with "dysfunctional families." Because "family" can mean so many different things to different people and in different contexts, various terms are used to specify types of families.

Types of Families

The members and structure of families differ according to the type of family being described. Various types of families include family of origin, family of procreation, nuclear family, binuclear family, and extended family.

Family of origin   Your family of origin is the family into which you were born or the family in which you were reared. It involves you, your parents, and your siblings. When you go to your parents' home for the holidays, you return to your *family of origin.* Your family of origin may be your biological family, adoptive family, or foster family. It may even be your grandparents if you were reared primarily by them.

Becoming independent of one's own family of origin is related to one's subsequent marital happiness. Rees et al. (1995) studied fifty-eight married couples and found that spouses who had terminated the hierarchical relationship between themselves and their parents and had begun to relate to their parents as peers reported happier marriages than those who were still controlled by their parents.

Family of procreation   The *family of procreation* represents the family that you will begin when you marry and have children. More than 90 percent of U.S. citizens living in the United States marry and establish their own family of pro-

Some children are reared by their grandmothers who become their family of origin.

INSIGHT

Politicians of both parties emphasize the theme of family values. The rhetoric implies that somehow the nuclear family of yesteryear with its working father, homemaking mother, and two children in suburbia was self-sufficient, patriotic, and God-fearing. Historian Stephanie Coontz argues that this is an inaccurate, nostalgic view of the family that never was. "Even at the height of this family form in the 1950s, only 60 percent of American children spent their entire childhoods in such a family" (1995, 12).

creation. Across the life cycle, individuals move from the family of origin to the family of procreation.

Nuclear family   The *nuclear family* may refer to one's family either of origin or of procreation. In practice, this means that your nuclear family consists of you, your parents, and siblings, or you, your spouse, and your children. Generally, one-parent households are not referred to as a nuclear family. They are a binuclear family if both parents are involved in the child's life or a single-parent family if only one parent is involved in the child's life and the other parent is either deceased or has no contact with the child.

Binuclear family   As previously noted, the *binuclear family* is a family that spans two households. It is created when spouses divorce and live separately so that the parents of the children set up two separate units, with the children remaining a part of each unit. Each of these units may also change again when the parents remarry and bring additional children into the respective units (*blended family*). Hence, some children go from living in a nuclear family with both parents to a binuclear unit with parents living in separate homes to a blended family when parents remarry and bring additional children into the respective units. Although we tend to think of blended families as a relatively recent family form, they are probably no more prevalent, as a percentage of total families, than they were 300 years ago. . . .

In colonial America, widowhood was frequent; life expectancy was limited by the risks of childbirth, war, disease, and inclement weather. Because of

This family represents the traditional nuclear family of father, mother, and children.

the harshness of their existence, people remarried because they needed helpmates to survive. Their remarriages created stepfamilies (Winton, 1995, 167).

Extended family   The *extended family* includes not only your nuclear family but other relatives as well. If you are a spouse, your extended family includes the parents, grandparents, aunts, uncles, and cousins of both you and your partner. A typical example of an extended family living together is a husband and wife, their children, and one of the parents of the spouses.

OTHER CULTURES   African-Americans, Hispanics, Asian-Americans, Native Americans, and Native Alaskan Innuits are more likely than Anglo-Americans to live with their extended families. Such extended families are sometimes the result of historical and religious influences. For example, the extended family model of Native Americans is related to the historical place of elders in the family. Not only are elderly men accorded respect because they provide spiritual guidance and maintenance of cultural heritage, but elderly women (now grandmothers) provide needed child care and help with household chores. In exchange, they are taken care of when they are too frail to care for themselves (Yee, 1992).

Among Asians, the status of the elderly in the extended family derives from religion. Confucian philosophy prescribes that all relationships are of the subordinate-superordinate type—husband-wife, parent-child, and teacher-pupil. "This implies that elders have more authority than younger members of the family" (Yee, 1992, 6). Abandoning the elderly rather than including them in larger family units would be unthinkable, although this may be changing as a result of the Westernization of Asian countries such as China, Japan, and Korea.

In addition to showing concern for the elderly, Asian-Americans are socialized to subordinate themselves to the group. Familism and group identity are valued over individualism and independence. Divorce is not prevalent, because Asians are discouraged from bringing negative social attention to the family. In addition, the relationship that is emphasized in Asian families (particularly among the Japanese) is the mother-child relationship, not the husband-wife relationship (Tamura & Lau, 1992).

African-American families are also characterized by their extended nature, multiple parenting and informal adoption practices, and child-centeredness. The family is regarded as the greatest source of life satisfaction for African-Americans.

African-Americans and Hispanics are more likely to have extended families for both economic and cultural reasons. As incomes are lower among these groups (because of racism and discrimination), extended-family living is a way of pooling resources. In addition, since many African-Americans and Hispanics have grown up in extended families, this type of family is perceived as culturally normative. ●

U.S. Census Definition of Family

The U.S. Census Bureau collects and publishes data on U.S. families. Most of these data focus on family households. A *household* consists of persons who share a housing unit, such as a house or apartment. For the U.S. Census, a *family household* consists of two or more persons who are related by blood, marriage, or adoption and who reside together in the same living quarters. Figure 1.1 shows the estimated percentage of various types of Census-defined family households for the year 2000.

Not all families constitute a household. A college student living in a dorm and that student's parent(s) and siblings constitute a family. Yet, these family members do not reside together and thus do not constitute a family household. Conversely, not all households are families; two unrelated dormmates reside in the same household, but are not a "family" (although they are members of different families). Indeed, the percentage of U.S. households that consist of nonfamily members is expected to increase from 19 percent in 1970 to 44 percent by 2000 (*Statistical Abstract of the United States: 1997*, Table 67). This increase in nonfamily households, which include cohabiting heterosexual and homosexual couples, represents one of many changes in our society that has affected legal and cultural definitions of the family. In the next section, we show how changes in society affect changes in marriage and family and vice versa.

American Families in Transition

Families are shaped by the social and cultural context in which they exist. In this section we discuss these influences on marriage and the family since the 1950s. We conclude by presenting new emerging legal and cultural conceptions of "family."

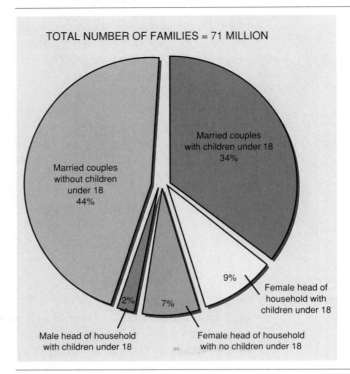

Source: *Statistical Abstract of the United States: 1997.* 117th ed. Washington, D.C.: U.S. Bureau of the Census, 1997. Table 67.

Social and Cultural Influences on Marriages and Families: Recent Changes

Things do not change, we change.

Henry David Thoreau
Social critic

From the early 1700s to the 1960s, the traditional nuclear family was regarded as the American culture's ideal family. This family form consisted of a husband and wife (who had not been previously married) and their biological and/or adopted children. Marital roles in the traditional nuclear family were clear: the husband earned the income and was dominant in the household; the wife took care of the children and the home and was economically dependent on her husband. Traditional nuclear families flourished in the 1950s, in part because this era of affluence and economic growth enabled many families to live comfortably on the income of one wage earner (Popenoe, 1996).

Changes began in the 1950s that continue as we move toward the twenty-first century. Table 1.2 reviews changes in marriage and family norms and values from the 1950s to 2000.

Other changes during this period have included greater social acceptability to remain single, to cohabit without being married, to divorce, to have children outside marriage, and to have an abortion. However, increased *tolerance* of these behaviors does not mean that there has been or is widespread *approval* of them. Thornton (1996) notes:

Table 1.2

Changes in Marriage and the Family from 1950s to 2000

1950s	2000

Family Relationship Values

Strong values for marriage and the family. Individuals who wanted to remain single or childless were considered deviant, even pathological. Husband and wife should not be separated by jobs or careers.

Individuals who remain single or childfree experience social understanding and sometimes encouragement. The single and childfree are no longer considered deviant or pathological but are seen as self-actuating individuals with strong job or career commitments. Husband and wife can be separated for job or career reasons and live in a commuter marriage. Married women in large numbers have left the role of full-time mother and housewife to join the labor market.

Sexual Values

Marriage was regarded as the only appropriate context for intercourse in middle-class America. Living together was unacceptable, and a child born out of wedlock was stigmatized. Virginity was sometimes exchanged for marital commitment.

For many, concerns about safer sex have taken precedence over the marital context for sex. Nonmarital sexual involvement is prevalent. Living together is regarded as not only acceptable but sometimes preferable to marriage. For some, unmarried single parenthood is regarded as a lifestyle option. It is certainly less stigmatized.

Mate Selection Criteria

Strong social pressure existed to date and marry within one's own racial, ethnic, religious, and social class group. Emotional and legal attachments were heavily influenced by obligation to parents and kin.

Dating and mating have become more heterogamous, with more freedom to select a partner outside one's own racial, ethnic, religious, and social class group. Attachments are more often by choice.

Media Portrayals of Intimate Relationships

Many aspects of intimate relationships were not discussed or portrayed in the media.

Talk shows, media events such as the O. J. Simpson trial, and magazine surveys are open about sexuality, violence, and relationships "behind closed doors."

Divorce

Society strongly disapproved of divorce. Familistic values encouraged spouses to stay married for the children. Strong legal constraints kept couples together. Marriage was forever.

Less stigma is associated with divorce. Individualistic values lead spouses to seek personal happiness. No-fault divorce allows for easy divorce. Marriage is tenuous. Increasing numbers of children are being reared in single-parent households apart from other relatives.

(continued)

Table 1.2 (continued)

Changes in Marriage and the Family from 1950s to 2000	
1950s	**2000**
Familism versus Individualism	
Families were focused on the needs of children. Mothers stayed home to ensure that the needs of their children were met.	The adult agenda of work and recreation has taken on increased importance, with less attention being given to children. Children are viewed as more sophisticated and capable of thinking as adults, which frees adults to pursue their own interests. Day care is used regularly.
Homosexuality	
Same-sex emotional and sexual relationships were a culturally hidden phenomenon. Gay relationships were not socially recognized.	Gay relationships are increasingly a culturally open phenomenon. Some definitions of the family include same-sex partners. Domestic partnerships are increasingly given legal status in some states. The legality of same-sex marriage is being considered in the courts.

While these previously proscribed behaviors have become increasingly accepted, there is no evidence suggesting that they have become positive goals to be achieved. The vast majority of Americans continue to value family life, plan to marry and have children, and are optimistic about achieving success in marriage (p. 76).

Indeed, in the 1980s and 1990s there has been a conservative backlash against changing family forms and family diversity. Religious groups and politicians have attributed many of society's problems to the breakdown in the traditional nuclear family and have called for a return to the two-parent family. Welfare reform laws passed in the late 1990s were aimed largely at discouraging out-of-wedlock childbearing. Legislators began proposing changes in divorce laws to make it harder for couples to divorce. Others have suggested bringing the shame back to nontraditional behaviors such as nonmarital sex, cohabitation, out-of-wedlock childbearing, and divorce.

Emerging Legal and Cultural Conceptions of the Family

If "family values" mean parents' togetherness with their natural children in their own house, these values have little to do with how the American people actually live.
James Hillman

As "nonfamily" (according to the U.S. Census) households and variations in family forms have become more common, many family scholars are advocating new cultural and legal definitions of the family. Geile (1996) suggests that "recognition as 'family' should . . . not be confined to the traditional two-parent unit connected by blood, marriage or adoption, but should be extended to include kin of a divorced spouse . . . same-sex partners, congregate households of retired persons, group living arrangements, and so on" (pp. 103–4). Family scholars recognize that couples who are not legally married but who live together and are economically, emotionally, and sexually interdependent are functioning as a family (Scanzoni & Marsiglio, 1993).

Increasingly, social and legal policy and court decisions are expanding the concept of family to include unmarried couples, heterosexuals and homosexuals who live together in an emotionally and financially interdependent relationship, and those who view themselves as a family. In some U.S. cities and counties, heterosexual cohabiting couples as well as gay couples (with and without children) are granted *domestic partnership* status and may receive legal entitlements such as health insurance benefits/inheritance rights that have traditionally been reserved for married couples. Domestic partnerships have also been granted to a single parent and his/her mother who resides with him/her to care for the children. In addition, some employers, including Disney, Levi Strauss, Ben & Jerry's, and IBM, have begun to extend employee benefits to cohabiting partners of employees.

Alternatives to Traditional Marriage and Family

Not all people get married, have children while married, or are heterosexual. In this section we examine alternatives to traditional marriage and family patterns.

Singlehood

NATIONAL DATA
In 1970, 19 percent of the men and 11 percent of the women aged 25–29 had never married; by 1996, these percentages had jumped to 52 percent and 38 percent, respectively (*Statistical Abstract of the United States: 1997*, Table 59).

Less than 5 percent of individuals who are 65 and older have never married (*Statistical Abstract of the United States*, 1997). Despite the fact that the vast majority of people eventually do marry, today more people are staying single longer.

One of the reasons for increased singlehood is increased social support:

At the most basic level, more people today can choose if, who, and when to marry. There is less pressure to marry, and less, if any stigma to living alone or cohabiting. Choosing not to marry does not prevent people from being sexually active, or indeed from becoming parents. As a society, we are far more willing to accept a range of possible living arrangements, than people a generation ago (Jones et al., 1995, 14).

Other reasons for remaining single are presented in Table 1.3, which lists the benefits of singlehood and the limitations of marriage.

I had often wondered why young women should marry, as they have so much more freedom, and so much more attention paid to them while unmarried, than when married.

Samuel Johnson
English poet

The primary perceived advantage of remaining single is freedom and control over one's life. Once a decision is made to involve another in one's life, it follows that one's options become vulnerable to the other's influence. The person who chooses singlehood may view such restrictions on freedom as constraining.

Rates of persons remaining single are higher in some ethnic groups than in others. Because African-American women outnumber African-American men (in some southern nonmetropolitan counties there are one hundred African-American women for seventy-eight African-American men), "marriage is simply not possible for many African American women" (Kiecolt & Fossett, 1997, 16).

NATIONAL DATA
In 1995, 36 percent of black women had never married, in contrast to 14 percent of white women who had never married (*Statistical Abstract of the United States: 1997,* Table 58).

Table 1.3

Reasons to Remain Single	
Benefits of Singlehood	**Limitations of Marriage**
Freedom to do as one wishes	Restricted by spouse or children
Responsible for one's self only	Responsible for spouse and children
Close friends of both sexes	Pressure to avoid close opposite-sex friendships
Spontaneous lifestyle	Routine, predictable lifestyle
Feeling of self-sufficiency	Potential to feel dependent
Spend money as wish	Expenditures influenced by needs of spouse/children
Freedom to move as career dictates	Restrictions on career mobility
Avoid being controlled by spouse	Potential to be controlled by spouse
Avoid emotional/financial stress of divorce	Possibility of divorce

Difficulty finding a suitable marriageable partner also contributes to singlehood among many black women. Not only are there fewer black men available, but those who are available may be victims of racial discrimination in hiring, firing, and salary advancement.

The deteriorating economic position of black men has been blamed for further discouraging the formation of married-couple families. Black men with low wages and little job security, have difficulty fulfilling the traditional role as the major breadwinner for a family (p. 18).

Singlehood is also affected by social approval in society. The stigma associated with singlehood in the United States has lessened, but not disappeared. Chasteen (1994) interviewed twenty-five single women between the ages of 27 and 54. A 34-year-old woman reported:

Society has an image of single women, a double standard. We're looked down on if we go to bars. We're not expected to do what the men do. We're supposed to be "ladies" . . . the perception people have of you is strange. You've got people on the one hand who admire what you've done and others who look down on you. It's a catch-22. You're in the middle (Chasteen, 323).

OTHER CULTURES The Netherlands is more positive about singlehood. Gibbons (1992) noted that the "new ideologies that emphasize individual fulfillment over the importance of commitment to others may be making some impact on this alternative lifestyle" (p. 63). Alternatively, singlehood in Japan is strongly stigmatized. ●

Single Parenthood

NATIONAL DATA

In 1990, 28 percent of U.S. births were to unmarried women. By 1995, this figure had increased to 32 percent (*Statistical Abstract of the United States: 1997*, Table 98).

I'm not going to discuss the father, the method or anything of that nature.

Jodie Foster

Actress

Single parenthood may be thought of as a relatively recent phenomenon but in fact has a long history. In prior centuries, high mortality rates of spouses ensured the prevalence of single-parent households (Coontz, 1995). Today such households are also numerous but for different reasons—unwed parenthood, divorce, widowhood, and choice. Blankenhorn notes that "unwed parenthood has . . . become, by far, the nation's fastest-growing family structure trend" (1995, 132). Jodie Foster and Rosie O'Donnell are among the unmarried celebrities who have chosen to be single parents.

Race is a variable related to unmarried parenthood. Black women are more likely than white women to give birth without being married (East, 1998). In 1994, of all U.S. births to unmarried women, only 25 percent were to white women; 70 percent were to black women (*Statistical Abstract*, 1997). Two researchers suggested why blacks are more likely to have single-parent families:

> The history of the Black family in the United States includes consistent and persistent economic strain. The result has been that Blacks are more likely than Whites to live in extended families. Children born out of wedlock and children from dissolved families are often informally adopted by others in the community, most often by grandparents, so that three-generation families are not uncommon (Heaton & Jacobson, 1994, 304).

Hence, black women experience less social pressure to be married before having a child. When they become pregnant out of wedlock, they are less likely than their white female counterpart to get an abortion or to marry.

OTHER CULTURES The prevalence of unmarried parenthood varies widely among minorities. Only 4 percent of Korean women have a child while unmarried, in contrast to 47 percent of Hawaiian women. Chinese and Asian Indian women also have low frequencies—6 percent and 8 percent, respectively (Martin, 1995). ●

Single-parent families and households develop from four avenues: involuntary unmarried pregnancy and childbirth, divorce, widowhood, and choice. Women are much more likely to be single parents for the first three reasons than by choice. And they are five times more likely than men to be rearing a family alone (Ahlburg & DeVita, 1992).

Regardless of the route individuals take to single parenthood, the experience tends to involve several challenges less often faced by parents in couple relationships. These include the strain of tending to the emotional and physical needs of one's children—alone and living on less money.

There may also be problems for children reared in single-parent homes. Researcher Sarah McLanahan, herself a single mother, set out to prove that children reared by single parents were just as well off as those raised by two parents. McLanahan's data on 35,000 children of single parents led her to a different conclusion—children in single-parent homes were twice as likely to drop out of high school, get pregnant before marriage, have drinking problems, and experience a

NATIONAL DATA

The mean income for female-headed, single-parent households is considerably less than the mean income for two-parent households ($21,348 versus $47,129) (*Statistical Abstract of the United States: 1997*, Table 719).

It is important to distinguish between a single-parent "family" and a single-parent "household." A single-parent family is one in which there is only one parent—the other parent is completely and totally out of the child's life through death, sperm donation, or complete abandonment, and no contact is ever made with the other parent. In contrast, a single-parent household is one in which one parent typically has primary custody of the child or children but the parent living out of the house is still a part of the child's family. This is also referred to as a binuclear family. In most divorce cases where the mother has primary physical custody of the child, the child lives in a single-parent household, since he or she is still connected to the father, who remains part of the child's family. In cases in which one parent has died, the child or children live with the surviving parent in a single-parent family, since there is only one parent.

I know what it feels like to try to blend in so that everybody else will think that you are OK and they won't hurt you.
Ellen DeGeneres
Actress, *Ellen*

We all have the potential to be bisexual.
Frank Pittman
Psychiatrist

RECENT RESEARCH

Prejudice against homosexuals continues. Matsuda and Harsel (1997) studied 365 secondary and university students and found that they maintained the greatest social distance from homosexuals (as opposed to persons of other races or the elderly).

host of other difficulties (including getting divorced themselves) as children reared by two married parents (McLanahan & Booth, 1989; McLanahan, 1991).

Nevertheless, most children reared in single-parent homes are happy, well adjusted, and successful. Benefits to the single parent include a sense of pride and self-esteem that results from being independent.

Homosexual Unions

In a landmark television event in the spring of 1997, the character of Ellen on the sitcom *Ellen* blurted (inadvertently) over a microphone in an airport, "I am gay." Ellen DeGeneres, the actress who plays Ellen, also made a public personal disclosure on ABC's *20/20* and *Primetime Live* that she was gay and in a stable love relationship ("This is forever") with Anne Heche. Ellen's "coming out" in her professional role and private life focused nationwide attention on homosexuality as a lifestyle.

Although singlehood and single parenthood are lifestyles individuals may decide to pursue, being homosexual may not be a choice, as some contend that there is no more choice in being homosexual than in being right-handed or left-handed or heterosexual. However, for those who are homosexual, living a homosexual or "gay" lifestyle is a choice (i.e., some who have a homosexual orientation may not choose to live a gay lifestyle).

Sexuality is much more than the biological act of procreation. It involves values, emotions, thoughts, lifestyles, identities, behaviors, and relationships. The term *sexual orientation* refers to the direction of one's thoughts, feelings, and sexual interactions—toward members of the same sex, the other sex, or both sexes.

Homosexuality refers to the predominance of cognitive, emotional, and sexual attraction to those of the same sex. The term *gay* is synonymous with the term *homosexual* and may refer to either males or females who have a same-sex orientation. More often the term *gay* is used to refer to male homosexuals, and the term *lesbian* is used to refer to homosexual women (Committee on Lesbian and Gay Concerns, 1991). *Heterosexuality* refers to the predominance of cognitive, emotional, and sexual attraction to those of the other sex.

Bisexuality refers to a sexual orientation that involves cognitive, emotional, and sexual attraction to members of both sexes. However, in a study of one hundred people who belonged to a bisexual organization, very few of the bisexuals were equally involved with men and women (Weinberg, Williams, & Pryor, 1994). Some of the respondents leaned toward homosexuality; others leaned toward heterosexuality. The majority had primary heterosexual relationships and secondary homosexual relationships. Some respondents reported that their sexual behavior leaned one way and their emotional feelings another.

In addition to being defined in terms of cognitive, emotional, and sexual attraction, sexual orientation is defined by one's sexual self-identity. This complicates the definition of homosexuality, heterosexuality, and bisexuality because attractions and sexual behavior are not always consistent with sexual self-identity. For example, in a study of fifty-two men who labeled themselves heterosexual, almost a quarter (23%) had had sex with both women and men in the past two years, and 6 percent had had sex exclusively with men (Doll et al., 1992). In another study (N = 6,982), of those men who reported having

All we wanted was a little baby and a house and a quiet little life.
Jan Holden, one of
Adam's two gay fathers

prior sexual experiences with both men and women, 69 percent described themselves as heterosexual, 29 percent as bisexual, and 2 percent as homosexual (Lever et al., 1992). These data indicate that adult bisexual experiences do not necessarily result in the acquisition of a bisexual self-identity. Also, labeling one's self as bisexual does not require having had prior bisexual experience. In the above study, 18 percent of those who labeled themselves bisexual reported no adult homosexual experiences.

Although most people view heterosexuality and homosexuality as discrete categories, Alfred Kinsey and his colleagues (1948) suggested that sexual orientation exists on a seven-point heterosexual-homosexual continuum with zero and six reflecting exclusively heterosexual and homosexual experience, respectively. They developed the scale after finding that many women and men reported having had sexual experiences involving both sexes. According to this continuum, few people are entirely either heterosexual or homosexual but are somewhere in between. In addition, individuals of either sexual orientation share more similarities than differences—they both have the capacity to love and to feel jealous. Although there remains a strong societal preference for defining marriage as a relationship between a man and a woman, legalization of homosexual marriages is now being considered in the courts (see Social Policy).

Theoretical Frameworks

Theoretical frameworks provide a set of interrelated principles designed to explain a particular phenomenon and provide a point of view. Common frameworks for viewing marriage and the family include structural-functional, conflict, symbolic interaction, family development, social exchange, and systems.

Structural-Functional Framework

Just as the human body is made up of different parts that work together for the good of the individual, society is made up of different institutions (family, education, economics, etc.) that work together for the good of society. Functionalists view the family as an institution with values, norms, and activities meant to provide stability for the larger society. Such stability is dependent on families serving various functions for society.

First, families serve to replenish society with socialized members. Since our society cannot continue to exist without new members, we must have some way of ensuring a continuing supply. But just having new members is not enough. We need socialized members—those who can speak the language and know the norms and roles of our society. The legal bond of marriage and the obligation to nurture and socialize offspring help to ensure that this socialization will occur.

Second, marriage and the family promote the emotional stability of the respective spouses. Society cannot provide enough counselors to help us whenever we have problems. Marriage ideally provides an in-residence counselor who is a loving and caring partner with whom a person shares his or her most difficult experiences.

Legalization of Same-Sex Marriage?

In late 1996, Hawaii Circuit Court Judge Kevin Chang made history when he ruled in *Baehr v Miike* that Hawaii's refusal to grant marriage licenses to same-sex couples violated the state's constitution. This decision made Hawaii the first state to recognize same-sex marriages. Judge Chang based his ruling on the state's failure to show that the well-being of children and families would be adversely affected by same-sex marriages. Although the matter has been sent back to the state supreme court for further deliberation, Hawaii is expected to be issuing marriage licenses to homosexual couples by the year 2000.

Advocates of same-sex marriage argue that banning same-sex marriage is a form of sex discrimination—a violation of the U.S. Constitution, which says that every person is entitled to equal protection under the law. Banning same-sex marriages or refusing to recognize gay marriages granted in other states denies gay and lesbian couples the many legal and financial benefits granted to heterosexual married couples. As noted earlier in this chapter, married couples have numerous "rights," including the right to inherit from a spouse who dies without a will, the right to make crucial medical decisions for a partner in the event of the partner's critical injury or illness, the right to Social Security survivor benefits, and the right to include a partner in his or her health insurance coverage. The extent to which public policies, entitlements, and protections are family-based is evidenced by the fact that in the United States Code, the term "family" appears over 2,000 times. In New York statutes, the word "family" also occurs over 2,900 times, and in California, over 4,000 times (Hartman, 1996).

Most Americans support the traditional idea that marriage is a legal heterosexual relationship and oppose the legalization of same-sex marriages. Two-thirds (67%) of a national sample of adults said that they would vote against a law making homosexual marriages legal in their state (Lawrence, 1996). Opponents who view homosexuality as unnatural and against our country's moral standards do not want their children to learn that homosexuality is an accepted, "normal" lifestyle. Indeed, the most common argument against same-sex marriage is that it subverts the stability and integrity of the heterosexual family. However, Sullivan (1997a) suggests that homosexuals are already part of heterosexual families by virtue of being "sons and daughters, brothers and sisters, and even mothers and fathers of heterosexuals" (p. 147). Opponents of gay marriage also express concern for the adjustment of children in lesbian and gay families. Over 8 million homosexuals are rearing children they had in previous heterosexual marriages. New Jersey was the first state to allow two gay partners to adopt a child (Parker, 1997). A summary of research findings suggests that children in these families are as happy and well adjusted as other children (Patterson, 1997).

Advocates of same-sex marriage argue that permitting such marriages might encourage many lesbians and gays to live within long-term, committed relationships. Hartinger (1994) suggests that "the result would be more people living more conventional lifestyles. . . . It's actually a conservative move, not a liberal one" (p. 239). Permitting gay marriage would, advocates argue, benefit not only homosexuals but the larger society as well. Marriage encourages monogamy. "In the wake of AIDS, encouraging gay monogamy is simply rational public health policy" (Hartinger, 1994, 239).

In sum, advocates suggest that gay marriage would strengthen, not weaken, the family. Indeed, individuals in lesbian pair-bonded relationships report more satisfaction than heterosexual couples. Individuals in gay male relationships report slightly less satisfaction than heterosexual married couples but decidedly more than divorced individuals (Green et al., 1996).

Hawaii's movement toward recognizing same-sex marriage has created a conservative backlash. Opponents used the Hawaii case as a rallying cry for support of the federal Defense of Marriage Act, which was passed by Congress and signed into law by President Clinton in September 1996. This law asserts that marriage is a "legal union between one man and one woman" and denies federal recognition of homosexual marriage. The law also permits states not to recognize such marriages licensed in other states; thus, if Hawaii does legalize same-sex marriages, other states will be able to ignore them. However, states wishing to recognize same-sex marriages may do so. As of December 1996, eighteen states have declared—either by law or through executive action—that they won't accept same-sex marriages granted in Hawaii or any other state.

The debate over the legalization of same-sex marriage in the United States raises questions about how society defines "marriage" and "family." Any movement toward redefining a social structure as fundamental as marriage and family is bound to be surrounded by heated debate that will not be easily resolved.

In July 1997 Hawaii passed legislation allowing unmarried couples (heterosexual or homosexual) to receive health and family leave benefits (Zimmerman, 1997). The law gives cohabiting couples, referred to as "reciprocal beneficiaries," or RBs for short, some of the same rights as married couples.

REFERENCES

Green, R. J., Bettinger, M., and Zacks, E. 1996. Are lesbian couples fused and gay male couples disengaged? In *Lesbians and gays in couples and families,* edited by Joan Laird and Robert-Jay Green. San Francisco: Jossey-Bass, 185–230.

Hartinger, Brent. 1994. A case for gay marriage: In support of loving and monogamous relationships. In *Taking sides: Clashing views on controversial issues in human sex-* *uality,* edited by Robert T. Francoeur. Guilford, Conn.: Dushkin Publishing Group, 236–41.

Hartman, Ann. 1996. Social policy as a context for lesbian and gay families. In *Lesbians and gays in couples and families,* edited by Joan Laird and Robert-Jay Green. San Francisco: Jossey-Bass, 69–85.

Lawrence, Jill. 1996. Gay issue sizzles in the Senate. *USA Today,* 15 July, 4A.

Parker, L. 1997. Adoption by N.J. gays sparks praise, criticism. *USA Today.* A1.

Patterson, C. 1997. Children of lesbian and gay parents: Summary of research findings. In *Same-sex marriage: Pro and con,* edited by A. Sullivan. New York: Vintage Books, 240–45.

Sullivan, A. 1997a. The conservative case. In *Same-sex marriage: Pro and con,* edited by A. Sullivan. New York: Vintage Books, 146–54.

Sullivan, A. 1997b. Introduction. In *Same-sex marriage: Pro and con,* edited by A. Sullivan. New York: Vintage Books.

Zimmerman, J. 1997. Hawaii OKs benefits to same-sex couples. *USA Today.* 8 July, 2A.

Children also need people to love them and to give them a sense of belonging. This need can be fulfilled in a variety of family contexts (two-parent family, single-parent family, extended family). The affective function of the family is one of its major offerings. No other institution focuses so completely on fulfilling the emotional needs of its members as do marriage and the family.

Third, families provide for the economic support of their members. Although modern families are no longer self-sufficient economic units, they provide food, shelter, and clothing for their members. One need only consider the homeless in our society to be reminded of this important function of the family.

While the replacement, emotional, and economic functions of the family are primary, other functions include

The institutional bond of marriage between biological parents, with the essential function of tying the father to the mother and the child, is found in virtually every society.

David Popenoe
Sociologist

- Physical care—Families provide the primary care for their infants, children, and aging parents. Other agencies—Day care, school, and nursing homes may help, but the family remains the primary caregiver.

- Regulation of sexual behavior—Spouses are expected to confine their sexual behavior to each other, which reduces the risk of having children who do not have socially and legally bonded parents and of spreading HIV and other sexually transmitted diseases.

- Status placement—Being born in a family provides social placement of the individual in society. One's social class, religious affiliation, and future occupation are largely determined by one's family of origin.

Conflict Framework

INSIGHT

Conflict theorists view conflict not as good or bad but as a natural and normal part of relationships. They also regard conflict as necessary for change and growth of individuals, marriages, and families.

One reason equality in marriage remains so elusive is that, in spite of the rhetoric of equality, men have more power and resources in American society than women, and this fact continues to structure marital relationships.

Carmen Knudson-Martin
Anne Rankin Mahoney
Family-life educators

Whereas functionalists look at family practices as good for the whole, conflict theorists recognize that not all family practices are good for every member of the family. Indeed, some activities that are good for one member are not good for others. For example, a woman who has devoted her life to staying home and taking care of the family may decide to seek full-time employment outside the home. This may be a good decision for her personally, but her husband and children may not like it. Conflict theorists recognize different goals and values among family members that cause disagreement and conflict. Conflict theory provides a lens through which to view these differences.

Cohabitation relationships, marriages, and families all have the potential for conflict. Cohabitants are in conflict about commitment to marry, spouses conflict about the division of labor, and parents are in conflict with their children over rules such as curfew and chores. These three units may also be in conflict with other systems. For example, cohabitants are in conflict with the economic institution for health benefits for their partners. Similarly, couples and parents are in conflict with their employers for flexible work hours, maternity/paternity benefits, and day-care facilities.

Karl Marx emphasized that conflict emanates from disputes over scarce resources and the struggle for power. Whereas Marxist theorists focus on conflict between the owners of production (bourgeoisie) and workers (proletariat), conflict also exists within relationships.

Symbolic Interaction Framework

For the most part we do not first see then define; we define and then see.

Walter Lippman
Social critic

Marriages and families represent symbolic worlds in which the various members give meaning to one another's behavior. Herbert Blumer (1969) used the term *symbolic interaction* to refer to the process of interpersonal interaction. Concepts inherent to this framework include the definition of the situation, the looking-glass self, and the self-fulfilling prophecy.

Definition of the situation   What happens between two people who have just spotted each other at a party depends on how they define the situation and respond to those definitions. Is the glance from the other person (a) an invitation to approach, (b) an approach, or (c) a misinterpretation—the other person was looking at someone behind the person? The definition a person arrives at will affect subsequent interaction (Thomas & Zanecki, 1958).

Individual, couple, and family identity grow out of meanings attributed to interactions in the family and the interpretations of those meanings. Symbolic interactionists focus on the process of interpretations and attribution of meaning that occurs in relationships and how these affect individuals and relationships.

Looking-glass self   The image people have of themselves is a reflection of what other people tell them about themselves (Cooley, 1964). Individuals develop their self-concept by the way others act toward them. Family members and intimate partners hold up social mirrors for one another into which the respective members look for definitions of self.

Parents (and other caregivers) have a tremendous influence on the self-concept of children. But adult partners also affect the self-concept of each other.

G. H. Mead (1934) believed that people are not passive sponges but reflect on the perceived appraisals of others, accepting some evaluations and not others. For example, if your partner criticizes you for spending too much time at the library and calls you "selfish," you can choose to reject your partner's negative evaluation of you and, instead, view yourself as "ambitious and conscientious" rather than "selfish."

Self-fulfilling prophecy   Once people define situations and the behaviors they are expected to engage in, they are able to behave toward one another in predictable ways. Such predictability of behavior also tends to exert influence on subsequent behavior. If you feel that your partner expects you to be faithful to him or her, your behavior is likely to conform to these expectations. The expectations thus create a self-fulfilling prophecy.

Symbolic interactionism as a theoretical framework helps to explain various choices in relationships. Individuals who decide to marry have defined their situation as a committed reciprocal love relationship. This choice is supported by the belief that the partners will view each other positively (looking-glass self) and be faithful spouses and cooperative parents (self-fulfilling prophecies).

Family Development/Family Life Cycle Framework

The family development/family life cycle framework emphasizes how families change over time by identifying the stages of the traditional family life cycle, the positions of various members of the family, and the developmental tasks associated with each stage (see Table 1.4). If developmental tasks at one stage are not accomplished, functioning in subsequent stages will be impaired. For example, one of the developmental tasks of early marriage is to emotionally and financially separate from one's family of origin. If such separation does not take place, independence as individuals and as a couple is impaired.

The eight-stage family life cycle presented in Table 1.4 helps to identify not only transitions and developmental tasks at each stage but also the choices with which many individuals are confronted throughout life. For example, the never-married are choosing partners, the newly married are considering when to begin their family, the soon-to-be-divorced are making decisions about custody/child support/division of property, and the remarried are making choices with regard to stepchildren and ex-spouses. Grandparents are making choices about their grandchildren, and widows/widowers are concerned with where to live (children, retirement home, with a friend, alone).

Social Exchange Framework

Chapter 4, Dating and Mate Selection, discusses the social exchange framework in detail as it relates to mate selection. Here, we point out that each interaction among spouses, parents, and children can be understood in terms of each individual seeking the most "benefits" at the least "cost" so as to have the highest

The world's a theatre, the earth a stage, which God and Nature do with actors fill.
Thomas Heywood
Apology for Actors

Table 1.4

Stage-Critical Family Development Tasks Throughout the Family Life Cycle

Stage of the Family Life Cycle	Positions in the Family	Stage-critical Family Developmental Tasks
1. Married couple	Wife Husband	Establishing a mutually satisfying marriage Adjusting to pregnancy and the promise of parenthood Fitting into the kin network
2. Childbearing	Wife-mother Husband-father Daughter-sister Son-brother	Having, adjusting to, and encouraging the development of infants Establishing a satisfying home for both parents and infant(s)
3. Preschool age	Wife-mother Husband-father Daughter-sister Son-brother	Adapting to the critical needs and interests of preschool children in stimulating, growth-promoting ways Coping with energy depletion and lack of privacy as parents
4. School age	Wife-mother Husband-father Daughter-sister Son-brother	Fitting into the community of school-age families in constructive ways Encouraging children's educational achievement
5. Teenage	Wife-mother Husband-father Daughter-sister Son-brother	Balancing freedom with responsibility as teenagers mature and emancipate themselves Establishing postparental interests and careers as growing parents
6. Launching center	Wife-mother-grandmother Husband-father-grandfather Daughter-sister-aunt Son-brother-uncle	Releasing young adults into work, military service, college, marriage, etc., with appropriate rituals and assistance Maintaining a supportive home base
7. Middle-aged parents	Wife-mother-grandmother Husband-father-grandfather	Rebuilding the marriage relationship Maintaining kin ties with older and younger generations
8. Aging family members	Widow/widower Wife-mother-grandmother Husband-father-grandfather	Coping with bereavement and living alone Closing the family home or adapting it to aging Adjusting to retirement

Source: *Marriage and Family Development*, 5th ed. by Evelyn Millis Duvall. Copyright © 1977, 1971, 1967, 1962, 1957 by Harper & Row Publishers, Inc. Reprinted by permission of Addison Wesley Educational Publishers Inc.

"profit" and avoid a "loss" (Homans, 1958; Blau, 1964). Teenagers are constantly aware of what they have to do ("costs") in such terms as cleaning their room and making good grades to get "benefits" in the form of freedom and money from their parents so that they have a "profit" in the social exchange with them.

Dating partners are in an exchange relationship, but the value one partner places on a behavior of the other partner may vary from the value placed on the same behavior by the other partner. In a study of 506 college students, Regan and Sprecher found that

When befriended, remember it.
When you befriend, forget it.
Benjamin Franklin
Philosopher

> women attached more value than men to contributions related to expressiveness (i.e. a warm and understanding disposition, sexual fidelity, love for the partner and demonstrating one's love and being a good com-municator) . . . men valued physical attractiveness more than women (1995, 234).

A social exchange view of marital roles emphasizes that spouses negotiate the division of labor on the basis of exchange. For example, he participates in child care in exchange for her earning an income, which relieves him of the total financial responsibility. Social exchange theorists also emphasize that power in relationships is the ability to influence, and avoid being influenced by, the partner. The various bases of power, such as money, the need for a partner, and brute force, may be expressed in various ways, including withholding resources, decreasing investment in the relationship, and violence. We discuss power in greater detail in Chapter 6, Communication and Conflict Resolution.

Family Systems Framework*

Whatever affects one directly, affects all indirectly. I can never be what I ought until you are what you ought to be. This is the interrelated structure of reality.

Dr. Martin Luther King, Jr.
Civil rights leader

Family members develop rules of interaction (Becvar & Becvar, 1993; Goldenberg & Goldenberg, 1991). These rules may be written (e.g., parents will write down what chores they expect their children to perform) or unwritten (e.g., spouses expect fidelity from each other). These rules serve various functions (Burr et al., 1993; Minuchin, 1974), such as allocating the resources (e.g., allowance), specifying the division of power (e.g., who decides how money is spent), and defining closeness and distance between systems (e.g., seeing or avoiding parents/grandparents) or subsystems (e.g., between parents and children). Rules are most efficient if they are flexible. For example, they should be adjusted over time in response to children's growing competence. A rule about not leaving the yard when playing may be appropriate for a 4-year-old but inappropriate for a 10-year-old.

Family members also develop boundaries that define the individual and the group and separate one system or subsystem from another (Constantine, 1991). Such boundaries may be physical, such as a closed bedroom door, or social, such as expectations that family problems will not be aired in public. Boundaries may also be emotional, such as communication, which maintains closeness or distance in a relationship.

In addition to rules and boundaries, family systems have roles or positions for the respective family members. Four such roles are leader, follower, opposer, and bystander (Kantor & Lehr, 1975). These roles may be shared by more than one person and may shift from person to person during an interaction or across time. In healthy families, individuals are allowed to alternate roles rather than being locked in one role. In problem families, one family member is often allocated the role of "scapegoat," or the cause of all the family's problems (e.g., alcoholic spouse).

Family systems may be open, in that they are open to information and interaction with the outside world, or closed, in that they feel threatened by such contact (Kantor & Lehr, 1976). The boundary between open family systems is

*Appreciation is expressed to Dr. Esther Devall of the Department of Family and Consumer Sciences at New Mexico State University for the development of this section.

permeable. Efforts are made to involve all family members in decision making and rule setting. The crew in *Star Trek Voyager* is an example of an open family system.

In closed family systems, the boundary between the family system and the outside world is rigid. The outside world is viewed as unpredictable and threatening, so the family closes rank. The Amish have a closed family system in that they want minimal contact with the outside world. Television viewing is prohibited.

The major theories are summarized in Table 1.5.

Other Frameworks

Though not formal theoretical frameworks, stratification and race provide ways of viewing marriage and the family. *Stratification* refers to the ranking of people according to their socioeconomic status, usually indexed according to income, occupation, and educational attainment. Passengers on the *Titanic* were stratified and assigned to different decks on the ship. Individuals who occupy a similar socioeconomic status are said to be in the same social class.

Marriages and families are also stratified into different social classes, such as the upper, middle, working, or lower social class. Families in these various social classes reflect differences in their attitudes, values, and behavior. For example, individuals from the lower class are more likely to divorce than individuals from the higher social classes. Parents in lower socioeconomic classes are also more likely to discuss personal and financial problems with their children than parents in higher socioeconomic groups. The former feel that the sooner their children become aware of the harsh realities of life, the better. Middle-class parents, on the other hand, tend to believe that they should protect their children from the realities that lie ahead. Social class is also related to gender role attitudes, with lower social status and less education being associated with more traditional views (Hoffman & Kloska, 1995).

Whether conflict exists in a relationship over fairness in household tasks is also related to social class. Middle-class wives who expect fairness with their husbands in household tasks report the greatest amount of conflict. In contrast, working-class wives who engage in a higher proportion of household tasks than middle-class wives (but who do not expect fairness) report the least amount of marital conflict. Hence, it is the socialization women receive in their respective social classes that influences their equity expectations, which, in turn, affect their level of conflict (Perry-Jenkins & Folk, 1994).

Social class is also affected by race. Racial minorities reflect a diverse range of family values and patterns, as noted throughout the text. Table 1.6 lists types of racial minorities in the United States.

The term *race* refers to a category of people who share biological traits that are deemed socially significant. Many of the differences between blacks and whites reflect differences in social class rather than race. Individuals in the lower class (whether white or black) have higher rates of unemployment, premarital pregnancies, divorce, and crime. In looking at the comparisons between blacks and whites throughout this text, it is important to keep in mind that

I'm from the projects—I became a movie star.
Whoopi Goldberg
Actress

I know my dad was somebody when Elvis came over for dinner.
Chris Mitchum
Son of Robert Mitchum

We are rapidly moving towards a multicultural and demographically diverse society.
Harriette Pipes McAdoo
Past President, National Council on Family Relations

INSIGHT

A person's membership in a particular racial or ethnic group is sometimes not as clear-cut as might be expected. For example, not all black people are African-Americans. Black immigrants from the Caribbean have a strong ethnic identity that is not African-American.

Table 1.5

Comparison of Theoretical Frameworks					
Theory	**Description**	**Concepts**	**Level of Analysis**	**Strengths**	**Weaknesses**
Structural-Functional	The family has several important functions within society; within the family, individual members have certain functions.	Structure Function	Institution	Emphasizes the relation of family to society, noting how families affect and are affected by the larger society.	Families with nontraditional structures (single-parent, same-sex couples) are seen as dysfunctional.
Conflict	Conflict in relationships is inevitable, due to competition over resources and power.	Conflict Resources Power	Institution	Views conflict as a normal part of relationships and as necessary for change and growth.	Sees all relationships as conflictual, and does not acknowledge cooperation.
Symbolic Interaction	People communicate through symbols and interpret the words and actions of others.	Definition of the situation Looking-glass self Self-fulfilling prophecy	Relationship	Emphasizes the perceptions of individuals, not just objective reality or the viewpoint of outsiders.	Ignores the larger social context and minimizes the influence of external forces.
Family Development	At each stage of the family life cycle, there are certain tasks that must be accomplished.	Family life cycle Developmental tasks	Family	Families are seen as dynamic, rather that static. Useful in working with families who are facing life cycle transitions.	Does not apply as easily to families without children or where divorce or remarriage has occurred.
Social Exchange	In their relationships, individuals seek to maximize their benefits and minimize their costs.	Benefits Costs	Individual	Provides explanations for human behavior based on outcome.	Assumes that people always act rationally and all behavior is calculated.
Family Systems	The family is a system of interrelated parts that function together to maintain the unit.	Subsystem Roles Rules Boundaries Open system Closed system	Family	Very useful in working with families who are having serious problems (violence, alcoholism). Describes the effect family members have on each other.	Based on work with troubled families and may not apply to nonproblem families.

Source: Esther L. Devall, Ph.D., CFLE, New Mexico State University. Developed specifically for this text. Used by permission.

Table 1.6

Racial Minorities of the United States in 2000 (est.)			
Group	**Total Number**	**Percentage of U.S. Pop.**	**Examples**
African-American	34 million	12	African, Caribbean
Hispanic	31 million	11	Mexican, Puerto Rican, Cuban, Central and South Americans
Asian/Pacific American	11 million	4	Chinese, Japanese, Korean, Vietnamese, Cambodian, Thai, Filipino, Laotian, Lao-Hmong, Samoan, Guamanian
American Indian Alaskan Native	2 million	1	Cherokee, Navajo, Sioux, Chippewa, Aleut, Innuit

Source: Statistical Abstract of the United States: 1997, 17th ed. Washington, D.C.: U.S. Bureau of the Census, 1997, Table 38.

many presumed racial differences are really those of social class. However, racism still exists in the form of discrimination against minorities in education, employment, and housing, and it affects spouses, parents, and children.

Research in Marriage and the Family: Some Cautions

My latest survey shows that people don't believe in surveys.
Lawrence Peter
Humorist

As you read this and other texts about marriage and the family, you should be critical of the research cited. Following are specific cautions.

Sampling

It is the sample that we observe, but it is the population which we seek to know.
George Snedecor
William Cochran
Sociologists

Some of the research on marriage and the family is based on random samples. *Random sampling* involves selecting individuals at random from an identified population. In a random sample, each individual in the population has an equal chance of being included in the sample. Studies that use random samples are based on the assumption that the individuals studied are similar to and therefore representative of the population that the researcher is interested in. For example, suppose you want to know the percentage of unmarried seniors (US) on your campus who are living together. Although the most accurate way to get this information is to secure an anonymous yes or no response from every US, doing so is not practical. To save yourself time, you could ask a sample of USs to complete your questionnaire and assume that the rest of them would say yes or no in the same proportion as those who did. To decide who would be in your sample, you could put the names of every US on campus on separate note cards, stir these cards in your empty bathtub, put on a blindfold, and draw one hundred cards. Because each US would have an equal chance of having his or her card drawn from the tub, you would obtain a random sample. After administering the questionnaire to this sample and adding the yes and no answers, you would have a fairly accurate idea of the percentage of USs on your campus who are living together.

INSIGHT

Even the term *random sample* may not always mean random. For example, in the study of unmarried seniors, not all the names you put in the bathtub to select from would have addresses and phone numbers. Hence, even if you drew the person's name, finding her or him to complete a questionnaire could be difficult. In addition, some people refuse to complete a questionnaire.

Because of the trouble and expense of obtaining random samples, many researchers study subjects to whom they have convenient access. This often means students in the researchers' classes. The result is an overabundance of research on "convenience" samples consisting of white, Protestant, middle-class college students. Because college students cannot be assumed to be similar to their noncollege peers or older adults in their attitudes, feelings, and behaviors, research based on college students cannot be generalized beyond the base population. Some research shows that nontraditional (over age 25) college students tend to have more strict values than traditional college-age (18–22) students (Bee, 1993). To provide a balance, where possible, we have attempted to include data in this text that describe people of different ages, marital statuses, racial backgrounds, lifestyles, religions, and social classes. When only data on college samples are presented, it is important not to generalize the findings too broadly.

In addition to having a random sample, it is important to have a large sample. The random study at 464 colleges and universities of over 250,000 students referred to in the first paragraph of this chapter represented a large national sample, which provides credible data on first-year college students. If only fifty college students were in the sample, the results would be very unreliable in terms of generalizing beyond that sample. Be alert to the sample size of the research you read. Many studies are based on small samples.

Control Groups

Any study concluding that divorce (or any independent variable) is associated with the fact that children in those homes make lower grades (or any dependent variable) must necessarily include two groups: (1) children whose parents are divorced and (2) children whose parents are still married. The latter would serve as a *control group*—the group not exposed to the independent variable you are studying. Hence, if you find that children in both groups make low grades, you know that divorce cannot be the cause. Be alert to the existence of a control group, which is not uncommon in research studies. Kunz (1995) reviewed sixty-five studies on the intellectual functioning of children from divorced homes, and each study included a control group of children from homes in which their parents were together. The level of confidence you can place in the findings of a study is vastly improved when researchers use a control group.

Age and Cohort Effects

In some research designs, different cohorts or age groups are observed and/or tested at one point in time. One problem that plagues such research is the difficulty—even impossibility—of discerning whether observed differences between the subjects studied are due to the research variable of interest, cohort differences, or some variable associated with the passage of time (e.g., biological aging). A good illustration of this problem is found in research on changes in marital satisfaction over the course of the family life cycle. In such studies, researchers may compare the level of marital happiness reported by couples who have been married for different lengths of time. For example, a researcher

INSIGHT

Be particularly cautious of research studies presented in popular magazines. In May of 1997, *Details'* magazine featured its "1997 College Sex Survey" and stated that 45 percent of the men and 42 percent of the women in their survey reported "cheating on a steady partner." However, of the 20,000 questionnaires that were sent out to college students, 90 percent were *not* returned; thus the information provided was based on the 10 percent who did return their questionnaires. Popular magazines exist to sell copies, not to provide credible research studies.

may compare the marital happiness of two groups of people—those who have been married for fifty years and those who have been married for five years. But differences between these two groups may be due to either (1) differences in age (age effect), (2) the different historical time period that the two groups have lived through (cohort effect), or (3) being married different lengths of time (research variable). It is helpful to keep these issues in mind when you read studies on marital satisfaction over time.

Terminology

What's in a name? that which we call a rose By any other name would smell as sweet.
Shakespeare
English playwright and poet

In addition to being alert to potential shortcomings in sampling, control groups, and age/cohort effects, you should consider how the phenomenon being researched is defined. For example, in a preceding illustration of unmarried seniors (US) living together, how would you define "living together"? How many people, of what sex, spending what amount of time, in what place, engaging in what behaviors will constitute your definition? Indeed, researchers have used more than twenty definitions of what constitutes "living together."

What about other terms? What is meant by marital satisfaction, commitment, interpersonal violence, and sexual fulfillment? Before accepting that most people report a high degree of marital satisfaction or sexual fulfillment, be alert to the definition used by the researcher. Exactly what is the researcher trying to measure?

Researcher Bias

But it is doubtless impossible to approach any human problem with a mind free of bias.
Simone de Beauvoir
The Second Sex

Although one of the goals of scientific studies is to gather data objectively, it may be impossible for researchers to be totally objective. Researchers are human and have values, attitudes, and beliefs that may influence their research methods and findings. It may be important to know what the researcher's bias is in order to evaluate that researcher's findings. For example, a researcher who does not support abortion rights may conduct research that focuses only on the negative effects of abortion. Conclusions of a study financed by a particular industry or special-interest group should also be scrutinized for bias.

In addition, some researchers present an interpretation of what other researchers have done. Two layers of bias may be operative here: (1) when the original data were collected and interpreted and (2) when the second researcher read the study of the original researcher and made his or her own interpretation. Much of this text is based on interpretations of other people's studies. As a consumer you should be alert to the potential bias in reading such secondary sources. To help control for this bias, we have presented references to the original source for your own reading.

Time Lag

Typically, a two-year lag exists between the time a study is completed and the study's appearance in a professional journal. Because textbooks (such as the one you are reading) take from three to five years to develop, by the time you read the results of a study, other studies may have been conducted that reveal different findings. Although we have tried to reduce this concern with our Recent Research feature, be aware that the research you read in this or any other text may not reflect current reality.

Distortion and Deception

Researchers in all fields may encounter problems of sampling, lack of a control group, age differences, terminology, researcher bias, and time lag, but other problems specific to social science research—particularly to marriage research—are distortion and deception. Marriage is a very private relationship that happens behind closed doors, and we have been socialized not to reveal to strangers the intimate details of our marriages. Therefore, we are prone to distort, omit, or exaggerate information, perhaps unconsciously, to cover up what we may feel is no one else's business. Thus, the researcher sometimes obtains inaccurate information. Marriage and family researchers know more about what people say they do than what they actually do.

An unintentional and probably more frequent form of distortion is inaccurate recall. Sometimes researchers ask respondents to recall details of their relationships that occurred years ago. Time tends to blur some memories, and respondents may not relate what actually happened but will relate only what they remember to have happened or worse, what they wish had happened. Table 1.7 summarizes potential weaknesses of research studies.

Table 1.7

Potential Weaknesses of Research Studies		
Weakness	**Consequences**	**Example**
Sample not random	Cannot generalize findings	Opinions of college students do not reflect opinions of other adults.
No control group	Inaccurate conclusions	Study on the effect of divorce on children needs control group of children whose parents are still together.
Age differences between groups of respondents	Inaccurate conclusions	Effect may be due to passage of time or to cohort differences.
Unclear terminology	Inability to measure what is not clearly defined	What is living together, marital happiness, sexual fulfillment, good communication, quality time?
Researcher bias	Slanted conclusions	Male researcher may assume that since men usually ejaculate each time they have intercourse, women should have an orgasm each time they have intercourse.
Time lag	Outdated conclusions	Often-quoted Kinsey sex research is over fifty years old.
Distortion	Invalid conclusions	Research subjects exaggerate, omit information, and/or recall facts or events inaccurately. Respondents may remember what they wish had happened.

The personal application section in each chapter may be used by both single and partnered individuals. While the former may regard it as an exercise in self-examination and values clarification, the latter may share their responses with their partners to facilitate self-disclosure and enable the partners to learn more about each other. This first application feature focuses on how well your family functions. You might complete the following scale when thinking about your family of origin or your family of procreation.

Family Functioning Scale

INSTRUCTIONS: Every family has strengths and capabilities, although different families have different ways of using their abilities. This questionnaire asks you to indicate whether your family is characterized by 26 different qualities. Please read each statement, then *circle* the response that is most true for your family (people living in your home). Please give your honest opinions and feelings. Remember that your family will not be like all the statements.

How is your family like the following statements:	Not at All Like My Family	A Little Like My Family	Sometimes Like My Family	Usually Like My Family	Almost Always Like My Family
1. We make personal sacrifices if they help our family.	1	2	3	4	5
2. We usually agree about how family members should behave.	1	2	3	4	5
3. We believe that something good always comes out of even the worst situations.	1	2	3	4	5
4. We take pride in even the smallest accomplishments of family members.	1	2	3	4	5
5. We share our concerns and feelings in useful ways.	1	2	3	4	5
6. Our family sticks together no matter how difficult things get.	1	2	3	4	5
7. We usually ask for help from persons outside our family if we cannot do things ourselves.	1	2	3	4	5
8. We usually agree about the things that are important to our family.	1	2	3	4	5
9. We are always willing to pitch in and help each other.	1	2	3	4	5
10. We find things to do that keep our minds off our worries when something upsetting is beyond our control.	1	2	3	4	5
11. We try to look at the bright side of things no matter what happens in our family.	1	2	3	4	5
12. We find time to be together even with our busy schedules.	1	2	3	4	5

How is your family like the following statements:	Not at All Like My Family	A Little Like My Family	Sometimes Like My Family	Usually Like My Family	Almost Always Like My Family
13. Everyone in our family understands the rules about acceptable ways to act.	1	2	3	4	5
14. Friends and relatives are always willing to help whenever we have a problem or crisis.	1	2	3	4	5
15. Our family is able to make decisions about what to do when we have problems or concerns.	1	2	3	4	5
16. We enjoy time together even if it is doing household chores.	1	2	3	4	5
17. We try to forget our problems or concerns for a while when they seem overwhelming.	1	2	3	4	5
18. Family members listen to both sides of the story during a disagreement.	1	2	3	4	5
19. We make time to get things done that we all agree are important.	1	2	3	4	5
20. We can depend on the support of each other whenever something goes wrong.	1	2	3	4	5
21. We usually talk about the different ways we deal with problems and concerns.	1	2	3	4	5
22. Our family's relationships will outlast our material possessions.	1	2	3	4	5
23. We make decisions like moving or changing jobs for the good of all family members.	1	2	3	4	5
24. We can depend upon each other to help out when something unexpected happens.	1	2	3	4	5
25. We try not to take each other for granted.	1	2	3	4	5
26. We try to solve our problems first before asking others to help.	1	2	3	4	5

(May be duplicated without permission with proper acknowledgment and citation.)

SCORING: Circling a 5 represents the most optimum family functioning response in terms of family strengths. Circling a 1 represents the least optimum family functioning response. The scale was administered to 206 mothers and 35 fathers of preschool children. The majority of items had average ratings between 3.00 and 4.00.

Source: Carol M. Trivette, Carl J. Dunst, Angela G. Deal, Deborah W. Hamby, and David Sexton. Family Functioning Style Scale, in Chapter 10, Assessing family strengths and capabilities. *Supporting and Strengthening Families: Volume 1—Methods, Strategies and Practices,* Carl J. Dunst, Carol M. Trivette, and Angela G. Deal, eds. Cambridge, Mass.: Brookline Books, 1994, 139. Permission to reproduce granted in text.

Other Research Problems

Nonresponse on surveys and the discrepancy between attitudes and behaviors are other research problems. With regard to nonresponse, not all individuals who complete questionnaires or agree to participate in an interview are willing to provide information about such personal issues as money, spouse abuse, family violence, rape, sex, and alcohol abuse. Such individuals leave the questionnaire blank or tell the interviewer they would rather not respond. Others respond but give only socially desirable answers. The implications for research are that data gatherers do not know the nature or extent to which something may be a problem because people are reluctant to provide accurate information.

The discrepancy between the attitudes people have and their behavior is another cause for concern about the validity of research data. It is sometimes assumed that if a person has a certain attitude (for example, extramarital sex is wrong), then his or her behavior will be consistent with that attitude (avoid extramarital sex). However, this assumption is not always accurate. People do indeed say one thing and do another. This potential discrepancy should be kept in mind when reading research on various attitudes.

Finally, most research reflects information provided by volunteers. The question we must ask is, Do volunteers answer questions the same way that nonvolunteers do? Some data suggest that the answer is no. Two sex researchers (Strassberg & Lowe, 1995) found that volunteers were significantly different from nonvolunteers. The former reported more positive attitudes toward sexuality, less sexual guilt, and more sexual experience than the latter. "Our data suggest that these differences (between volunteers and nonvolunteers) may be even larger than previously recognized" (p. 378).

In view of the research cautions identified here, you might ask, "Why bother to report the findings?" However, the quality of some family science research is excellent. For example, articles published in the *Journal of Marriage and the Family* (among other journals) reflect the high level of methodologically sound articles that are being published. Even less sophisticated journals provide useful information on marital, family, and other relationship data. The alternative to gathering data is to rely on personal experience and observation alone, and this is unacceptable to social scientists.

GLOSSARY

binuclear family   A family that spans two households, as when children whose parents divorce have a family with their mother and a family with their father.

bisexuality   Sexual orientation involving cognitive, emotional, and sexual attractions to both women and men.

blended family   A stepfamily consisting of remarried spouses in which one or both partners have children from a previous relationship.

common-law marriage   A marriage by mutual agreement between a cohabiting man and woman without a marriage license or ceremony (recognized in some but not all, states). A common-law marriage may require a legal divorce if the couple breaks up.

control group   In a research study, the group not exposed to the independent variable being studied.

domestic partnership   A relationship in which individuals who live together are emotionally and financially interdependent and are given some kind of official recognition by a city or corporation allowing them to receive partner benefits such as health insurance.

extended family   A nuclear family plus relatives including grandparents, aunts, uncles, cousins, and in-laws.

extradyadic sexuality   Having sexual intercourse with someone other than one's partner, or outside the dyad.

family   Defined by the U.S. Census Bureau as a group of two or more persons related by blood, marriage, or adoption. New cultural definitions of the family include non-related persons who reside together and who are economically, emotionally, and sexually interdependent (e.g., cohabiting heterosexual or homosexual couples.).

family household   Two or more persons who are related by blood, marriage, or adoption and who reside together in the same living quarters.

family of origin   The family into which an individual is born or reared, usually including a mother, father, and children.

family of procreation   The family one begins, which includes a spouse and children.

heterosexuality   Sexual orientation in which the predominance of cognitive, emotional, and sexual attraction is toward those of the other sex.

homosexuality   Sexual orientation in which the predominance of cognitive, emotional, and sexual attraction is toward those of the same sex.

household   Individuals who share a housing unit, such as a house or apartment.

lifestyle   The overall pattern of living that people evolve to meet their biological, social, and emotional needs.

marriage (in the U.S.)   A legal contract entered into voluntarily between two unmarried adults who meet the legal requirements (age, mental competence, other sex).

nuclear family   The family of origin or procreation in which the spouses and their children represent the core relationships, with other relatives being peripheral.

polyandry   A marriage in which one wife has two or more husbands.

polygamy   A marriage in which there are more than two spouses.

polygyny   A marriage in which one husband has two or more wives.

race   Category of individuals who share biological traits that are deemed socially significant.

random sample   For research purposes, a sample in which each individual in the population being studied has an equal chance of being included in the study.

sexual orientation   The direction of one's emotional and sexual interests—toward members of the same sex, the other sex, or both sexes.

stratification   The ranking of people according to socioeconomic status, usually indexed according to income, occupation, and educational attainment.

theoretical framework   A set of interrelated principles designed to explain a particular phenomenon and provide a point of view.

SUMMARY

Marriage

Although individuals are waiting until they are older to marry, over 95 percent eventually marry and regard it as an important goal. When compared with singles, marrieds are happier, have better health, live longer, and have more economic resources.

Marriage in the United States is a legal contract between two unmarried adults of different sexes that allows the state in which they reside to regulate their economic and sexual relationship. Marriages performed in religious contexts are not legal until the couple, witnesses, and person who conducts the ceremony sign the marriage license issued by the state. Common-law marriages (recognized by thirteen states and the District of Columbia) are marriages by mutual agreement between a cohabiting man and woman without a marriage license or ceremony. In common-law states, a woman and man who live together and consider themselves "married" by common law may inherit from each other and receive health and Social Security benefits. They are also responsible for each other's debts.

Family

Society's definition of family is changing. The U.S. Census Bureau defines family household as two or more persons who are related by blood, marriage, or adoption and who reside together in the same living quarters. But divorced parents may not live with their children although they are still considered "family." Contemporary definitions of family include nonrelated persons who reside together and who are economically, emotionally, and sexually interdependent. This definition encompasses cohabiting heterosexual or homosexual couples. Various types of families include family of origin, family of procreation, nuclear family, binuclear family, and extended family. Although the legality of denying same-sex marriages is now being challenged in the courts, there remains a strong cultural preference to define marriage in terms of a man-woman relationship.

American Families in Transition

American families have undergone considerable change in the last fifty years. It is now more socially acceptable to remain single, to cohabit without being married, to divorce, to have children outside marriage, and to have an abortion. However, increased *tolerance* of these behaviors does not necessarily imply increased *approval* of these same behaviors. Our society remains committed to marriage and to rearing children within marriage.

Alternatives to Traditional Marriage and Family

Alternative life paths to marriage and the traditional two-parent childrearing unit include being single, rearing children as a single parent, and/or having a homosexual lifestyle. A major attraction of singlehood is the freedom to do as one wishes. Although single parenthood is increasing, a major disadvantage is lack of money. Benefits include a sense of pride and self-esteem that results from being independent. Homosexuals are challenged by the lack of social acceptance. Indeed, conflict theorists point out the political power exercised by heterosexuals over homosexuals as the dominant group in society. Yet homosexuals continue to gain recognition and benefits. These alternatives to traditional marriage and the family serve society by providing diverse outlets for a variety of needs and interests. Since most people are reared in families and socialized to marry and have children, marriage will continue to be the dominant lifestyle pattern.

Theoretical Frameworks

Marriage and the family may be viewed from various theoretical perspectives. Structural-functionalists emphasize how marriage and the family replenish society with socialized members and provide for the emotional/sexual needs of its members. Conflict theorists view spouses and children competing for scarce resources of space, time, and money. Symbolic interactionists focus on the definitions, meanings, and interpretations of roles and behaviors as spouses and children interact. Developmentalists emphasize the various stages through which couples and families pass and the developmental tasks of each stage. Social exchange theorists identify the rewards and costs of various interpersonal choices. Spouses are continually negotiating rules and behaviors and making interpersonal exchanges. Family systems theorists emphasize that the family is a system of interrelated parts that function together to maintain the unit.

Research in Marriage and the Family: Some Cautions

Research in marriage and the family should be read cautiously. Methodological problems include the lack of random samples, lack of control groups, and vague terminology. Magazine surveys often report data from nonrandom samples where no control group was used and the terminology was poorly defined. Professional journals such as the *Journal of Marriage and the Family* provide superb examples of solid research.

REFERENCES

Ahlburg, Dennis. A., and Carol J. De Vita. 1992. New realities of the American family. *Population Bulletin* 47: 2–44.

American Council on Education and University of California. 1997. *The American freshman: National norms for Fall, 1997*. Los Angeles: Higher Education Research Institute, U.C.L.A. Graduate School of Education and Information Studies.

Becvar, S. D., and R. J. Becvar. 1993. *Family therapy: A systemic integration*. Boston: Allyn and Bacon.

Bee, Richard H. 1993. Age differentiation on ethical issues. *College Student Journal* 27: 490–97.

Blankenhorn, D. 1995. *Fatherless America: Confronting our most urgent social problem*. New York: HarperPerennial.

Blau, P. M. 1964. *Exchange and power in social life*. New York: Wiley.

Blumer, H. G. 1969. The methodological position of symbolic interaction. In *Symbolic interactionism: Perspective and method*. Englewood Cliffs, N.J.: Prentice-Hall.

Burr, W. R., R. D. Day, and K. S. Bahr. 1993. *Family science*. Pacific Grove, Calif.: Brooks/Cole.

Carter, S. A., and M. L. Morris. 1997. The marriage and family quiz: College students' beliefs in selected myths about marriage, remarriage, and parenting. Paper presented at 59th Annual Conference of the National Council on Family Relations, Crystal City, Va, November.

Chasteen, Amy L. 1994. The world around me: The environment and single women. *Sex Roles* 31: 309–28.

Committee on Lesbian and Gay Concerns. 1991. Avoiding heterosexual bias in language. *American Psychologist* 46: 973–74.

Constantine, L. 1991. *Family paradigms: An overview*. Pacific Grove, Calif.: Brooks/Cole.

Cooley, C. H. 1964. *Human nature and the social order*. New York: Schocken Books.

Coontz, Stephanie. 1995. The way we weren't: The myth and reality of the "traditional" family. *Phi Kappa Phi Journal* 75: 11–14.

Crook, J. H., and S. J. Crook. 1988. Tibetan polyandry: Problems of adaptation and fitness. In *Human reproductive behavior: A Darwinian perspective*, edited by Laura Betzig, Monique B. Bulder, and Paul Turke. New York: Cambridge University Press, 97–114.

Doll, Linda S., Lyle R. Petersen, Carol R. White, Eric S. Johnson, John W. Ward, and the Blood Donor Study Group. 1992. Homosexual and nonhomosexual identified men: A behavioral comparison. *Journal of Sex Research* 29: 1–14.

East, P. L. 1998. Racial and ethnic differences in girls' sexual, marital, and birth expectations. *Journal of Marriage and the Family* 60: 150–162.

Gage-Brandon, Anastasia J. 1992. The polygyny-divorce relationship: A case study of Nigeria. *Journal of Marriage and the Family.* 54: 285–92.

Geile, Janet Z. 1996. Decline of the family: Conservative, liberal, and feminist views. In *Promises to keep: Decline and renewal of marriage in America*, edited by D. Popenoe, J. B. Elshtain, and D. Blankenhorn. Lanham, Md.: Rowman & Littlefield, 89–115.

Geraghty, M. 1997. Finances are becoming more crucial in students' college choice, survey finds. *Chronicle of Higher Education*, 17 January, A41–44.

Gibbons, J. A. 1992. Alternative lifestyles: Variations in household forms and family consciousness. In *Family and marriage: Cross-cultural perspectives*, edited by K. Ishwaran. Toronto, Ontario: Thompson Educational Publishing, 61–74.

Glenn, N. D. 1997. *Closed hearts, closed minds: The textbook story of marriage.* New York: Institute for American Values.

Goldenberg, I., and H. Goldenberg. 1991. *Family therapy: An overview.* Pacific Grove, Calif.: Brooks/Cole.

Gose, B. 1998. More freshmen than ever appear disengaged from their studies, survey finds. *The Chronicle of Higher Education*, 16 January, A37. The complete research appears in American Council on Education and University of California. 1997. *The American freshman: national norms for Fall, 1997.* Los Angeles: Higher Education Research Institute, U.C.L.A. Graduate School of Education and Information Studies.

Green, R. J., Bettinger, M., and Zacks, E. 1996. Are lesbian couples fused and gay male couples disengaged? In *Lesbians and gays in couples and families*, edited by Joan Laird and Robert-Jay Green. San Francisco: Jossey-Bass, 185–230.

Gwanfogbe, P. N., W. R. Schumm, M. Smith, and J. L. Furrow. 1997. Polygyny and marital life satisfaction: An exploratory study from rural Cameroon. *Journal of Comparative Family Studies* 28: 55–71.

Heaton, Tim B., and Cardell K. Jacobson. 1994. Race differences in changing family demographics in the 1980s. *Journal of Family Issues*, 15: 290–308.

Hoffman, Lois W., and Deborah D. Kolska. 1995. Parent's gender-based attitudes toward marital roles and child rearing: Development and validation of new measures. *Sex Roles*, 32: 273–96.

Homans, G. 1958. *Social behavior: Its elementary forms.* New York: Harcourt Brace.

Hunt, Janet G. 1996. Elephants in the living room: Why we can't have a coherent national debate on family policy. Paper presented at the 58th Annual Conference of the National Council on Family Relations, Kansas City.

Jensen, Gerald, and Larry Jensen. 1997. *Social policies to strengthen the family.* Denver: Jenex Corporation.

Jones, Charles L., Lorne Tepperman, and Susannah J. Wilson. 1995. *The futures of the family.* Englewood Cliffs, N.J.: Prentice-Hall.

Kantor, D., and W. Lehr. 1975. *Toward a theory of family process.* San Francisco: Jossey-Boss.

Kiecolt, K. J., and M. A. Fossett. 1997. Mate availability and marriage among African Americans. *African American Research Perspectives* 3:12–18.

Kinsey, A. C., W. B. Pomeroy, and C. E. Martin. 1948. *Sexual behavior in the human male.* Philadelphia: Saunders.

Knox, D., and M. E. Zusman. 1998. Unpublished data collected for this text.

Kunz, Jenifer. 1995. The impact of divorce on children's intellectual functioning: A meta-analysis. *Family Perspective* 29: 75–101.

Lawrence, Jill. 1996. Gay issue sizzles in the Senate. *USA Today,* 15 July, 4a.

Lever, J., D. E. Kanouse, W. H. Rogers, S. Carson, and R. Hertz. 1992. Behavior patterns and sexual identity of bisexual males. *Journal of Sex Research* 29: 141–67.

Levine, Robert, Suguru Sato, Tsukasa Hashimoto, and Jvoti Verma. 1995. Love and marriage in eleven cultures. *Journal of Cross-Cultural Psychology* 26: 554–71.

Martin, J. A. 1995. Birth characteristics for Asian or Pacific Islander subgroups. Monthly vital statistics report 43, no. 10. Hyattsville, Md.: National Center for Health Statistics.

Matsuda, Y., and S. Harsel. 1997. Factors influencing social distance from ethnic minorities, homosexuals, and the aged. *Australian Journal of Social Research* 3: 37–56.

McLanahan, Sara S. 1991. The long term effects of family dissolution. In *When families fail: The social costs,* edited by Bryce J. Christensen. New York: University Press of America for the Rockford Institute, 5–26.

McLanahan, Sara, and Karen Booth. 1989. Mother-only families: Problems, prospects and politics. *Journal of Marriage and the Family* 51: 557–80.

Mead, G. H. 1934. *Mind, self and society.* Chicago: University of Chicago Press.

Michael, Robert T., John H. Gagnon, Edward O. Laumann, and Gina Kolata. 1994. *Sex in America: A definitive survey.* Boston: Little, Brown.

Minuchin, S. 1974. *Families and family therapy.* Cambridge, Mass.: Harvard University Press.

National Center for Health Statistics. 1998. Births, marriages, divorces, and deaths for June 1998. Monthly vital statistics report 46, no. 6. Hyattsville, Md.: National Center for Health Statistics, 28 January.

Perry-Jenkins, Maureen, and Karen Folk. 1994. Class, couples, and conflict: Effects of the division of labor on assessments of marriage in dual-earner families. *Journal of Marriage and the Family* 56: 65–180.

Popenoe, David. 1996. Modern marriage: Revising the cultural script. In *Promises to keep: Decline and renewal of marriage in America,* edited by D. Popenoe, J. B. Elshtain, and D. Blankenhorn. Lanham, Md.: Rowman & Littlefield, 247–70.

Rees, Fred L., Toni S. Zimmerman, and Brooke Jacobsen. 1995. Personal authority and its relationship to marital satisfaction. *Journal of Couples Therapy* 5: 117–39.

Regan, Pamela C., and Susan Sprecher. 1995. Gender differences in the value of contributions to intimate relationships: Egalitarian relationships are not always perceived to be equitable. *Sex Roles* 33: 221–38.

Resnick, Michael D., Peter S. Bearman, Robert W. Blum, Karl E. Bauman, Kathleen M. Harris, Jo Jones, Joyce Tabor, Trish Beuhring, Renee E. Sieving, Marcia Shew, Marjorie Ireland, Linda H. Bearinger, and Richard Udry. 1997. Protecting adolescents from harm: Findings from the National Longitudinal Study on Adolescent Health. *Journal of the American Medical Association* 278: 823–32.

Risman, Barbara J., and Myra Marx Ferree. 1995. Making gender invisible. *American Sociological Review* 60: 775–81.

Rodriquez, M., R. Young, S. Renfro, M. Asencio, and D. W. Haffner. 1997. Teaching our teachers to teach: A study on preparation for sexuality education and HIV/AIDS prevention. *Journal of Psychology and Human Sexuality* 9: 121–141.

Rogers, R. G. 1995. Marriage, sex, and mortality. *Journal of Marriage and the Family* 57: 515–526.

Rogers, S. J., and P. R. Amato. 1997. Is marital quality declining? The evidence from two generations. *Social Forces* 75, 3: 1089–1100.

Ross, C. E., J. Mirowsky, and K. Goldsteen. 1990. The impact of the family on health. *Journal of Marriage and the Family* 52: 1059–78.

Saluter, Arlene F. *Marital status and living arrangements: March 1994.* 1996. U.S. Bureau of the Census, Current population reports, Series P 20-484. U.S. Government Printing Office, Washington, D.C.

Scanzoni, J., and W. Marsiglio. 1993. New action theory and contemporary families. *Journal of Family Issues* 14: 105–32.

Statistical Abstract of the United States: 1997, 117th ed. Washington, D.C.: U.S. Bureau of the Census.

Strassberg, Donald S., and Kristi Lowe. 1995. Volunteer bias in sexuality research. *Archives of Sexual Behavior* 24: 369–82.

Tamura, T., and A. Lau. 1992. Connectedness versus separateness: Applicability of family therapy to Japanese families. *Family Process* 31: 319–40.

Thomas, W. I., and F. Zanecki. 1958. *The Polish peasant in Europe and America.* Vols. 1 and 2. New York: Dover.

Thornton, Arland. 1996. Comparative and historical perspectives on marriage, divorce, and family life. In *Promises to keep: Decline and renewal of marriage in America,* edited by D. Popenoe, J. B. Elshtain, and D. Blankenhorn. Lanham, Md.: Rowman & Littlefield, 69–87.

U.S. Bureau of the Census. 1995. *Population Profile of the U.S. 1995*. Washington, D.C.: U.S. Government Printing Office.

Waite, L. J. 1995. Does marriage matter? *Demography* 32: 483–507.

Weinberg, Martin, Colin Williams, and Douglas Pryor. 1994. *Dual attraction: Understanding bisexuality*. Oxford University Press.

Winton, Chester A. 1995. *Frameworks for studying families*. Guilford, Conn.: Dushkin Publishing Group.

Yee, B. W. K. 1992. Gender and family issues in minority groups. In *Cultural diversity and families*, edited by K. G. Arms, J. K. Davidson, Sr., and N. B. Moore. Dubuque, Iowa: Brown and Benchmark, 5–10.

CHAPTER 2 Gender

I hope the twenty-first century sees an end to the nature-nurture argument. . . . We need to move forward and investigate how nature and culture interact.

Helene Fisher, Anthropologist

*No value has gotten modern
man into more trouble than
that of "being a man."*

Ken Druck

The Secrets Men Keep

When a video segment that depicted U.S. Marines in Camp LeJune, North Carolina, pounding spiked "wings" into the chests of paratrooper cadets was aired on national television, the American public watched with horror. The scenes in the video were so graphic and violent that some television viewers, disgusted and sickened by what they saw, turned their heads or closed their eyes. Americans asked why any self-respecting individual would allow himself to be subjected to such a painful and brutal ritual. They also asked why any sane individual would inflict such unnecessary pain on another person. Defending the practice, James Perry, a U.S. Marine Corps First Lieutenant, said that the "blood wings" are symbols of masochism reflecting membership in an elite unit that has to go out and do dangerous jobs. He felt such a ritual was important and added that he couldn't even remember the puncture he received when he got his "wings" (Associated Press, 1997). Although public reaction to the televised video included shock and disbelief, another sentiment was "well, boys will be boys." The wing-pounding ritual is consistent with traditional male gender role socialization, which teaches men to be tough and stoic.

The influence of gender and gender roles is evident not only in dramatic examples, such as the wing-pounding ritual, but in the everyday lives of women and men. In this chapter we focus on how gender influences one's development and relationships. We begin by defining basic terms.

Terminology

In common usage the terms *sex* and *gender* are often interchangeable. But to sociologists, family/consumer science educators, human development specialists, and health educators, these terms are not synonymous. After clarifying the distinction between sex and gender, we discuss other relevant terminology, including gender identity, gender role, and gender role ideology.

Sex

Sex refers to the biological distinction between being female and being male. The primary sex characteristics that differentiate women and men include external genitalia (vulva and penis), gonads (ovaries and testes), sex chromosomes (XX and XY), and hormones (estrogen, progesterone, and testosterone). Secondary sex characteristics include the larger breasts of women and the deeper voice and growth of facial hair in men.

Gender

Gender refers to the social and psychological characteristics associated with being female or male. For example, characteristics typically associated with the female gender include being gentle, emotional, and cooperative; characteristics typically associated with the male gender include being aggressive, rational, and competitive. In popular usage, gender is dichotomized as an either/or concept (feminine or masculine). But gender may also be viewed as existing along a continuum of femininity and masculinity.

RECENT RESEARCH

Cole et al. (1997) studied 435 trans-sexuals (318 male to female; 117 female to male) to assess the presence of major psychopathology. They found that "gender dysphoric individuals appear to be relatively 'normal' in terms of an absence of diagnosable, comorbid psychiatric problems" (p. 21).

RECENT RESEARCH

Bullough and Bullough (1997) surveyed 372 cross-dressers: 67 percent identified themselves as heterosexual, 11 percent as bisexual, and 2 percent as homosexual.

I don't think boys in general watch the emotional world of relationships as closely as girls do. Girls track that world all day long, like watching the weather.

Carol Gilligan

Gender researcher

Is this child a boy or a girl and what are the implications for the child's future life?

Traditional gender expectations are changing. Almost 2,000 (1,842) undergraduate women at a large metropolitan university said that their ideal man had both compassion and intellect. Over 1,000 (1,148) undergraduate men at the same university identified compassion, intellect, sexuality, and power as the most important preferences for their ideal woman (Street, Kimmel, & Kromrey, 1995).

There is an ongoing controversy about whether gender differences are innate as opposed to learned or socially determined. Just as sexual orientation (discussed in Chapter 1, Marriage and the Family: An Overview) may be best explained as an interaction of biological and social/psychological variables, gender differences may also be explained as a consequence of both biological and social/psychological influences. However, social scientists tend to emphasize the role of social influences in gender differences.

Gender Identity

Gender identity is the psychological state of viewing one's self as a girl or boy and later as a woman or man. Some individuals experience *gender dysphoria*, a condition in which one's gender identity does not match one's biological sex. These individuals, known as *transsexuals,* have the genetic and anatomical characteristics of one sex but the self-concept of the other. "I am a woman trapped in a man's body" (or the reverse) reflects the feeling of being a transsexual. Transsexuals are to be distinguished from transvestites. *Transvestites,* also known as cross-dressers, dress in the clothing of the other sex and may experience erotic stimulation from wearing such clothing, but they are not interested in sex-reassignment surgery.

Gender Role

Gender roles are the social norms that dictate what is socially regarded as appropriate female and male behavior. All societies have expectations of how boys and girls, men and women "should" behave. Gender roles influence women and men in virtually every sphere of life, including family and occupation. For example, traditional gender roles have influenced women to be housekeepers. In a random phone survey of 947 individuals in relationships, women reported doing more housework than men (Stanley & Markman, 1997). Women have also been socialized to enter "female" occupations such as elementary school teacher, day-care worker, and nurse. Traditional gender roles have influenced men to be the primary breadwinners and to enter "male" occupations such as engineering, construction, and mechanical repair work. The concentration of women in certain occupations and men in other occupations is referred to as *occupational sex segregation.*

Gender Role Ideology

Gender role ideology refers to beliefs about the proper role relationships between women and men in any given society.

All human societies consist of men and women who must interact with one another, usually on a daily basis, and who have developed customs embracing prescriptive beliefs about the manner in which men and women are to relate to one another (Williams & Best, 1990a, 87).

Traditional American gender role ideology has perpetuated and reflected male dominance and male bias in almost every sphere of life. Even our language reflects this male bias. For example, the words *man* and *mankind* have traditionally been used to refer to all humans. There has been a growing trend away from using male-biased language. Dictionaries have begun to replace "chairman" with "chairperson" and "mankind" with "humankind."

While traditional heterosexual relationships have reflected male dominance, lesbian and gay male relationships tend to be more equal, with greater gender role flexibility. When gender role ideology in homosexual relationships is assessed, lesbian relationships tend to be more egalitarian and flexible than gay male relationships (Green, Bettinger, & Zacks 1996).

Transgendered

Transgendered is a generic term that refers to a broad spectrum of individuals who express characteristics other than those of their assigned gender. Transgendered individuals may be cross-dressers who enjoy dressing in the clothes of the other gender or transsexuals who feel that they are so like the other sex that they want to have their genitals altered to match that sex. A *transgenderist* is an individual who lives in a gender role that does not match his or her biological sex, but has no desire to surgically alter his or her genitalia. The example is a genetic male who prefers to live as a female but does not want surgery.

Theories of Gender Role Development

Various theories attempt to explain why women and men exhibit different characteristics and behaviors.

Sociobiology

Sociobiology emphasizes that there are biological explanations for social behavior. Sociobiological explanations of gender roles acknowledge biological differences between women and men and suggest that these differences account for the different roles. Women and men have different sex chromosomes: women have the XX chromosome; men the XY. Though each sex has hormones of the other, women have higher levels of estrogen and progesterone; men have higher levels of testosterone. These chromosomes and hormones produce such markedly different physical characteristics as breasts in women and beards in men.

Sociobiologists emphasize that the different genetic and hormonal make-ups of women and men account for differences in their social behavior. Consider the following differences in sexual behavior of men and women:

INSIGHT

The view that at least some gender role
behaviors are biologically based is
deeply ingrained in our culture. For example, we tend to think of mothers as
naturally equipped and inclined to perform the role of primary child caregiver. Chodorow (1978) argued that
the role of child caregiver has been assigned to women, although there is no
biological reason why fathers cannot
be the primary caregivers or participate
in child-rearing activities.

- In a national sample (3,432) of adults in the United States, 17 percent of men, compared with 3 percent of women, reported having had more than twenty-one sexual partners since age 18 (Michael et al., 1994).

- Men are also more likely to engage in casual sex. Twenty-five percent of men, compared with 48 percent of women, reported that their first intercourse experience involved "affection for the partner" (Michael et al., 1994).

- Most acts of sexual aggression (e.g., rape and sexual harassment) are perpetrated by men (see Chapter 10).

Sociobiologists have attempted to explain these differences in sexual behavior on the basis of the different hormonal makeups of women and men. For example, the hormone testosterone has been associated with male aggressive behavior. Before puberty, male and female testosterone levels are about the same. At maturity, these levels have increased by a factor of ten or twenty in males, but they only double in females (Udry, Talbert, & Morris, 1986). Progesterone has been associated with female nurturing behavior. When female rats are given large doses of testosterone, they become aggressive; similarly, when male rats are given large doses of progesterone, they become nurturers (Arnold, 1980). The same hormonal reversal changes have been found in monkeys (Goy & McEwen, 1980).

In mate selection, heterosexual men tend to value women who are youthful and physically attractive, while heterosexual women tend to value men with higher incomes. The pattern of men seeking physically attractive young women and women seeking economically ambitious men was observed in thirty-seven groups of women and men in thirty-three different societies (Buss, 1989). This pattern is also evident in courtship patterns in the United States (Davis, 1990). An evolutionary explanation for this pattern argues that men and women have different biological agendas in terms of reproducing and caring for offspring (Symons & Ellis, 1989; Symons, 1987).

The term *parental investment* refers to any investment by a parent that increases the offspring's chance of surviving and hence increases reproductive success. Parental investment requires time and energy. Women have a great deal of parental investment in their offspring (nine months' gestation, taking care of dependent offspring) and tend to mate with men who have high status, economic resources, and a willingness "to invest their resources in a given female and her offspring" (Ellis & Symons, 1990, 533). Men, on the other hand, focus on the importance of "health and youth" in their selection of a mate because young, healthy women are more likely to produce healthy offspring (Ellis & Symons, 1990, 534). Men also "have an aversion to invest in relationships with females who are sexually promiscuous," since men want to ensure that the offspring is their own (Grammer, 1989, 149).

The sociobiological explanation for mate selection is extremely controversial. Critics argue that women may show concern for the earning capacity of a potential mate because women have been systematically denied access to similar economic resources, and selecting a mate with these resources is one of their remaining options. In addition, it is argued that U.S. women and men, when selecting a mate, think more about their partners as emotional companions than as future parents of their offspring. Finally, the sociobiological perspective fails to acknowledge the degree to which social and psychological factors influence behavior.

Identification

Freud was one of the first researchers to study gender role development. Freud suggested that children acquire the characteristics and behaviors of their same-sex parent through a process of identification. Boys identify with their fathers; girls identify with their mothers.

In *The Reproduction of Mothering*, Nancy Chodorow uses Freudian identification theory as a basis for her theory that gender role specialization occurs in the family because of the "asymmetrical organization of parenting" (1978, 49).

> Women, as mothers, produce daughters with mothering capacities and the desire to mother. These capacities and needs are built into and grow out of the mother-daughter relationship itself. By contrast, women as mothers (and men as not-mothers) produce sons whose nurturant capacities and needs have been systematically curtailed and repressed (Chodorow, 1978, 7).

In other words, all activities associated with nurturing and child care are identified as female activities because women are the primary caregivers of young children. This one-sidedness (or asymmetry) of nurturing by women increases the likelihood that females, because they identify with their mothers, will see their own primary identity and role as mother.

Social Learning

As evidence accumulates that social processes may be critical for genetic expression, all notions of biological determinism should be laid to their eternal rest.

David Reiss

Psychiatrist

Derived from the school of behavioral psychology, social learning theory emphasizes the roles of reward and punishment in explaining how a child learns gender role behavior. For example, two young brothers enjoyed playing "lady." Each of them would put on a dress, wear high-heeled shoes, and carry a pocketbook. Their father came home early one day and angrily demanded that they "take those clothes off and never put them on again. Those things are for women," he said. The boys were punished for playing "lady" but rewarded with their father's approval for playing "cowboys," with plastic guns and "Bang! You're dead!" dialogue.

Reward and punishment alone are not sufficient to account for the way in which children learn gender roles. Direct instruction ("girls wear dresses," "a man walks on the outside when walking with a woman") by parents or peers is another way children learn. In addition, many of society's gender rules are learned through modeling. In modeling, the child observes another's behavior and imitates that behavior. Gender role models include parents, peers, siblings, and characters portrayed in the media.

The impact of modeling on the development of gender role behavior is controversial. For example, a modeling perspective implies that children will tend to imitate the parent of the same sex, but children in all cultures are usually reared mainly by women. Yet this persistent female model does not seem to interfere with the male's development of the behavior that is considered appropriate for his gender. One explanation suggests that boys learn early that our society generally grants boys and men more status and privileges than girls and women. Therefore, boys devalue the feminine and emphasize the masculine aspects of themselves.

OTHER CULTURES The fact that gender roles differ across societies provides evidence that gender roles are learned. Mead (1935) visited three New Guinea tribes in the early 1930s and observed that the Arapesh socialized both men and women to be feminine by Western standards. The Arapesh person was taught to be cooperative and responsive to the needs of others. In contrast, the Tchambuli were known for dominant women and submissive men—just the opposite of our society. And both of these societies were unlike the Mundugumor, which socialized only ruthless, aggressive, "masculine" personalities. The inescapable conclusion of this cross-cultural study is that human beings are products of their social and cultural environment and that gender roles are learned. ●

Male and female personalities are socially produced.
Margaret Mead
Anthropologist

OTHER CULTURES Modern-day cross-cultural examples of gender roles include subservient Hindu and Muslim women. In the presence of others, a wife must not speak to her husband or stare at him. At meals, a woman eats only after the men have been served. A wife walking with her husband is expected to follow a few steps behind him. Though we tend to think of the United States as having very liberal gender roles, it was only forty years ago that women were "refused the right to serve on juries, sign contracts, take out credit cards in their own names, or establish legal residence" (Coontz, 1995, 13). ●

In Islam, the most male-oriented of the modern religions, a woman is nothing but a vehicle for producing sons.
Joseph Campbell
Anthropologist

Cognitive Developmental Theory

The cognitive-developmental theory of gender role development reflects a blend of biological and social learning views. According to this theory, the biological readiness, in terms of cognitive development, of the child influences how the child responds to gender cues in the environment (Kohlberg, 1966, 1976). For example, gender discrimination (the ability to identify social and psychological characteristics associated with being female or male) begins at about age 30 months. At that age, toddlers are able to assign a "boy's toy" to a boy and a "girl's toy" to a girl (Etaugh & Duits, 1990). However, at this age, children do not view gender as a permanent characteristic. Thus, while young children may define people who wear long hair as girls and those who never wear dresses as boys, they also believe they can change their gender by altering their hair or changing clothes.

Not until age six or seven does the child view gender as permanent (Kohlberg, 1966, 1969). In Kohlberg's view, this cognitive understanding involves the development of a specific mental ability to grasp the idea that certain basic characteristics of people do not change. Once children learn the concept of gender permanence, they seek to become competent and proper members of their gender group. For example, a child standing on the edge of a school playground may observe one group of children jumping rope while another group is playing football. That child's gender identity as either a girl or a boy connects with the observed gender-typed behavior, and the child joins one of the two groups. Once in the group, the child seeks to develop the behaviors that are socially defined as appropriate for her or his gender.

Agents of Gender Role Socialization

Three of the four theories discussed in the preceding section emphasize that gender roles are learned through interaction with the environment. Indeed, while biology may provide a basis for some roles (being 7'5" is helpful for a basketball player), cultural influences in the form of various socialization agents (parents, peers, teachers, religion, and the media) shape the individual toward various gender roles. These powerful influences in large part dictate what a person thinks, feels, and does in his or her role as a woman or as a man.

Parents

Parents are one of the most important influences in a child's life. This influence usually varies by whether the parent is the mother or the father. In a nationally representative sample of about 2,000 children and parents (one parent for each child), Starrels (1994) found that fathers spend more time with sons than with daughters and provide more instrumental attention (such as giving gifts or money) to their children than do mothers. In contrast, mothers provide more

RECENT RESEARCH

Whether one's gender role socialization is traditional or egalitarian may be influenced by the type of family structure in which one is reared. Leve and Fagot (1997) compared 67 two-parent households, 32 single-mother households, and 13 single-father households and found that single-parent households and mothers provided less traditional gender-role socialization than two-parent households and fathers.

What is this child learning about gender role behavior?

affective nurturance in the form of love/interest, affection, and verbal praise to daughters and sons equally.

Both parents tend to assign different chores to their children depending on whether the child is a boy or a girl (McHale et al., 1990). Boys tend to be assigned maintenance chores, such as mowing lawns, while girls tend to be assigned domestic tasks, such as cooking, laundry, and taking care of siblings. Such differential chore assignment may encourage the development of different personal qualities. For example, girls required to take care of their younger siblings may be more likely to develop nurturing behaviors than boys who are required to cut the grass (Basow, 1992). Parents might be aware that the chores they assign to their children will affect what they learn.

Peers

Though parents are usually the first socializing agents that influence a child's gender role development, peers become increasingly important during the school years. The gender role messages from adolescent peers are primarily traditional.

> For adolescent boys, such traits include being tough (through body building or athletic achievement), being cool (not showing emotions, not fearful of danger, staying reasonable under stress), being interested in girls and sex, being good at something, being physically attractive and having an absence of any trait or characteristic that is female or feminine (Harrison & Pennell, 1989, 32).

NATIONAL DATA   Twenty percent of a random sample of all first year undergraduate women and 32 percent of a random sample of all first-year undergraduate men in 464 colleges and universities in the United States agreed that "the activities of married women are best confined to the home and family" (American Council on Education and University of California, 1997).

Female adolescents are under tremendous pressure to be physically attractive (thin), popular, and achievement-oriented. The latter may be traditional (cheerleading) or nontraditional (sports or academics). Adolescent females are sometimes in conflict because high academic success may be viewed as being less than feminine.

Teachers and Educational Materials

Although parents have the earliest influence and adolescent peers have increasing influence during the teen years, teachers are another important socialization influence. Research suggests that

> elementary and secondary teachers give far more active teaching attention to boys than girls. They talk to boys more, ask them more lower- and higher-order questions, listen to them more, counsel them more, give them more extended directions, and criticize and reward them more frequently.

This pattern of more active teacher attention directed at male students continues at the post-secondary level. . . . In general, women are rarely called on; when female students do participate, their comments are more likely to be interrupted and less likely to be accepted or rewarded (Sadker & Sadker, 1990, 177).

Gender role stereotypes are also conveyed through educational materials. In a study of 1,883 stories used in schools, Purcell and Stewart (1990) found that males were more often presented as clever, brave, adventurous, and income-producing, while females were more often presented as passive and as victims. Females were also more likely to be in need of rescue and were depicted in fewer occupational roles than were males.

Children's books more often show women holding domestic household artifacts (skillet), while men are more often shown holding nondomestic production products (wrench) (Crabb & Bielawski, 1994). Gender stereotyping also occurs in textbooks at the college level. Ferree and Hall (1990) found that women were underrepresented in sociology textbooks in terms of their participation in American society and their contributions to sociological theory and research.

Finally, the models children are exposed to in the educational institution reflect an unequal ratio of men to women in power positions. At the elementary and secondary levels, men are more often principals and administrators. At the university level, men are more often presidents, deans, and departmental chairs.

Religion

Traditional and conservative interpretations of the Bible reflect the patriarchal nature of family roles. Basow found that "to the extent that a child has any religious instruction, he or she receives further training in the gender stereotypes" (1992, 156).

> Wives be subject to your husband, as to the Lord. . . . Husbands, love your wives, even as Christ also loved the church (Ephesians 5:22–25). (See also Colossians 3:18-19.)

While the Bible has been interpreted in both sexist and nonsexist terms, male dominance is indisputable in the hierarchy of religious organizations, where power and status have been accorded mostly to men. Basow (1992) noted that the Catholic church does not have female clergy, and men dominate the nineteen top positions in the U.S. dioceses.

Male bias is also reflected in terminology used to refer to God in Jewish, Christian, and Islamic religions. For example, God is traditionally referred to as "He," "Father," "Lord," and "King." Two researchers observed that individuals who attend religious services frequently are more likely to have traditional gender role ideologies than individuals who do not attend church frequently (Willetts-Bloom & Nock, 1994).

Media

In reference to media, the phrase "you are what you eat" might be replaced with "you are what you see and read" (Doyle, 1995, 96). Media such as movies,

---

The Christian Right undermines the potential for a more mature husband-wife relationship by consistently . . . suggesting that women are both different and inferior.
Kenneth Chafin
Theologian

INSIGHT

In response to an interest in removing sexist language from religious works, some hymns are being rewritten. The 1865 hymn "Rejoice, You Pure in Heart" no longer speaks of "strong men and maidens meek" but speaks of "strong souls and spirits meek." And "The Father, Son, and Holy Ghost" has been changed to "Praise God the Spirit, Holy Fire."

What Melrose, the soaps, and talk shows fail to address are the political, economic, cultural, and social structures that limit women's roles to ensure their continued dependence (Baxter & Kane, 1995). The real issues of limited female representation in Congress and sexism in employment and wages are avoided. Rather, the media exist to sell products, not to press political agendas. And men, who control media content, would not allow it anyway. As Rapping asserts, "we are not allowed to rock the political or economic boat of television by suggesting that things could be different. That would rightly upset the sponsors and network heads" (1994, 196).

Men are also victims of abusive stereotyping that has implications for their family life. Men, particularly fathers, have been presented by the media as "buffoons whose only contribution was to bring home a paycheck or hand out weekly allowances. Society views fathers, then, as simply the providers" (Doyle, 1995, 13). Such a view suggests that men are not emotional, nurturing caretakers for their children, which sometimes influences judges in custody disputes to give mothers custody and give fathers limited visitation rights.

television, magazines, newspapers, books, and music both reflect and shape gender roles. Gender role relationships in movies reflect male dominance (Hedley, 1994). Media images of women and men typically conform to traditional gender stereotypes, and media portrayals depicting the exploitation, victimization, and sexual objectification of women are common. Women also have fewer movie roles than men. For example:

- In Hollywood films, women play only 34 percent of the roles and earn 33 percent less than male counterparts for comparable industry jobs (Rapping, 1994).

- On television, such programs as *Melrose Place* and *Beverly Hills 90210* are known for their "excessive attention to physical beauty and fashion. Everyone is gorgeous, in perfect shape, and up to the minute on what's pictured in this month's *Elle* and *GQ*" (Rapping, 1994, 159). These shows, and others like them, promise that "good looks will buy more than is realistic, because there is no other optimistic message to offer" (p. 160).

    Among the daytime soaps, "all the men are doctors and lawyers, with an occasional police chief or millionaire thrown in for plot variety. And the women are wives and mothers" (p. 177).

    Although Oprah, Montel Williams, and Geraldo discuss an array of women's troubles from bulimia, to drug abuse, to prostitution, the programs are presented as "frivolous and trashy" and "are not taken seriously by those in power" (p. 194).

The cumulative effect of family, peers, education, religion, and mass media is to perpetuate gender stereotypes. Each agent of socialization reinforces gender roles that are learned from other agents of socialization, thereby creating a gender role system that is deeply embedded in our culture.

Consequences of Traditional Gender Role Socialization

This section discusses different consequences, both negative and positive, for women and men, of traditional female or male socialization in the United States.

Consequences of Traditional Female Role Socialization

Table 2.1 summarizes some of the negative and positive consequences of being socialized as a woman in our society.

Each consequence may or may not be true for a specific woman. For example, while women in general have less education and income, a particular woman may have more education and a higher income than a particular man.

Negative consequences

Less Education/Income Ross and Van Willigen (1997) emphasized the value of education to one's subjective quality of life—lower levels of depression, anxiety,

Women tend to expect disadvantages in education and in their careers regardless of their ethnic heritage. McWhirter (1997) examined a sample of 1,139 Mexican-American and Euro-American high school juniors and seniors and found that female students, regardless of ethnic background, anticipated more educational and career barriers than male students.

I need to cut back on my work schedule and take a very aggressive approach to having a baby.

Connie Chung
TV journalist

Women who elect to have children experience an income penalty. Waldfogel (1997) analyzed national data on mothers and other women and found a 4 percent "family penalty" for one child and a 12 percent penalty for two or more children in terms of lower wages. Whether this lower income was due to discrimination from employers or employee adjustments (e.g., changing jobs after childbirth) is not clear.

This woman's place is in the House—the House of Representatives.

Bella Abzug
Representative

Table 2.1

Consequences of Traditional Female Role Socialization	
Negative Consequences	**Positive Consequences**
Less education/income (more dependent)	Longer life expectancy
High HIV infection risk	Closer mother-child bond
Negative self-concept	Greater emotionality
Value defined by youthfulness and beauty	Identity not tied to job
Less marital satisfaction; no "wife" at home	Greater relationship focus

and anger and lower levels of physical problems. Women earn fewer advanced degrees beyond a master's degree than do men. For the year 1996, women earned 40 percent of the Ph.D.s (Henderson and Woods, in press).

The strongest explanation for why women earn fewer advanced degrees than men is that women are socialized to value motherhood over career preparation. Two researchers compared the pronatalist attitudes of a sample of adult men and women over the age of 18 (n = 13,017) and found that women associate greater rewards with parenthood than do men (Seccombe & Warner, 1994). When 821 undergraduate women were asked to identify their lifestyle preference, less than 1 percent selected being unmarried and working full time. In contrast, 53 percent selected "graduation, full-time work, marriage, children, stop working at least until youngest child is in school, then pursue a full-time job" as their preferred lifestyle sequence (Schroeder et al., 1993, 243). Only 6 percent of 535 undergraduate men selected this same pattern. This lack of career priority on the part of women influences the lack of priority given to education to prepare for a career.

Less education is associated with lower income. Women still earn about two-thirds of what men earn, even when the level of educational achievement is identical.

NATIONAL DATA Women's and Men's Median Income with Similar Education

	Bachelor's Degree	Master's Degree	Doctorate Degree
Men	$39,040	$49,076	$57,356
Women	$24,065	$33,509	$39,821

Source: *Statistical Abstract of the United States: 1997.* 117th ed. Washington, D.C.: U.S. Bureau of the Census. Table 732.

OTHER CULTURES Economic inequality is not unique to the United States but also exists in Canada, Australia, Norway, and Sweden (Baxter & Kane, 1995). Women earn about 75 percent of what men earn in Australia, Norway, and Sweden. In China, women earn 60 percent of what men earn (Cox, 1995).●

Another factor that contributes to both the educational and salary differentials of women and men is occupational sex segregation. Many workers are em-

INSIGHT

Women who do not pursue higher levels of education or do not pursue higher-paying occupations may find themselves economically dependent on their partners or spouses. By choosing to pursue advanced education and well-paying occupations, women create more interpersonal choices and will not be pressured to stay in unsatisfying relationships because they are economically dependent.

With a 50 percent chance of divorce for marriages begun in the nineties, and the likelihood of being a widow for seven or more years, a woman without education and employment skills is often left high and dry. As one widowed mother of four said, "The shock of realizing you have children to support and no skills to do it is a worse shock than learning that your husband is dead." In the words of a divorced, 40-year-old mother of three, "If young women think it can't happen to them, they are foolish."

ployed in gender-segregated occupations, that is, occupations in which workers are either primarily male or primarily female. Female-dominated occupations tend to require less education, have lower status, and pay lower salaries than male-dominated occupations. If the role is typically occupied by men, it tends to pay more. For example, the job of child-care attendant requires more education than the job of dog pound attendant. However, dog pound attendants are more likely to be male and earn more than child-care attendants, who are more likely to be female.

High HIV Infection Risk   Since men are the primary sexual partners of women, and since men are at greater risk for HIV infection and other STDs as a result of a higher number of sexual partners, women are more likely to contract HIV from men than men are from women. Women have a greater vulnerability to HIV infection because semen, which may carry HIV, is deposited in the woman's body. Women also report that their partners are less likely to use a condom, and women feel relatively powerless to influence them to do so (Catania et al., 1992).

Negative Self-concept   Some research suggests that women have more negative self-concepts than men. In a study of first-year students at a southeastern university, significantly more women students reported feeling less intellectual and socially self-confident than men students (Smith, 1995). Schvaneveldt et al. (1997) also found women undergraduates reporting lower levels of self-esteem than men undergraduates.

Women in the United States live in a society that devalues them. "Women's natures, lives, and experiences are not taken as seriously, are not valued as much as those of men" (Walker, Martin, & Thomson, 1988, 18). *Sexism* is defined as an attitude, action, or institutional structure that subordinates or discriminates against an individual or group because of one's sex. Sexism against women reflects the tradition of male dominance and presumed male superiority in our society.

OTHER CULTURES   Sexist attitudes exist not only in the United States but also in the former Soviet Union, China, India, Japan, and Latin America (Lindsey, 1990). ●

Not all research demonstrates that women have more negative self-concepts than men. Indeed, young (8 to 10) boys and girls tend to view their own gender more positively than they view the other (Powlishta, 1995). In another study focusing on the self-esteem of 824 college students, the authors found no differences between women and men (Mitchell & Fandt, 1995). And when Williams and Best summarized their research on the self-concepts of women and men in the United States, they noted "no evidence of an appreciable difference" (1990b, 153).

OTHER CULTURES   Williams and Best (1990b) found no consistency in the self-concepts of women and men in fourteen different countries. "In some of the countries the men's perceived self was noticeably more favorable than the women's, whereas in others, the reverse was found" (p. 152). ●

RECENT RESEARCH

Dissatisfaction with one's body image is not unique to American women. Tang, Lai, Phil, and Chung (1997) found in their study of 305 Chinese college students that the women were significantly more likely to have negative body images than the men.

RECENT RESEARCH

Two researchers conducted in-depth interviews with 12 newly married dual-career couples and found that all the couples talked about their relationships using a "language of equality" (Knudson-Martin & Mahoney, 1998). However, none of the couples met the researchers' criteria for equal marriages, and most couples in the study avoided conscious confrontation of gender and equality issues in their marriages. These findings suggest that couples may develop a "myth of equality," convincing themselves that their relationship is equal when, in reality, it is not equal.

Negative body image   The murder of JonBenet Ramsey Christmas night of 1996 brought into cultural focus the degree to which even very young women (she was 6) are socialized to emphasize beauty and appearance (there are 3,800 beauty pageants annually). The effect for many women who do not match the cultural ideal is to have a negative body image. Koenig and Wasserman (1995) found that college women were more likely to have a negative body image than men. Women are also more likely than men to have an eating disorder (Seid, 1994) and to report going on a diet. In regard to the latter, 96 percent of college women in a national sample reported dieting, in contrast to 59 percent of college men (Elliott & Brantley, 1997).

Less marital satisfaction   Researchers disagree over whether wives are less happy than husbands. Basow (1992) found that wives reported less marital satisfaction than husbands and attributed this to role overload: women are expected to keep their husbands, children, parents, and employers happy and to be good homemakers (cook, clean, wash dishes/laundry). The results of coping with these unrealistic expectations internalized by or imposed on women are, by contrast with men, more frequent nervous breakdowns, greater psychological anxiety, increased self-blame for not living up to these expectations, and poorer physical health (Bird & Fremont, 1991). Women also are more likely than men to give up their own goals and become selfless as they try to please their husbands and children (Heyn, 1997).

In direct contrast, Aldous and Woodberry (1994) analyzed General Social Survey data covering nineteen years and found that women reported being happier than men. However, the researchers attributed this finding to the tendency of women to give socially desirable responses.

 INSIGHT

If the role overload hypothesis is accurate, more women than men are frustrated. To decrease their frustration, some wives reject the expectation that they be all things to all people and alert their partners to the need for shared housework/child care. Two researchers (Blaisure & Allen, 1995) interviewed one or both partners in twenty-three marriages in which the wife held a feminist ideology. One of the wives stated, "Feminism gives me the strength of demand: Either I get the treatment I think is adequate partnering or I leave" (p. 10).

No "wife" at home   As we will discuss in Chapter 9, Work, Leisure and the Family, both spouses may need a "wife" at home to take care of the house and children to free them to pursue their job or career. Since wives typically do more housework and child care, they have no help ("wife"), as do husbands. One exception is Sally Jessy Raphael, whose husband (Karl Soderlund) has adopted the role of the wife. Of their relationship Karl said, "If either of us is going to be rich, it'll probably be Sally. Therefore, I'm going to take jobs and give them up, and she'll have the career. . . ." (Rader, 1993, 5).

Successful professional career women are more likely to be single than their male counterparts. In her book *The Third Sex: The New Professional Woman*, anthropologist Pat McBroom (1992) notes that 60 percent of top female executives live single lives, versus 5 percent of top male executives.

NATIONAL DATA

Life expectancy for U.S. women born in 2000 is projected to be 79.7; for men, 73.0 (*Statistical Abstract of the United States: 1997*, Table 117).

INSIGHT

Close emotional relationships may provide enormous life satisfaction for both women and men. Personal inter-views with 2,374 adults revealed that satisfaction with family life (rather than work life) is the most important deter-minant of overall life satisfaction (Carl-son & Videka-Sherman, 1990).

Before leaving this section on negative consequences of being socialized as a female we look at the issue of female genital mutilation in this chapter's So-cial Policy feature. This is more of an issue for females born in some African, Middle Eastern, and Asian countries than for women in the United States. But the practice continues even here.

Benefits of traditional female role socialization   We have discussed the negative consequences of being born and socialized as a female. But there are also de-cided benefits. Three of these are the potential to live longer, a stronger rela-tionship focus, and a closer emotional bond with children.

Being embedded in a network of relationships may be related to longevity. Women are more involved in relationships than are men. Two primary groups are their parents and same-sex peers. This pattern occurs worldwide.

OTHER CULTURES   Williams and Best (1990a) collected data from university students in twenty-eight countries (including England, Australia, Nigeria, Japan, and Brazil) and observed that women were more likely than men to be associ-ated with having certain relationship characteristics:

- Succoring—soliciting sympathy, affection, or emotional support from others.
- Nurturance—engaging in behavior that extends material or emotional ben-efits to others.
- Affiliation—seeking and sustaining numerous personal relationships.

Other research supports the idea that women place more importance on rela-tionships (Hammersla & Frease-McMahan, 1990), have more close friends (Jones et al., 1990), and are more cooperative than men (Garza & Borchert, 1990).

In contrast, men were viewed as being more dominant, autonomous, ag-gressive (Williams & Best, 1990a), and competitive (Garza & Borchert, 1990). These qualities are counter to developing close relationships with others.

Another advantage of being socialized as a woman is the potential to have a closer bond with children. In general, women tend to be more emotionally bonded with their children than men. Although the new cultural image of the father is of one who engages emotionally with his children, most fathers con-tinue to be content for their wives to take care of their children, with the result that mothers, not fathers, become more bonded with their children.

OTHER CULTURES   The mother-child bond is particularly strong in African-American, Asian-American, and Hispanic families (Mindel, Habenstein, & Wright, 1998).

Consequences of Traditional Male Role Socialization

Male socialization in our society is associated with its own set of conse-quences. The women's movement has given widespread visibility to the re-strictions imposed upon women by the traditional female role. However, there

Although female genital operations, also called female genital mutilation (FGM), occur primarily in African and some Middle Eastern and Asian countries, the practice also occurs in the United States among immigrant families who bring their cultural traditions with them. Each year, there are about 7,000 immigrants to the United States from countries that practice FGM who undergo the procedure (cutting off the clitoris) in the United States or during a visit to their homelands (Burstyn, 1995).

Prior to 1996, only a handful of states had laws banning FGM. In the fall of 1996 Congress passed a federal law banning the practice of genital cutting of females under age 18 in the United States. This law also directs federal authorities to inform new immigrants from countries where FGM is practiced that parents who arrange for the genital cutting of their female children in the United States (as well as people who perform the cutting) face up to five years in prison (Dugger, 1996). France, the United Kingdom, Sweden, and the Netherlands have already outlawed the practice; in France over thirty immigrant families have been prosecuted for violating the ban. Such cases are usually reported to the police by doctors, who detect the practice while examining girls.

Some U.S. health care providers fear that the ban on FGM will not eliminate the practice but instead will cause African parents to withhold medical care for their daughters to avoid detection and prosecution. After African mothers repeatedly requested that their daughters have genital operations in the hospital, physicians at a Seattle hospital proposed offering a largely symbolic form of the ritual in which they would nick the tip of a girl's clitoris, with her consent, under a local anesthetic. No tissue would be removed. It is not clear whether this procedure would violate the U.S. law against female genital mutilation (Dugger, 1996).

Another policy issue concerning FGM involves giving asylum to refugee women who come to the United States to avoid genital cutting. In 1994, 17-year-old Fauziya Kasinga fled from her home in Togo to escape having her clitoris cut off. After escaping to the United States, Kasinga turned herself in to immigration officials, asking for asylum. To win asylum, refugees must show that because of their race, nationality, religion, politics, or membership in a particular social group, they either have been persecuted or

have a "well-founded fear" of persecution. Kasinga's claim for asylum was denied by a judge with the Executive Office for Immigration Review. While awaiting an appeal of the decision, she was held in prison for more than a year, part of it in a maximum security wing with serious criminals. When asked what she would do if her appeal was denied, Kasinga replied, "I'd prefer to stay in jail rather than go back and face what would be done" (McCarthy, 1996). In the summer of 1996, the Board of Immigration Appeals granted political asylum to Kasinga, recognizing FGM as a form of persecution against women. This ruling sets a binding precedent for all U.S. immigration judges. Some U.S. citizens are worried, that in the current anti-immigrant climate, this precedent may result in the influx of millions of women into the United States as refugees.

Finally, some U.S. policies attempt to discourage FGM in the countries that practice this ritual. For example, the federal ban on FGM requires U.S. representatives to the World Bank and other international financial institutions to oppose loans to countries that have not carried out educational programs to prevent it. However, Western government efforts to eradicate FGM in Africa or Asia are often perceived by members of these societies as a continuation of the racism, discrimination, and cultural imperialism that these countries have endured historically (Lane & Rubinstein, 1996).

Changing a country's deeply held beliefs and values concerning this practice cannot be achieved by denigration. If Western countries continue to denigrate those who practice female genital cutting, we may be creating "a backlash in which the custom is viewed as intrinsic to the group's threatened identity" (Lane & Rubinstein, 1996, 38). More effective approaches to discouraging the practice include the following (Lane & Rubinstein, 1996):

1. Respect the beliefs and values of countries that practice female genital cutting. Calling the practice "genital mutilation" and referring to it as a form of "child abuse" and "torture" conveys disregard for the beliefs and values of the cultures where it is practiced.

2. Use less inflammatory or derogatory language to describe the practice. Rather than call it "genital mutilation," we might call it "female circumci-

sion," "female genital cutting," "female genital operations," or "female genital surgeries."

3. Remember that the practice is "arranged and paid for by loving parents who deeply believe that the surgeries are for their daughters' welfare. Parents fear . . . that leaving their daughters uncircumcised will make them unmarriageable. Parents worry about their daughters during the procedures and care for their wounds afterward to help them recover. . . . Parents who do this are not monsters, but are ordinary, decent, caring persons" (p. 38).

4. Finally, it is important to recognize and support the efforts of individuals and groups in the countries that practice FGM who are trying to abolish this practice through education and policy change.

REFERENCES

Burstyn, Linda. 1995. Female circumcision comes to America. *Atlantic Monthly*, October, 28–35.

Dugger, Celia W. 1996. New law bans genital cutting in the United States. *New York Times*, 12 October, 1, 28.

Lane, Sandra D., and Robert A. Rubinstein. 1996. Judging the other: Responding to traditional female genital surgeries. *Hastings Center Report* 26, no. 3, 31–40.

McCarthy, Sheryl. 1996. Fleeing mutilation, fighting for asylum. *Ms. Magazine*, July/August, 12–16.

I love Mickey Mouse more than any woman I've ever known.
Walt Disney
(1901–1966)

is a growing recognition that the traditional male gender role is also restrictive. Sociologist Edwin Schur stated:

> There is no denying that the gender system controls men too. Unquestionably, men are limited and restricted through narrow definitions of "masculinity." . . . They too face negative sanctions when they violate gender prescriptions. There is little value in debating which sex suffers or loses more through this kind of control; it is apparent that both do (1984, 12).

The following subsections discuss some of the potential negative and positive consequences of being socialized as a male. These are summarized in Table 2.2. As with women, each consequence may or may not be true for a specific man.

Anything that gets in the way of work, personal or otherwise, I have to let go of it.
Martin Scorsese
Film director

Negative consequences *Identity Synonymous with Occupation* Ask men who they are, and many will tell you what they do. Society tends to equate a man's identity with his occupational role. Male socialization toward greater involvement in

Table 2.2

Consequences of Traditional Male Role Socialization	
Negative Consequences	**Positive Consequences**
Identity tied to work role/income	Higher occupational status
Limited emotional expression	More positive self-concept
Less time with children	Less job discrimination
Custody disadvantages	Higher income
Shorter life expectancy	Less marital stress

I was drunk with travel . . . indifferent to thoughts of home and family . . . .
Charles Kuralt
A Life on the Road

INSIGHT

A dilemma may await U.S. men. In England, the British labor market has become entrenched with part-time low-pay/low-skilled jobs, making it ever more difficult to find a permanent full-time job. As economies slow and fewer full-time jobs are available, men have a more limited opportunity to fulfill the stereotype of the full-time working male. Yet the definition of masculinity remains tied to the work sector. "This has had a catastrophic impact on men's self-identity" (Goodwin, 1990, 2).

Therapists report that the most common complaint of women in distressed marriages is that their husbands are too withdrawn and don't share openly enough.
Howard Markman
Marriage therapist

INSIGHT

Despite traditional male socialization, some men *do* talk about their feelings and value emotional intimacy. In a study of same-sex friendships, Walker interviewed men who "reported that they spoke intimately about spouses, other family members and their feelings" (1994, 253). For example, one lawyer reported that he and his friends had talked repeatedly about fertility problems they and their wives were having. Seventy-five percent of Walker's respondents reported talking about such feelings.

the labor force is evident in adolescence. Data from 1,481 rural high school students showed that males were more likely to be employed, to begin work earlier, to receive higher pay, and to work longer hours than females (Clifford & Shoffner, 1992).

Work is the principal means by which men confirm their masculinity and success. Black and Mexican-American men, primarily because of racism and discrimination, are more likely to be unemployed than white men. Hence, they may suffer more emasculation because of the connection between identity and occupation. Goodwin (1990), using a national sample of men in the United Kingdom, observed that men who work part time are more than twice as likely as men who work full time to suffer from some form of malaise or mental illness. The danger increases if the man's partner or close family members hold attitudes that are close to the traditional perceptions of masculinity. Men are also more likely than women to value working more hours to make more money. Indeed, 76 percent of all first-year undergraduate male students in contrast to 70 percent of female students said that "making more money" was an "important" reason for going to college (American Council on Education and University of California, 1997).

Limited Expression of Emotions   Some men feel caught between society's expectations that they be competitive, aggressive, and unemotional and their own desire to be more cooperative, passive, and emotional. Not only are men less likely to cry than women, they are also less able to express their feelings of depression, anger, fear, and sadness. According to Derlega et al.:

> Men raised in North American culture (emphasis on white, Protestant, Northern European background) are more likely to believe that task accomplishment is an important goal and that emotional control is one general strategy for facilitating that goal. Women, however, are more likely to believe that social-emotional closeness is an important goal and that emotional expression is one general strategy for facilitating that goal (1993, 45).

Custody Disadvantages   Courts are sometimes biased against divorced men who want custody of their children. Because divorced fathers are typically regarded as career focused and uninvolved in child care, some are relegated to seeing their children on a limited basis, such as every other weekend or four evenings a month. Although longer visitation times are afforded them on holidays and summer vacations, they are sometimes effectively removed from regular, meaningful interaction with their children. Even if divorced fathers see their children a month in the summer, a week at Christmas, a week during spring break, and every other weekend, they still end up being with their children less than 25 percent of the time. Some bitter ex-wives seize this time advantage supported by the courts to turn their children against their fathers. Warren Farrell, a strong advocate for father's rights, said at the 1996 National Congress for Fathers and Children that the right of men to parent their children will be the twenty-first-century equivalent of the twentieth-century rights of women in the workplace.

Shorter Life   Men die about seven years earlier than women. Although biological factors may play a role in greater longevity for women than for men, tradi-

NATIONAL DATA

Black men are the most vulnerable, as they die about 13 years earlier than white women (67 vs. 80) (*Statistical Abstract of the United States: 1997,* Table 118).

tional gender roles play a major part. For example, the traditional male role emphasizes achievement, competition, and suppression of feelings, all of which may produce stress. Not only is stress itself harmful to physical health, but it may also lead to compensatory behaviors, such as smoking, alcohol and other drug abuse, and dangerous risk-taking behavior.

The degree to which males are socialized to be violent and aggressive is another factor in their shorter life expectancy. Race is also a variable, as the rate of being murdered for young black men is ten times higher than that of young white men. In fact, homicide is the leading cause of death of young black men (Press, McCormick, & Wingert 1994). Luckenbill and Doyle (1989) emphasized that the high rate of homicide among young, lower-income males is related to the culturally transmitted willingness to settle disputes, especially those perceived as a threat to their masculinity, by using physical force.

That minority men have poorer health and a shorter life expectancy than white men may also be explained by the stress they experience from racism, prejudice, discrimination, and poverty. The frustration and anger that may result from the social and economic disadvantages of minority status may lead to hazardous compensatory behaviors, such as those previously described.

In sum, the traditional male gender role is hazardous to men's physical health. However, as women have begun to experience many of the same stresses and behaviors as men, their susceptibility to stress-related diseases has increased (Rodin & Ickovics, 1990). For example, since the 1950s, male smoking has declined while female smoking has increased, resulting in an increased incidence of lung cancer in women.

Adapting to New Expectations from the Modern Woman   Men can no longer dictate the role relationships between men and women because of the growing

Bachelors know more about women than married men; if they didn't they'd be married too.
H. L. Menken
Philosopher

This preadolescent male is more likely to display egalitarian role behaviors as an adult.

equality of those relationships (Schroeder et al., 1993). Today's modern heterosexual woman (according to 600 adult women) expects her male partner to be caring and nurturing and open with his thoughts and feelings. He is also expected to be ambitious and to participate in domestic chores (Rubenstein, 1990). Husbands in dual-earner marriages who do not share the workload of house and children are deemed "unfair" by their employed wives, and marital satisfaction drops (Blair, 1993).

Benefits of traditional male socialization   Though men may have a limited identity independent of occupational role, have a limited range of emotional expression, be disadvantaged in custody disputes, die earlier, and have to cope with learning how to be sensitive without being called a wimp, they also have a number of benefits. Men tend to have a more positive self-concept, have greater confidence in themselves, and enjoy higher incomes and occupational status than women. They also have a decided advantage in pursuing corporate careers and rarely are confronted with sexual harassment.

We have been discussing the respective ways in which women and men are affected by traditional gender role socialization. Table 2.3 summarizes twelve implications that traditional gender role socialization has for the relationships of women and men.

In spite of the widespread goal of equality, numerous studies of married couples suggest that few couples actually achieve it.
Carmen Knudson-Martin and
Anne Rankin Mahoney
Gender researchers

Beyond Gender Roles

Imagine a society in which women and men each develop characteristics, lifestyles, and values that are independent of gender role stereotypes. Characteristics such as strength, independence, logical thinking, and aggressiveness are no longer associated with maleness, just as passivity, dependence, showing emotions, intuitiveness, and nurturing are no longer associated with femaleness. Both sexes are considered equal and women and men may pursue the same occupational, political, and domestic roles. Some gender scholars have suggested that persons in such a society would be neither feminine nor masculine but would be described as androgynous, or a blend of feminine and masculine. The next subsection discusses androgyny and the related concept of gender role transcendence.

The splitting of female homemaker and male provisioner no longer provides an adequate long-term basis for a marriage enterprise, but no clear alternative has yet been established.
Daniel J. Levinson
The Seasons of a Woman's Life

Androgyny

Androgyny refers to a blend of traits that are stereotypically associated with masculinity and femininity. Street et al. (1995) reported a preference on the part of over 1,800 university women and over 1,000 university men for the ideal woman to be "androgynous." In reality, Twenge (1995) observed that U.S. college women are becoming more "masculine" in terms of being assertive, action-oriented, and goal-driven but that U.S. college men are no more "feminine" in terms of being nurturing and empathic than they were twenty years ago.

Androgyny also implies flexibility of traits; for example, an androgynous individual may be emotional in one situation, logical in another, assertive in another, and so forth.

I think we all have androgyny within us; we have both sides. It's just a matter of realizing that and liking both roles.
Grace Slick
Rock singer

Table 2.3

Relationship Consequences of Traditional Gender Role Socialization

Women	Men
1. A woman who is not socialized to pursue advanced education (which often translates into less income) may feel pressure to stay in an unhappy relationship with someone on whom she is economically dependent.	1. Men who are socialized to define themselves in terms of their occupational success and income and less in terms of positive individual qualities leave their self-esteem and masculinity vulnerable should they become unemployed.
2. Women who are socialized to play a passive role and not initiate relationships are limiting interactions that could develop into valued relationships.	2. Men who are socialized to restrict their experience and expression of emotions are denied the opportunity to discover the rewards of emotional interpersonal sharing.
3. Women who are socialized to accept that they are less valuable and important than men are less likely to seek or achieve egalitarian relationships with men.	3. Men who are socialized to believe it is not their role to participate in domestic activities (childrearing, food preparation, housecleaning) will not develop competencies in these life skills. Domestic skills are often viewed by potential partners as desirable qualities.
4. Women who internalize society's standards of beauty and view their worth in terms of their age and appearance are likely to feel bad about themselves as they age. Their negative self-concept, more than their age or appearance, may interfere with their relationships.	4. Heterosexual men who focus on cultural definitions of female beauty overlook potential partners who might not fit the cultural beauty ideal but who would nevertheless be wonderful life companions.
5. Women who are socialized to accept that they are solely responsible for taking care of their parents, children, and husbands are likely to experience role overload. Potentially, this could result in feelings of resentment in their relationships.	5. Men who are socialized to view negatively women who initiate relationships are restricted in their relationship opportunities.
6. Women who are socialized to emphasize the importance of relationships in their lives will continue to seek relationships that are emotionally satisfying.	6. Men who are socialized to be in control of relationship encounters may alienate their partners, who may desire equal influence in relationships.

> He had that rare weird electricity about him—that extremely wild and heavy presence that you only see in a person who has abandoned all hope of ever behaving "normally."
>
> Hunter S. Thompson
> Novelist

Thus, each androgynous individual has the opportunity to develop his or her potential to its fullest, without the restriction that only gender-appropriate behaviors are allowed (Basow, 1992, 326).

While some studies have found that androgyny is associated with high self-esteem, social competence, self-disclosure, flexibility, and fewer psychological problems, others have found it to be associated with increased work stress and less overall emotional adjustment (Harrison & Pennell, 1989).

Androgyny may be viewed as an alternative to traditional gender roles, but gender scholars have noted several problems with the concept. One problem is that "androgyny has come to be seen as a combination of the traits of the two sexes rather than as a transcendence of gender categorization itself" (Unger, 1990, 112).

This chapter has been about the gender roles of women and men. The following scale allows you to assess your attitudes toward feminism.

Attitudes toward Feminism Scale

Following are statements on a variety of issues. Left of each statement is a place for indicating how much you agree or disagree. Please respond as you *personally* feel and use the following letter code for your answers:

A—Strongly Agree   B—Agree   C—Disagree   D—Strongly Disagree
+2   +1   -1   -2

1. It is naturally proper for parents to keep a daughter under closer control than a son.

2. A man has the right to insist that his wife accept his view as to what can or cannot be afforded.

3. There should be no distinction made between woman's work and man's work.

4. Women should not be expected to subordinate their careers to home duties to any greater extent than men.

5. There are no natural differences between men and women in sensitivity and emotionality.

6. A wife should make every effort to minimize irritation and inconvenience to her husband.

7. A woman should gracefully accept chivalrous attentions from men.

8. A woman generally needs male protection and guidance.

9. Married women should resist enslavement by domestic obligations.

10. The unmarried mother is more immoral and irresponsible than the unmarried father.

11. Married women should not work if their husbands are able to support them.

12. A husband has the right to expect that his wife will want to bear children.

13. Women should freely compete with men in every sphere of economic activity.

14. There should be a single standard in matters relating to sexual behavior for both men and women.

15. The father and mother should have equal authority and responsibility for discipline and guidance of the children.

16. Regardless of sex, there should be equal pay for equal work.

17. Only the very exceptional woman is qualified to enter politics.

18. Women should be given equal opportunities with men for all vocational and professional training.

19. The husband should be regarded as the legal representative of the family group in all matters of law.

20. Husbands and wives should share in all household tasks if both are employed an equal number of hours outside the home.

21. There is no particular reason why a girl standing in a crowded bus should expect a man to offer her his seat.

22. Wifely submission is an outmoded virtue.

23. The leadership of a community should be largely in the hands of men.

24. Women who seek a career are ignoring a more enriching life of devotion to husband and children.

25. It is ridiculous for a woman to run a loco-motive and for a man to darn socks.

26. Greater leniency should be adopted toward women convicted of crime than toward male offenders.

27. Women should take a less active role in courtship than men.

28. Contemporary social problems are crying out for increased participation in their solution by women.

29. There is no good reason why women should take the name of their husbands upon marriage.

30. Men are naturally more aggressive and achievement-oriented than women.

 31. The modern wife has no more obligation to keep her figure than her husband to keep down his waistline.

 32. It is humiliating for a woman to have to ask her husband for money.

 33. There are many words and phrases which are unfit for a woman's lips.

34. Legal restrictions in industry should be the same for both sexes.

 35. Women are more likely than men to be devious in obtaining their needs.

36. A woman should not expect to go to the same places or to have quite the same freedom of action as a man.

 37. Women are generally too nervous and high-strung to make good surgeons.

 38. It is insulting to women to have the "obey" clause in the marriage vows.

 39. It is foolish to regard scrubbing floors as more proper for women than mowing the lawn.

 40. Women should not submit to sexual slavery in marriage.

 41. A woman earning as much as her male date should share equally in the cost of their common recreation.

 42. Women should recognize their intellectual limitations as compared with men.

Reproduced by permission of Bernice Lott, Department of Psychology, University of Rhode Island.

SCORING: Score your answers as follows: A = +2, B = +1, C = −1, D = −2. Because half the items were phrased in a pro-feminist and half in an antifeminist direction, you will need to reverse the scores (+2 becomes −2, etc.) for the following items: 1, 2, 6, 7, 8, 10, 11, 12, 17, 19, 21, 23, 25, 26, 27, 30, 33, 35, 36, 37, and 42. Now sum your scores for all the items. Scores may range from +84 to −84.

INTERPRETING YOUR SCORE: The higher your score, the higher your agreement with feminist (Lott used the term "women's liberation") statements. You may be interested in comparing your score, or that of your classmates, with those obtained by Lott (1973) from undergraduate students at the University of Rhode Island. The sample was composed of 109 men and 133 women in an introductory psychology class, and 47 additional older women who were participating in a special Continuing Education for Women (CEW) program. Based on information presented by Lott (1973), the following mean scores were calculated: Men = 13.07, Women = 24.30, and Continuing Education Women = 30.67.

More recently, Biaggio, Mohan, and Baldwin (1985) administered Lott's questionnaire to 76 students from a University of Idaho introductory psychology class and 63 community members randomly selected from the local phone directory. Although they did not present the scores of their respondents, they reported they did not find differences between men and women. Unlike Lott's students, in Biaggio et al.'s sample, women were not more pro-liberation than men. Biaggio et al. (1985, p. 61) stated, "It seems that some of the tenets of feminism have taken hold and earned broader acceptance. These data also point to an intersex convergence of attitudes, with men's and women's attitudes toward liberation and child rearing being less disparate now than during the period of Lott's study." It would be interesting to determine if there are differences in scores between members of each sex in your class.

Biaggio, M. K., Mohan, P. J., & Baldwin, C. 1985. Relationships among attitudes toward children, women's liberation, and personality characteristics, *Sex Roles* 12: 47–62.
Lott, B. E. 1973. Who wants the children? Some relationships among attitudes toward children, parents, and the liberation of women. *American Psychologist* 28: 573–82.

The androgyny model continues to acknowledge and even depend on the conventional concepts of femininity and masculinity. Thus, in spite of its emancipatory promise, the model retains the classic dualism and, hence, the assumption of some real gender difference (Morawski, 1990, 154).

In other words, though androgyny represents a broadening of gender role norms, it still implies two differing sets of gender-related characteristics (i.e., masculine = active-instrumental; feminine = expressive-nurturant). One solution to this problem is to simply describe characteristics such as active-instrumental and expressive-nurturant without labeling these traits as masculine or feminine. Another solution is to go beyond the concept of androgyny and focus on gender role transcendence.

Gender Role Transcendence

As noted earlier, we tend to impose a gender-based classification system on the world. Thus, we associate many aspects of our world, including colors, foods, social/occupational roles, and personality traits, with either masculinity or femininity. The concept of *gender role transcendence* involves abandoning gender schema (i.e., becoming "gender aschematic," Bem, 1983) so that personality traits, social/occupational roles, and other aspects of our life become divorced from gender categories.

GLOSSARY

androgyny   A blend of traits that are stereotypically associated with masculinity and femininity.

gender   The social and psychological characteristics associated with being female or male.

gender dysphoria   The condition in which one's gender identity does not match one's biological sex.

gender identity   The psychological state of viewing one's self as a girl or boy, and later as a woman or man.

gender role   A set of social norms that dictate what is considered appropriate female and male behavior.

gender role ideology   The socially defined role relationships between women and men in a society.

gender role transcendence   Abandoning gender schema and viewing personality traits, social/occupational roles, and other aspects of our world without labeling them as "feminine" or "masculine."

parental investment   Any investment by a parent that increases the offspring's chance of surviving and hence increases reproductive success of the parent.

sex   The biological distinction between being female and being male—for example, having XX or XY chromosomes.

sexism   An attitude, action, or institutional structure that subordinates or discriminates against an individual or group because of their biological sex.

sociobiology   A theoretical perspective that emphasizes biological explanations for social behavior.

transgendered A generic term that refers to a broad spectrum of individuals who express characteristics other than those of their assigned gender.

transgenderist An individual who lives in a gender role that does not match his or her biological sex, but has no desire to surgically alter his or her genitalia.

transsexual An individual who has the anatomical and genetic characteristics of one sex but the self-concept of the other.

transvestite A person who enjoys dressing in the clothes of the other sex.

SUMMARY

Terminology

In common usage the terms *sex* and *gender* are often interchangeable. But these terms are not synonymous. *Sex* refers to the biological distinction between females and males. *Gender* refers to the social and psychological characteristics often associated with being female or male. Job and credit card applications ask for the sex of an individual. Other terms related to sex and gender include *gender identity, gender role, gender role ideology, transgendered,* and *transgenderist.*

Theories of Gender Role Development

Most family life specialists acknowledge an interaction effect of biology and environment in gender role behaviors. Sociobiology emphasizes biological sources of social behavior such as higher sex aggression on the part of males due to higher levels of testosterone. Identification theory focuses on the importance of the same-sex parent for learning gender roles. Social learning theory discusses how children are rewarded and punished for expressing various gender role behaviors and how gender is learned through direct instruction and modeling. Mead's study of three societies displaying different gender roles emphasizes the impact of culture on gender role behavior. Cognitive-developmental theorists are concerned with the developmental ages at which children are capable of learning social roles.

Agents of Gender Role Socialization

Whereas biological differences may predispose people to behave in certain ways, agents of socialization (parents, peers, teachers, religion, and media) influence what people learn. Each of these agents may have more influence at different times throughout the family life cycle. Though parents are the first and most enduring influence, peers may become particularly influential during adolescence.

Consequences of Traditional Gender Role Socialization

Traditional female role socialization may result in less education, less income, greater dependence, lower marital satisfaction, a longer life, a closer emotional bond with children, and a larger number of quality relationships for women. Traditional male role socialization may result in the fusion of self and occupation, a more limited expression of emotion, disadvantages in child custody disputes, a shorter life, higher income, and higher status.

Beyond Gender Roles

Androgyny refers to the blend of traits stereotypically associated with masculinity and femininity. *Gender role transcendence* involves abandoning gender schema so that the world is not divided into feminine and masculine categories. *Transgenderists* are individuals who express the social role of the other gender—for example, a genetic male living as a female. Cross-dressers are often heterosexual men who enjoy expressing the feminine side of themselves.

REFERENCES

Aldous, J., and R. Woodberry. 1994. Gender, marital status, and the pursuit of happiness. Paper presented at the 56th annual meeting of the National Council on Family Relations, Minneapolis, MN.

American Council on Education and University of California. 1997. The American freshman: National norms for Fall, 1997. Los Angeles: Higher Education Research Institute.

Arnold, A. P. 1980. Sexual differences in the brain. *American Scientist*, March-April, 165–73.

Associated Press. 1997. Hazing video gets range of reactions. *Daily Reflector*. 2 February, 1.

Basow, Susan A. 1992. *Gender: Stereotypes and roles*. 3d ed. Pacific Grove, Calif.: Brooks/Cole.

Baxter, Janeen, and Emily W. Kane. 1995. Dependence and independence: A cross-national analysis of gender inequality and gender attitudes. *Gender and Society* 9: 193–215.

Bem, S. L. 1983. Gender schema theory and its implications for child development: Raising gender-aschematic children in a gender-schematic society. *Signs* 8: 596–616.

Bird, C. E., and A. M. Fremont. 1991. Gender, time, use, and health. *Journal of Health and Social Behavior* 32: 114–29.

Blair, S. L. 1993. Employment, family, and perceptions of marital quality among husbands and wives. *Journal of Family Issues* 14: 189–212.

Blaisure, Karen R., and Katherine R. Allen. 1995. Feminists and the ideology and practice of marital equality. *Journal of Marriage and the Family* 57: 5–19.

Bullough, B., and V. Bullough. 1997. Are transvestites necessarily heterosexual? *Archives of Sexual Behavior* 26: 1–12.

Buss, D. M. 1989. Sex differences in human mate preferences: Evolutionary hypotheses tested in 37 cultures. *Behavioral and Brain Sciences* 12: 1–13.

Carlson, B. E., and Lynne Videka-Sherman. 1990. An empirical test of androgyny in the middle years: Evidence from a national survey. *Sex Roles* 23: 305–24.

Catania, J. A., T. J. Coates, R. Stall, H. Turner, J. Peterson, N. Hearst, M. M. Dolcini, E. Hudes, J. Gagnon, J. Wiley, and R. Groves. 1992. Prevalence of AIDS-related risk factors and condom use in the United States. *Science* 258: 1101–06.

Chodorow, N. 1978. *The reproduction of mothering*. Berkeley: University of California Press.

Clifford, D. M., and S. M. Shoffner. 1992. Gender-based differences in high school employment: Is there different socialization? In *Family and work*. Vol. 2 of *Proceedings*. 54th Annual Conference of the National Council on Family Relations, no. 1, 30.

Cole, C. M., M. O'Boyle, L. E. Emory, and W. J. Meyer. 1997. Comorbidity of gender dysphoria and other major psychiatric diagnoses. *Archives of Sexual Behavior* 26: 13–26.

Coontz, Stephanie. 1995. The way we weren't: The myth and reality of the "traditional" family. *Phi Kappa Phi Journal* 75: 11–14.

Cox, James. 1995. China's women make small strides. *USA Today*, 29 August, A1.

Crabb, Peter B., and Dawn Bielawski. 1994. The social representation of material culture and gender in children's books. *Sex Roles* 30: 69–79.

Davis, Simon. 1990. Men as success objects and women as sex objects: A study of personal advertisements. *Sex Roles* 23: 43–50.

Derlega, V. J. S., S. Metts, S. Petronio, and S. T. Margulis. 1993. *Self-disclosure*. Newbury Park, Calif.: Sage.

Doyle, James A. 1995. *The Male Experience*. Madison, Wis.: WCB Brown and Benchmark Publishers.

Elliott, L., and Brantley, C. 1997. *Sex on campus: The naked truth about the real sex lives of college students*. New York: Random House.

Ellis, Bruce J., and Donald Symons. 1990. Sex differences in sexual fantasy: An evolutionary psychological approach. *Journal of Sex Research* 27: 527–56.

Etaugh, Claire, and Terri Duits. 1990. Development of gender discrimination: Role of stereotypic and counterstereotypic gender cues. *Sex Roles* 23: 215–22.

Ferree, M. M., and E. J. Hall. 1990. Visual images of American society: Gender and race in introductory sociology textbooks. *Gender and Society* 4: 500–33.

Garner, D. M. 1997. The 1997 body image survey results. *Psychology Today*. February, 30 et passim.

Garner, P. W., S. Robertson, and G. Smith. 1997. Preschool children's emotional expressions with peers. *Sex Roles* 36: 675–691.

Garza, R. T., and J. E. Borchert. 1990. Maintaining social identity in a mixed-gender setting: Minority/majority status and cooperative/competitive feedback. *Sex Roles* 22: 679–91.

Goodwin, Robin. 1990. Sex differences among partner preferences: Are the sexes really very similar? *Sex Roles* 23: 501–14.

Goy, R. W., and B. S. McEwen. 1980. *Sexual differentiation of the brain.* Cambridge, Mass.: MIT Press.

Grammer, Karl. 1989. Human courtship behavior: Biological basis and cognitive processes. In *Sociobiology of sexual reproductive strategies,* edited by A. E. Rasa, C. Vogel, and E. Voland. London: Chapman and Hall, 147–69.

Green, R. J., M. Bettinger, and E. Zacks. 1996. Are lesbian couples fused and gay male couples disengaged? In *Lesbians and gays in couples and families,* edited by Joan Laird and Robert-Jay Green. San Francisco: Jossey-Bass, 185–230.

Hammersla, Joy Fisher, and Lynne Frease-McMahan. 1990. University students' priorities: Life goals vs. relationships. *Sex Roles* 23: 1–14.

Harrison, D. F., and R. C. Pennell. 1989. Contemporary sex roles for adolescents: New options or confusion? *Journal of Social Work and Human Sexuality* 8: 27–45.

Hedley, Mark. 1994. The presentation of gendered conflict in popular movies: Affective stereotypes, cultural sentiments, and men's motivation. *Sex Roles* 31: 721–40.

Henderson, P. H., and C. Woods. In press. *Summary report 1996: Doctorate recipients from United States universities.* Washington, D.C.: National Academy Press.

Heyn, Dalma. 1997. *Marriage shock: The transformation of women into wives.* New York: Villard.

Jones, D. C., N. Bloys, and M. Wood. 1990. Sex roles and friendship patterns. *Sex Roles* 23: 133–45.

Knudson-Martin, C., and Mahoney, A. R. 1998. Language and processes in the construction of equality in new marriages. *Family Relations* 47: 81–91.

Koenig, Linda J., and Erika L. Wasserman. 1995. Body image and dieting failure in college men and women: Examining links between depression and eating problems. *Sex Roles* 32: 225–49.

Kohlberg, L. 1966. A cognitive-developmental analysis of children's sex-role concepts and attitudes. In *The development of sex differences,* edited by E. E. Macoby. Stanford, Calif.: Stanford University Press.

Kohlberg, L. 1969. State and sequence: The cognitive-developmental approach to socialization. *Handbook of socialization theory and research,* edited by D. A. Goslin. Chicago: Rand McNally, 347–480.

Kohlberg, L. 1976. Moral stages and moralization: The cognitive-developmental approach. In *Moral development and behavior,* edited by T. Lickona. New York: Holt, Rinehart, & Winston.

Leve, L. D., and B. I. Fagot. 1997. Gender-role socialization and discipline processes in one and two parent families. *Sex Roles* 36: 1–21.

Lindsey, L. L. 1990. *Gender roles: A sociological perspective.* Englewood Cliffs, N.J.: Prentice-Hall.

Luckenbill, D. F., and D. P. Doyle. 1989. Structural position and violence: Developing a cultural explanation. *Criminology* 27: 419–33.

McBroom, P. A. 1992. *The third sex: The new professional woman.* New York: Paragon House.

McHale, S. M., W. T. Bartko, A. C. Crouter, and M. Perry-Jenkins. 1990. Children's housework and psychosocial functioning: The mediating effects of parents' sex-role behaviors and attitudes. *Child Development* 61: 1413–26.

McWhirter, E. H. 1997. Perceived barriers to education and career: Ethnic and gender differences. *Journal of Vocational Behavior* 50: 124–140.

Mead, Margaret. 1935. *Sex and temperament in three primitive societies.* New York: William Morrow.

Michael, Robert T., John H. Gagnon, Edward O. Laumann, and Gina Kolata. 1994. *Sex in America: A definitive survey.* Boston: Little, Brown.

Mindel, C. H., R. W. Habenstein, and R. Wright, Jr., eds. 1998. *Ethnic families in America: Patterns and variations.* 4th ed. New York: Elsevier.

Mitchell, Grace, and Patricia M. Fandt. 1995. Examining the relationship between role-defining characteristics and self-esteem of college students. *College Student Journal* 29: 96–102.

Morawski, J. G. 1990. The troubled quest for masculinity, femininity, and androgyny. In *Sex and gender*, edited by P. Shaver and C. Hendrick. Newbury Park, Calif.: Sage.

Morrow, Frances. 1991. *Unleashing our unknown selves: An inquiry into the future of femininity and masculinity.* New York: Praeger.

Purcell, P., and L. Stewart. 1990. Dick and Jane in 1989. *Sex Roles* 22: 177–85.

Powlishta, Kimberly K. 1995. Gender bias in children's perceptions of personality traits. *Sex Roles* 32: 17–28.

Press, Aric, John McCormick, and Pat Wingert. 1994. A Crime as American as a colt .45. *Newsweek,* 15 August, 22–23.

Rader, D. 1993. How to live without answers. *Parade Magazine,* 25 April, 4–5.

Rapping, Elayne. 1994. *Media-tions: Forays into the culture and gender wars.* Boston, Mass: South End Press.

Risman, B. J., and K. Myers. 1997. As the twig is bent: Children reared in feminist households. *Qualitative Sociology* 20: 229–252.

Rodin, J., and J. R. Ickovics. 1990. Women's health: Review and research agenda as we approach the twenty-first century. *American Psychologist* 45: 1018–34.

Ross, C. E., and M. Van Willigen. 1997. Education and the subjective quality of life. *Journal of Health and Social Behavior* 38: 275–97.

Rubenstein, C. 1990. A brave new world. *New Woman,* October, 158–64.

Saad, Lydia. 1995. Children, hard work taking their toll on baby boomers. *Gallup Poll Monthly,* April, 21–24.

Sadker, Myra, and David Sadker. 1990. Confronting sexism in the college classroom. In *Gender in the classroom: Power and pedagogy,* edited by Susan L. Gabriel and Isaiah Smithson. Chicago: University of Illinois Press, 176–87.

Schroeder, K. A., L. L. Blood, and D. Maluso. 1993. Gender differences and similarities between male and female undergraduate students regarding expectations for career and family roles. *College Student Journal* 27: 237–49.

Schvaneveldt, Paul L., Jennifer L. Kerpelman, and Guy Cunningham. 1997. Anticipated identity commitments of young adults to career, marital, and parental roles. Paper presented at the 59th Annual Conference of the National Council on Family Relations, Crystal City, Va. November.

Schur, Edwin. 1984. *Labeling women deviant: Gender, stigma, and social control.* New York: Random House.

Seccombe, Karen, and Rebecca L. Warner. 1994. The influence of gender roles and socioeconomic status on pronatalism: A comparison of men and women. In *Families and justice: From neighborhoods to nations. NCFR Vol. 4 of Proceedings.* 56th Annual Conference of the National Council on Family Relations, 41.

Seid, R. P. 1994. Too "close to the bone": The historical context for women's obsession with slenderness. In *Feminist perspectives on eating disorders,* edited by P. Fallon, M. A. Katzman, and S. C. Wooley. New York: Guilford Press, 3–16.

Smith, Kris M. 1995. First-year student survey. Report 9596-1, East Carolina University, Research, Assessment, and Testing. Greenville, N.C.

Stanley, S. M., and H. J. Markman. 1997. *Marriage in the 90s: A nationwide random phone survey.* Denver: PREP, Inc. (303-759-9931).

Starrels, Marjorie E. 1994. Gender differences in parent-child relations. *Journal of Family Issues* 15: 148–65.

Statistical Abstract of the United States: 1997. 117th ed. Washington, D.C.: U.S. Bureau of the Census.

Street, Sue, E. B. Kimmel, and J. D. Kromrey. 1995. Revisiting university student gender role perceptions. *Sex Roles* 33: 183–201.

Symons, D. 1987. An evolutionary approach: Can Darwin's view of life shed light on human sexuality? In *Theories of human sexuality,* edited by J. H. Greer and W. T. O'Donohue. New York: Plenum.

Symons, D., and B. Ellis. 1989. Human male-female differences in sexual desire. In *Sociobiology of sexual and reproductive strategies,* edited by A. E. Rasa, C. Vogel, and E. Voland. London: Chapman and Hall, 131–46.

Tang, C. S., F. D. Lai, M. Phil, and T. K. H. Chung. 1997. Assessment of sexual functioning for Chinese college students. *Archives of Sexual Behavior* 26: 79–90.

Twenge, J. M. 1995. Twenty years of change: Differences in Bem-Sex Role Inventory means across time, regions, and schools. Poster session, annual meeting of American Psychological Association, New York, August 11–15.

Udry, R. J., L. M. Talbert, and N. M. Morris. 1986. Biosocial foundations for adolescent female sexuality. *Demography* 23: 217–27.

Unger, Rhoda K. 1990. Imperfect reflections of reality: Psychology constructs gender. In *Making a difference: Psychology and the construction of gender,* edited by R. T. Hare-Mustin and J. Marecek. New Haven, Conn.: Yale University Press, 102–49.

Waldfogel, Jane. 1997. The effect of children on women's wages. *American Sociological Review* 62: 209–17.

Walker, Alexis J., Sally S. Kees Martin, and Linda Thomson. 1988. Feminist programs for families. *Family Relations* 37: 17–22.

Walker, Karen. 1994. Men, women, and friendship: What they say, what they do. *Gender and Society* 8: 246–65.

Williams, John E., and Deborah L. Best. 1990a. *Measuring sex stereotypes: A multination study.* London: Sage.

―――― . 1990b. *Sex and psyche: Gender and self viewed cross-culturally.* London: Sage.

Williams, Lindy, and Teresa Sobieszczyk. 1997. Attitudes surrounding the continuation of female circumcision in the Sudan: Passing the tradition to the next generation. *Journal of Marriage & the Family* 59: 966–81.

Willetts-Bloom, Marion C., and Steven L. Nock. 1994. The influence of maternal employment on gender role attitudes of men and women. *Sex Roles* 30: 371–89.

There is no happiness in the world comparable to loving another person, totally and unqualifiedly, and finding that love returned.

Richard Taylor, Philosopher

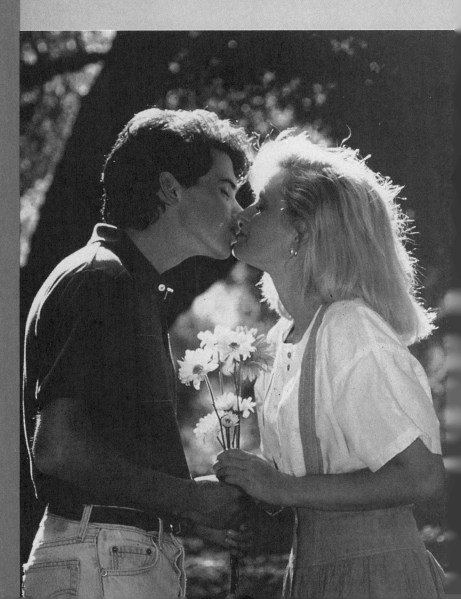

*T*itanic is steaming toward being the most commercially successful film of all time. Although the film is a technical marvel, its box-office success may have more to do with its content as a love story between Rose and Jack, who are forever separated by the icy waters of the North Atlantic. But love is not only the stuff of films. It is a common experience among university students. Eighty-three percent of 620 university students reported that they had experienced being in love with a dating partner (Knox & Zusman, 1998). In this chapter we examine this elusive phenomenon we call love—its conceptions, theories, and conditions for development. We begin by looking at three conceptions of love.

Conceptions of Love

Love may be conceptualized as a triangle of three basic elements, as a style of interpersonal interaction, and as a continuum from romanticism to realism.

Sternberg's Triangular View of Love

According to Sternberg (1986), love consists of various degrees of intimacy, passion, and commitment. Intimacy includes disclosing personal feelings and thoughts, giving/receiving emotional support to/from the beloved, and being able to count on the beloved in times of need. Passion involves sexual longing and sexual needs. Commitment is characterized by a decision to maintain the love relationship with the partner both now and in the future (see Figure 3.1). Erich Fromm (1963) also emphasized that love is a commitment, a decision:

> To love somebody is not just a strong feeling—it is a decision, it is a judgment, it is a promise. If love were only a feeling, there would be no basis for the promise to love each other forever. A feeling comes and it may go (p. 47).

Using these three basic elements of love, Sternberg identified

1. Nonlove: Absence of all three components.
2. Liking: Intimacy without passion or commitment.
3. Infatuation: Passion without intimacy or commitment
4. Romantic love: Intimacy and passion without commitment
5. Companionate love: Commitment and intimacy without passion
6. Fatuous love: Passion and commitment without intimacy
7. Empty love: Commitment without passion or intimacy
8. Consummate love: Combination of intimacy, passion, and commitment

There is some overlap among these types of love. For example, some level of commitment is felt between romantic lovers (romantic love), and some level of passion is felt between companionate lovers (companionate love). However, the predominant focus of romantic love is passion, and the predominant quality of companionate love is commitment.

Figure 3.1

Sternberg's Triangle of Love

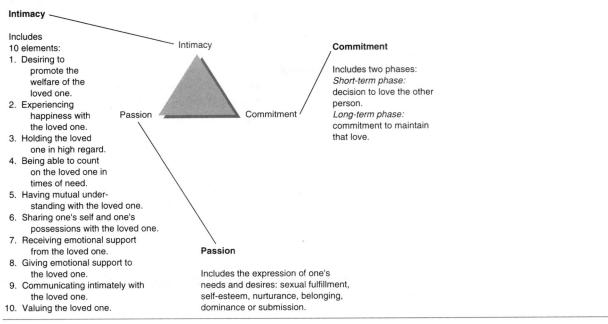

Intimacy

Includes
10 elements:
1. Desiring to promote the welfare of the loved one.
2. Experiencing happiness with the loved one.
3. Holding the loved one in high regard.
4. Being able to count on the loved one in times of need.
5. Having mutual understanding with the loved one.
6. Sharing one's self and one's possessions with the loved one.
7. Receiving emotional support from the loved one.
8. Giving emotional support to the loved one.
9. Communicating intimately with the loved one.
10. Valuing the loved one.

Commitment

Includes two phases:
Short-term phase: decision to love the other person.
Long-term phase: commitment to maintain that love.

Passion

Includes the expression of one's needs and desires: sexual fulfillment, self-esteem, nurturance, belonging, dominance or submission.

Source: Based on R. J. Sternberg. 1986. A Triangular Theory of Love. *Psychological Review* 93:119–35.

Love is a process, not a destination.

Daphne Rou Kingma

Therapist

We always believe our first love our last, and our last love our first.

George Whyte-Melville

Scottish author

The only reason Romeo and Juliet's love is eternal is because they died. If they had married, their love would not have lasted. People didn't expect lasting passion from marriage.

John Gray

Psychologist

Individuals bring different love triangles to their relationship. One lover may bring a predominance of passion with some intimacy but no commitment (romantic love) while the other person may bring intimacy but no passion or commitment (liking). The triangular theory of love allows lovers to see the degree to which they are matched in terms of the three basic elements of passion, intimacy, and commitment.

Lee's Love Styles

Theorist John Lee (1973, 1988) identified a number of styles of love that describe the way lovers relate to each other:

1. Ludus. The ludic lover views love as a game, as play, as fun, and refuses to become dependent on any one person and does not encourage another's intimacy. Two essential skills of the ludic lover are to juggle several people at the same time and to manage each relationship so that no one is seen too often. These strategies help to ensure that the relationship does not deepen into an all-consuming love. Don Juan represented the classic ludic lover.

2. Pragma. This is the love of the pragmatic who is logical and rational. The pragma lover assesses his or her partner on the basis of assets and liabilities. Economic security may be regarded as very important. The pragma lover does not seek interracial, long-distance, or age-discrepant partners, because logic argues against doing so.

The love of this mother for her child is an example of agape love whereby the mother expects nothing in return for her love and care for her baby.

3. Eros. Just the opposite of pragma, the erotic lover is consumed by passion and romance. Intensity of both emotional and sexual feelings dictates one's love involvements. The *Titanic* love between Rose and Jack was of the eros style. Inman-Amos et al. (1994) observed that eros was the most common love style of women and men in a sample of college students.

4. Mania. The person with manic love feels intense emotion and sexual passion but is out of control. The person is possessive and dependent and "must have" the beloved. Persons who are extremely jealous and controlling reflect manic love. Prosecutors in the O. J. Simpson trial implied that Simpson's love for Nicole Brown was of the manic variety.

5. Storge. Storge love is a calm, soothing, nonsexual love devoid of intense passion. Respect, friendship, commitment, and familiarity are characteristics that help to define the relationship. The partners care deeply about each other but not in a romantic or lustful sense. Their love is also more likely to endure than fleeting romance.

6. Agape. Agape is selfless and giving, expecting nothing in return. The nurturing and caring partners are concerned only about the welfare and growth of each other. The love parents have for their children is often described as agapic love.

Romantic versus Realistic Love

Love may also be described as being on a continuum from romanticism to realism. For some people, love is romantic; for others, it is realistic. Romantic love is characterized by such beliefs as love at first sight, there is only one true love, love conquers all, and the beloved will meet one's highest ideals. The symptoms of romantic love include drastic mood swings, palpitations of the heart, and intrusive thoughts about the partner. Fantasy thoughts make the individual soar. When Cinderella thinks of her prince, she says, "In the arms of my love I am flying over mountains and meadows and glen."

A 28-year-old woman in the blossom of romantic love (she met and married the man in six weeks) wrote:

> The love feelings I have for my partner are truly unlike anything I've felt before—they are beyond what I experienced with my first love at 16, more passionate than anything I ever shared with my husband during the eight years we were married, and more secure than anything I ever shared with my first lover after my divorce. With my new love, I feel safe and secure and yet he is able to continually surprise me. . . . He makes me feel complete . . . as if we are two gears in which the teeth come closer and closer together, finally locking into place and the machine is able to move forward as the gears shift. I truly feel that I've waited my whole life to meet this person, that everything up to this point has been in preparation for the time when he would appear in my life. I feel that I have known him forever and that we will never be parted no matter what happens to either of us at this point (authors' files).

Romantic love is associated with Valentine's Day, named after Valentinus, a priest in third-century Rome. One legend holds that Valentinus was a Christian imprisoned for refusing to worship pagan gods. On the day of his execution, February 14, Valentinus sent a farewell note to the jailor's blind daughter, whom he had befriended. He signed the note, "Your Valentine" (Ackerman, 1994).

Romantic love is embedded in the fabric of our culture as expressed in the media. Notice that the bulk of popular and country music is about either becoming involved in a romantic relationship or getting over a broken one. Romance novels, steeped in passion, have the same dual themes. Movies also celebrate the beginning and ending of love, and television soaps and sitcoms regularly feature romantic themes.

OTHER CULTURES We have already noted the importance of romantic love in the United States. But what about other societies? Jankowiak and Fischer (1992) found evidence of passionate love in 147 out of 166 (88.5%) of the societies they studied. Passionate love was evidenced by the presence of at least one of the following: accounts depicting personal anguish and longing, love songs, elopement due to mutual affection, native accounts affirming the existence of passionate love, or the researchers' affirmation that romantic love was present. The researchers' study stands "in direct contradiction to the popular idea that romantic love is essentially limited to or the product of Western culture. Moreover, it suggests that romantic love constitutes a human universal, or at the least a near-universal" (p. 154). In their cross-cultural study of love, Hatfield and Rapson concluded, "People throughout the modern world seem to be equally susceptible to love and to experience it with the same fervor" (1996, 88).

Although romantic love is an experience most people throughout the world are capable of, its importance as a prerequisite for marriage varies throughout the world. ●

NATIONAL DATA

A Roper Starch Worldwide Survey found that 40 percent of the respondents believed that "love at first sight" happens; 28 percent believed that "it could happen"; and 32 percent said that there is "no way" that it happens (Hall & Staimer, 1995).

INTERNATIONAL DATA In a survey of 497 male and 673 female college students in eleven countries, the following percentages reported that they would marry a person even though they were not in love (the lower the percentage, the more important romantic love is as a prerequisite for marriage): Japan (2.3%), United States (3.5%), Brazil (4.8%), Mexico (10.2%), Philippines (11.4%), India (49%), Pakistan (50.4%) (Levine et al., 1995, 561).

RECENT RESEARCH

Two-thirds (67%) of 620 never-married undergraduates reported that they would divorce if they fell out of love with their spouse (Knox & Zusman, 1998).

Infatuation is sometimes regarded as synonymous with romantic love. Infatuation comes from the same root word as *fatuous*, meaning "silly" or "foolish," and refers to a state of passion or attraction that is not based on reason. Infatuation is characterized by the tendency to idealize the love partner. People who are infatuated magnify their lovers' positive qualities ("My partner is always happy") and overlook or minimize their negative qualities ("My partner doesn't have a problem with alcohol; he just likes to have a good time"). Differences between infatuation (unrealistic love) and mature love (realistic love) include speed of development, denial of any relationship problems, and inability to cope with separation or a crisis. Infatuated love happens quickly, denies problems, and is associated with poor coping strategies.

In contrast to romantic love, realistic love, or *conjugal love*, tends to be characteristic of stable marriages. Conjugal (married) love is less emotional, passionate, and exciting than romantic love and is characterized by companionship, calmness, comfort, and security. The Love Attitudes Scale at the end of the chapter allows you to assess the degree to which your love attitudes are romantic or realistic.

Are men or women more romantic in their attitudes toward love? When 436 men and 447 women completed an abridged version of the love attitudes inventory, men were (though not significantly) more romantic than women (Knox & Schacht, 1995). Similarly, Shepard (1993) collected data from 535 students who completed the love attitudes inventory at Southeast Missouri State University and found that men were more romantic than women.

OTHER CULTURES In a study of sixty-eight unmarried Caucasians in South Africa, the women scored higher on conjugal (realistic) love than men. Conjugal love was defined as "the belief that love demands serious thought and careful consideration" (Stone & Philbrick, 1991, 220). ●

Hendrick and Hendrick (1991) suggested that women are more practical in their attitudes toward love than men because they are seeking men who would be good parents/providers. In effect, the woman must delay her romantic investment until she is sure that the man has an interest in providing for and protecting her offspring.

Finally, Silliman and Schumm (1995) found that women were much more interested in analyzing premarital relationships, suggesting a more realistic perspective.

Don't part with your illusions. When they are gone you may still exist, but you have ceased to live.

Mark Twain

INSIGHT

The answer to the question, Who is more romantic, men or women? depends on how one defines and measures romanticism. If one conceptualizes romanticism as a tendency to fall in love quickly, men are more romantic, as they tend to fall in love more quickly than women. But if one conceptualizes romanticism as a tendency to form intense and lasting bonds, women are more romantic (Walsh, 1991).

Theories on the Origins of Love

Various theories have been suggested to explain the origins of love. These include evolutionary, learning, psychosexual, ego-ideal, ontological, biochemical, and attachment theories.

Evolutionary Theory

Love has an evolutionary purpose by providing a bonding mechanism between the parents during the time their offspring are dependent infants. Love's strongest bonding lasts about four years, the time when children are most dependent and two parents can cooperate in handling their new infant. "If a woman was carrying the equivalent of a 12-lb bowling ball in one arm and a

Romantic love is seen by exchange theorists as a person perceiving that he or she is getting a high ratio of rewards at very little cost.

Chester Winton
Sociologist

Fall in love, fall into disgrace.

Chinese proverb

pile of sticks in the other, it was ecologically critical to pair up with a mate to rear the young" observes anthropologist Helene Fisher (Toufexis, 1993). The "four year itch" is Fisher's term that points to the time that parents with one child are most likely to divorce—the time when the woman can more easily survive without parenting help. If the couple has a second child, doing so resets the clock so that "the seven year itch" is the next most vulnerable time.

Learning Theory

Unlike evolutionary theory, which views the experience of love as innate, learning theory emphasizes that love feelings develop in response to certain behaviors occurring in certain contexts. Individuals on a date who look at each other, smile at each other, compliment each other, touch each other endearingly, do things for each other, and do enjoyable things together are engaging in behaviors that make it easy for love feelings to develop. In contrast, individuals who avoid, criticize, and reject each other are engaging in behaviors that make it difficult for love feelings to develop.

During a developing relationship, couples also have a high frequency of reinforcing each other for behaviors they enjoy in each other. The continuation of love feelings depends on each partner's continuing to reinforce the desirable behaviors of the other so that these behaviors (and the love feelings they elicit) continue.

Love is learned in a context. Sternberg and Beall noted that "one can become attracted to someone else not because of who he or she is, but because one just happens to experience positive reinforcement in the presence of the person—reinforcement that may have nothing to do with the person to whom one is attracted" (1991, 264). For example, doing a lot of fun things with a partner can positively influence the development of love feelings, but it may be the enjoyable events and not the person that elicit or facilitate the development of the love feelings.

Psychosexual Theory

According to psychosexual theory, love results from blocked biological sexual desires. In the sexually repressive mood of his time, Sigmund Freud ([1905] 1938) referred to love as "aim-inhibited sex." Love was viewed as a function of the sexual desire a person was not allowed to express because of social restraints. In Freud's era, people would meet, fall in love, get married, and have sex. Freud felt that the socially required delay from first meeting to having sex resulted in the development of "love feelings." By extrapolation, Freud's theory of love suggests that love dies with marriage (presumed access to the sex partner).

Ego-Ideal Theory

Love—it's everything I understand and all the things I never will.

Mary Chapin Carpenter
Songwriter

Theodore Reik (1949) suggested that love springs from a state of dissatisfaction with one's self and represents a vain urge to reach one's "ego-ideal." He believed that love is a projection of one's ideal image of himself/herself on another person. For example, suppose you are a shy, passive, dependent person but wish that you were assertive, outgoing, and independent. According to

Reik's theory, you will probably fall in love with a person who has the qualities you admire but lack in yourself.

Researcher David Lewis (1985) invited his readers to make a list of their qualities, both favorable and unfavorable. Such a list describes one's "real self." He then asked the readers to make another list of the qualities that they don't possess but would like to. This list describes the "ideal self." Lewis suggested that many people seek mates who possess the qualities they put on their ideal-self list. He noted that by attaching ourselves to and loving a person who possesses the qualities of our ideal self, we compensate for our inadequacies. We vicariously love our ideal self through loving someone who has the qualities we admire.

Ontological Theory

Ontology is a branch of philosophy that is concerned with being. Love from an ontological perspective arises from a lack of wholeness in our being. Such lack of wholeness is implied by the division of humans into males and females. From an ontological perspective, love represents women's desire to be united with their other half (i.e., men) and men's desire to be united with their other half (i.e., women). Eric Fromm (1963) viewed love as a means of overcoming the "separateness" of an individual and of quelling the anxiety associated with being lonely. When men and women develop a love relationship, they become whole.

Biochemical Theory

Ackerman (1994) emphasized that hormones and chemicals are an important basis for love. Oxytocin is a hormone that encourages contractions during childbirth and endears the mother to the suckling infant. It has been referred to as the "cuddle chemical" because of its significance in bonding. Oxytocin also seems operative in the development of love feelings between lovers during sexual arousal. Oxytocin may be responsible for the fact that more women than men prefer to continue cuddling after intercourse.

Phenylethylamine (PEA) is a natural amphetamine-like substance that makes the lovers feel euphoric and energized. The "high" that they report feeling just being with each other is the PEA released by their brain into their bloodstream. Walsh said:

> Stimulant drugs such as cocaine and amphetamine have much the same effect as love—love is a natural high (1991, 188).

Attachment Theory

The attachment theory of love emphasizes that a primary motivation in life is to be connected with other people "because it is the only security we ever have. Maintaining closeness is a bona fide survival need" (Johnson & Marano, 1994, 34). Attachment theorists point to the infant who becomes upset when separated from his or her caretaker and the behavior designed to regain proximity—crying, protesting, reaching out. When closeness is achieved, "they do all they can to maintain it: They hug, they coo, they make eye contact, they cling—and, that all-time pleaser, they smile" (p. 35).

RECENT RESEARCH

Lamm and Wiesmann (1997) asked 99 German students (mostly women) how they could tell that they were in love with someone. Sixty-two percent reported arousal as the most important indicator.

INSIGHT

The natural chemical high associated with love may explain why the intensity of passionate love decreases over time. As with any amphetamine, the body builds up a tolerance to PEA, and it takes more and more to produce the special kick. Hence, lovers develop a tolerance for each other. "Love junkies" are those who go from one love affair to the next in rapid succession to maintain the "high." In effect, they are addicted to love (Peele & Brodsky, 1976). Alternatively, some lovers break up and get back together frequently as a way of making the relationship new again and keeping the high going.

Not everything that can be
counted counts, and not
everything that counts can be
counted.

Albert Einstein

Physicist

Table 3.1

Criticisms of Love Theories	
Theory	**Criticism**
Evolutionary	Assumption that women and children need men for survival is not necessarily true today. Also, does not account for love between partners who do not want children or who are beyond their child-bearing years.
Learning	Does not account for (1) why some people will share positive experiences yet will not fall in love and (2) why some people stay in love despite negative behavior.
Psychosexual	Does not account for people who report intense love feelings yet are having sex regularly.
Ego-Ideal	Does not account for the fact that people of similar characteristics fall in love.
Ontological	Does not account for homosexual love; implies that people need a love relationship in order to be "whole."
Biochemical	Chemicals alone cannot create the state of love; cognitions are also important. Does not explain the continuation of love after the "high" wears off.
Attachment	Not all people feel the need to be emotionally attached to others. Some prefer to be detached and alone.

*To realize itself, love must
violate all the rules of our
world.*

Octavio Paz

Nobel Prize-winning author

As adults, we seek "close physical proximity to a partner, and rely on their continuing affections and availability, because it is a survival need. What satisfies the need for attachment in adults is what satisfies the need in the young: Eye contact, touching, stroking, and holding a partner deliver the same security and comfort" (p. 36).

Each of the theories on the origin of love has been criticized. In Table 3.1 we identify the criticism of each theory.

Conditions for the Development of Love

Certain social, psychological, physiological, and cognitive conditions contribute to the development of love.

Social Conditions

Love is a grave mental illness.

Plato

Greek philosopher

The society in which an individual is socialized influences the development of his or her love relationships. Societal norms also determine the degree to which individuals value or require love as a condition for marriage.

Social pressures influence the development of love relationships by encouraging some relationships and discouraging others. Although miscegenation laws (laws that prohibited whites and blacks from marrying) no longer exist in the United States, interracial love relationships are not socially encouraged. Less than 5 percent of U.S. marriages are interracial (*Statistical Abstract of the United States: 1997*, Table 62). Social pressures also discourage love relationships between individuals who have different religions, or who are "too young" or "too old," age-discrepant, already married, or who are of the same sex.

OTHER CULTURES In India and other countries (e.g., Palestine) that practice arranged marriages, the development of romantic love relationships is tightly controlled. For example, parents select the mates for their children in an effort to prevent any potential love relationship from forming with the "wrong person" and to ensure that the child marries the "right person." Such a person must belong to the desired social class and have the economic resources desired by the parents. These traditional child marriages are regarded as the linking of two families; the love feelings of the respective partners are less important. Love is expected to follow marriage, not precede it. ●

Psychological Conditions

Two psychological conditions associated with the development of healthy love relationships are a positive self-concept and the ability to self-disclose.

Self-concept   A positive self-concept is important for developing healthy love relationships because it enables an individual to feel worthy of being loved. Feeling good about yourself allows you to believe that others are capable of loving you. Indeed, a direct positive relationship exists between self-esteem on the part of either or both partners and relationship stability (Christensen & Busby, 1992).

Having self-esteem provides other benefits:

1. It facilitates the ability to be open and honest with others, about both strengths and weaknesses.
2. It enables one to feel generally equal to others.
3. It strengthens the ability to take responsibility for one's own feelings, ideas, mistakes, and failings.
4. It facilitates the acceptance of self and others, both in strengths and in weaknesses.
5. It enables one to validate one's self and not to expect the partner to do this.
6. It encourages empathy—a very important skill in relationships.
7. It allows separateness and interdependence, as opposed to fusion and dependence.

Positive physiological and general success outcomes also follow from high self-esteem.

Therefore love moderately;
long love doth so.
Friar Laurence in Shakespeare's
Romeo and Juliet

I was not wanted by my
parents (they wanted a boy),
by my husband (he had
lovers) or the Court (I was an
embarrassment).
Diana
Princess of Wales

Although it helps to have a positive self-concept going into a love relationship, sometimes this develops after becoming involved in the relationship. "I've always felt like an ugly duckling," said one woman. "But once I fell in love with him and he with me, I felt very different. I felt very good about myself then because I knew that I was somebody that someone else loved." A positive self-concept, then, is not a prerequisite for falling in love. People who have a very negative self-concept may fall in love with someone else as a result of feeling deficient (ego-ideal theory of love). The love they perceive the other person has for them may compensate for the deficiency and improve their self-esteem.

Partners who spend a lot of time disclosing themselves to each other are likely to report experiencing love feelings.

People who feel good about themselves get fewer ulcers, have fewer sleepless nights, deal with anxiety better, and show more persistence with difficult tasks than those who don't. High levels of self-esteem may also be characteristic of effective leaders and those perceived as successful in society (Mitchell & Fandt, 1995, 96).

In contrast, a negative self-concept has devastating consequences for individuals and the relationships in which they become involved. Not feeling loved as a child and, worse, feeling rejected and abandoned encourage a negative self-concept. Comedian Rodney Dangerfield reported that he felt his mother never cared about him—"She wouldn't even look at my report card," he said. People who have never felt loved and wanted may require constant affirmation from their partners as to their worth and may cling desperately to those persons so as to not feel alone. Such dependence (the modern term is *codependency*) can result in unhealthy relationships.

One characteristic of individuals with a negative self-concept is that they may "love too much" and be addicted to "unhealthy love relationships." Petrie and colleagues studied fifty-two women who reported that they were involved in unhealthy love relationships in which they had selected men with problems (such as alcohol/other drug addiction) they attempted to solve at the expense of neglecting themselves. "Their preoccupation with correcting the problems of others may be an attempt to achieve self-esteem" (Petrie et al., 1992, 17).

Negative self-concepts often have their origin in the family in which one was reared. The fifty-two women in unhealthy relationships with low self-concepts reported that they had grown up "in nonsupportive, nonaffectionate childhood homes in which their emotional needs were not met" (Petrie et al., 1992, 16).

Self-disclosure   Self-disclosure is also necessary if one is to love and be loved by another. Disclosing yourself is a way of investing yourself in another. Once the other person knows some of the intimate details of your life, you will tend to feel more positively about that person because a part of you is now a part of him or her. Open communication tends to foster the development of an intense love relationship. Rodriguez and Devall (1997) found that increased self-disclosure is related to marital satisfaction.

It is not easy for some people to let others know who they are, what they feel, or what they think. Some have been reared in homes in which parents and siblings were models of nondisclosure. Or, they may fear that if others really know them, they will be rejected as a friend or lover. To guard against this possibility, they may protect themselves and their relationships by allowing only limited access to their past behaviors and present thoughts and feelings.

Trust is the condition under which people are often willing to disclose themselves. When people trust someone, they tend to feel that whatever feelings or information they share will not be judged and will be kept safe with that person. If trust is betrayed, a person may become bitterly resentful and vow never to disclose herself or himself again. One woman said, "After I told my partner that I had had an abortion, he told me that I was a murderer and he never wanted to see me again. I was devastated and felt I had made a mistake telling him about my past. You can bet I'll be careful before I disclose myself to someone else" (authors' files).

In order for this couple to experience being in love, they must label their feelings for each other as love.

Physiological and cognitive conditions Physiological and cognitive variables are also operative in the development of love. The individual must be physiologically aroused and interpret this stirred-up state as love (Walster & Walster, 1978).

> Suppose, for example, that Dan is afraid of flying, but his fear is not particularly extreme and he doesn't like to admit it to himself. This fear, however, does cause him to be physiologically aroused. Suppose further that Dan takes a flight and finds himself sitting next to Judy on the plane. With heart racing, palms sweating, and breathing labored, Dan chats with Judy as the plane takes off. Suddenly, Dan discovers that he finds Judy terribly attractive, and he begins to try to figure out ways that he can continue seeing her after the flight is over. What accounts for Dan's sudden surge of interest in Judy? Is Judy really that appealing to him, or has he taken the physiological arousal of fear and mislabeled it as attraction (Brehm, 1992, 44)?

Since most people are not aroused in this way yet develop love feelings, they may be aroused or anxious about other issues (being excited at a party, feeling apprehensive about meeting new people) and mislabel these feelings as those of attraction when they meet someone.

In the absence of one's cognitive functioning, love feelings are impossible. Individuals with brain cancer who have had the front part of the brain between the eyebrows removed are incapable of love. Indeed, emotions are not present in them at all (Ackerman, 1994).

Problems Associated with Love

Philosopher Bertrand Russell once said, "To fear love is to fear life, and those who fear life are already three parts dead." His statement reflects the belief that the experience of being in love is worth whatever problems befall the lovers. Though love can be a positive force (the greater the love, the greater the desire to share the responsibility for tasks—Ripke & Huston, 1997), it is also associated with problems. In this section we discuss problems of love, including simultaneous loves, being in love with an abusive partner, and love as a context that encourages risky choices.

Simultaneous Loves

Country songs abound on the theme of one person with two lovers (e.g., "Torn between Two Lovers", "We Ain't Done Nothing Wrong"). While romantics believe that such an experience is not possible (that there is only one true love), realists view individuals as being capable of being in love with several different people at the same time.

Some people involved in one relationship develop love feelings for a new person at work. How the business world responds to office romances is the subject of our Social Policy section.

Love in the Workplace

With an increase in women in the workforce, an increase in the age at first marriage, and longer work hours, the workplace has become a common place for romantic relationships to develop. According to a Bureau of National Affairs study, more future spouses will meet at work than at school, social, or neighborhood settings (Loftus, 1995).

Pros and cons of office romances   The energy that both fuels and results from intense love feelings can also fuel productivity on the job. And, if the coworkers eventually marry or enter a nonmarital but committed and long-term relationship, they may be more satisfied with and committed to their jobs than spouses whose partners work elsewhere. Working at the same location enables married couples to commute together, go to company-sponsored events together, and "talk shop" together.

However, workplace romances can also be problematic for the individuals involved as well as for their employers. When a workplace romance involves a supervisor/subordinate relationship, other employees might make claims of favoritism or differential treatment. In a typical differential-treatment allegation, an employee (usually a woman) claims that the company denied her a job benefit because her supervisor favored a female coworker—who happens to be the supervisor's girlfriend.

If a workplace relationship breaks up, it may be difficult to continue to work in the same environment. A breakup that is less than amicable may result in efforts by the partners to sabotage each other's work relationships and performance, incidents of workplace violence, harassment, and/or allegations of sexual harassment.

Policies regarding intimate relationships in the workplace   Although some companies have tried to impose restrictions on romantic relationships between employees, such policies are rare. Wal-Mart used to fire any employee who committed adultery with a coworker. In 1992, the company fired a New York woman who had a legal separation and her boyfriend, who was single. However, the state attorney general successfully sued Wal-Mart for violating a New York law that prohibits employers from interfering with employees' private lives as long as the employees are not doing something illegal. Other states have also passed workplace privacy laws designed to protect any lawful activities conducted off the employers' premises and outside working hours.

Wal-Mart's current policy, consistent with that of most companies, does not prohibit romantic relationships among employees. However, the new policy does prohibit open displays of affection between employees in the workplace and romantic relationships between supervisor and subordinate. In the early 1990s, a survey of members of the Society for Human Resource Management found that 92 percent of the companies surveyed had no policy regarding love relationships at work. Over 70 percent said that they "permit and accept" it, and only 1.5 percent had policies that banned it (Fisher, 1995, 255). In a 1994 survey of 200 CEOs, three-quarters said that romances between workers are "none of the company's business" (Loftus, 1995, 28).

Recognizing the potential benefits of increased job satisfaction, morale, productivity, creativity, and commitment, some companies even look favorably upon love relationships among employees. Apple Computers, in Cupertino, California, encourages socializing among employees by sponsoring get-togethers every Friday afternoon with beer, wine, food, and, on occasion, live bands. The company also has ski clubs, volleyball clubs, and Frisbee clubs, providing employees with opportunities to meet and interact socially.

However, there are some exceptions to the generally permissive policies regarding workplace romances. Many companies have written policies prohibiting intimate relationships when one member of the couple is in a direct supervisory position over the other. These policies may be enforced by transferring, or dismissing, employees who are discovered in romantic relationships. However, companies can be subject to discrimination suits if they do not enforce their dating policies evenhandedly. For example, a clerk was fired for dating her manager, but nothing happened to the manager. The clerk sued for discrimination and was awarded financial compensation (Employment Law Resource Center, 1997). A New Jersey company implemented an antidating policy among married employees but not singles. However, the courts ruled that this violated New Jersey's law against marital status discrimination. Adulterous relationships are prohibited among military personnel. Violation of this policy may result in a court-martial or, as in the 1997 case of former Air Force Lieutenant Kelly Flinn, resignation to avoid a court-martial.

Many colleges and universities also have policies regarding dating among students and teachers in order to avoid actual or perceived conflicts of interest and potential abuses of authority. Some university policies prohibit any dating and sexual relationships between a faculty member and a student enrolled in that faculty member's course or employed as that faculty member's teaching or research assistant. Other universities do not ban professor-student consensual relationships but make it mandatory for professors to inform their department heads of any intimate relationship they have with a student. Once the department head is informed of the situation, an in-dependent committee evaluates the student's class work to avoid a conflict of interest and decrease the potential for grade retaliation if the student wants to end the relationship.

REFERENCES

Employment Law Resource Center. 1997. Dating among employees. Website: http://www.ahipubs.com/problem-solvers/dating.html.

Fisher, Anne B. 1995. Is the office the right place for romance? *Cosmopolitan*, May, 253–55.

Loftus, Mary. 1995. Romance in the workplace. *USA Today*, November, 28–30.

INSIGHT

Some people do not like the feeling of being in love with two people at once and try to reduce their feelings for one of them. This may be accomplished by deciding to see only one of the persons and by thinking negative thoughts about the other. For example, Jan, who was in love with both her husband and her colleague, decided to stop seeing her colleague socially. When she did think of him, she made herself think only of what she perceived to be the negative aspects of being involved with him: he was married, he drank heavily, he smoked, he was twelve years older than she, he had little ambition, and he had three teenage children he would bring to live with them if they eventually married. She soon lost her "love feelings."

Abusive or Unfulfilling Love Relationship

Another problem associated with love is being in love with someone who is emotionally or physically abusive (Chapter 10, Abuse in Relationships, discusses abusive relationships in detail). "I love him/her but I can't live with him/her" is the common expression. Someone who is an alcoholic and beats you, who criticizes you continually ("you're ugly, stupid, pitiful"), or who is dishonest with you (sexually unfaithful) may create a great deal of pain, stress, and disappointment. Nevertheless, you might love that person and feel emotionally drawn to him or her. Most marriage therapists would suggest that you examine why you love and continue to stay with this person. Do you feel that you deserve this treatment because you are "no good"? Do you feel pity for the person and feel responsible for rescuing him or her? Do you feel you would not be able to find a better alternative? Do you feel you would rather be with a person who treats you badly than be alone?

Another explanation for why some people who are abused by their partners continue to be in love with them is that the abuse is only one part of the relationship. When the partner is not being abusive, he or she may be kind, loving, and passionate. It is these latter qualities that overshadow the occasional abuse and keep the love alive. Love stops when the extent of the abuse is so great that there are insufficient positive behaviors to counteract the abusive behavior.

Love relationships that do not involve emotional or physical abuse may be unfulfilling for other reasons. Partners in love relationships may experience lack of fulfillment if they have radically different values, religious beliefs, role expectations, recreational or occupational interests, sexual needs, or desires concerning family size.

Context for Risky Choices

The negative power of love is illustrated in the study "What I Did for Love" (Knox, Zusman, & Nieves, 1998), in which college students identified the most risky or dangerous behaviors they had engaged in while they were in love. Topping the list was "having sex without protection," followed by "giving up who I was" (e.g., dropping out of or changing schools for the beloved), lying to the partner, and driving drunk. In addition, one researcher observed in his sample of 181 university students that the more in love they were with their partners, the less concerned they were about AIDS and the less likely they were to use condoms (Pilkington et al., 1994).

Jealousy

Jealousy is another major problem associated with love and can be defined as an emotional response to a perceived or real threat to an important or valued relationship. Although jealousy may not occur in all cultures (polyandrous societies value cooperation, not sexual exclusivity; Cassidy & Lee, 1989), it is a part of relationships in our society. Thirteen percent of 620 university students reported that jealousy was the most frequent problem that they encountered in their current or most recent relationship (Knox & Zusman, 1998).

OTHER CULTURES Over 2,000 college students from seven industrialized countries (Hungary, Ireland, Mexico, the Netherlands, the former Soviet Union, the former Yugoslavia, and the United States) were asked about what makes them jealous. Jealousy existed in all countries, with flirting and sexual involvement eliciting the most negative reactions. Students from the Netherlands and the former Soviet Union tended to be the least and most jealous, respectively. Students from the United States were mid-range (Buunk & Hupka, 1987). ●

Causes of Jealousy

Jealousy can be triggered by external or internal factors.

External causes External factors refer to behaviors the partner engages in that are interpreted as (1) an emotional and/or sexual interest in someone (or something) else or (2) a lack of emotional and/or sexual interest in the primary partner. Examples of behavior that may be interpreted as threatening to the partner and the relationship include flirting, spending a lot of time with someone else, or saying positive things about someone else. External causes may also include an interpretation that the partner is spending too little time with or showing little interest in being with the beloved.

Internal causes Jealousy may also exist even when there is no external behavior that indicates the partner is involved or interested in an extradyadic relationship. Internal causes of jealousy refer to characteristics of individuals that predispose them to jealous feelings, independent of their partner's behavior.

The person who is more involved in or dependent on the relationship not only is more likely to experience jealousy but also may intentionally induce jealousy in the partner. Such attempts to induce jealousy may involve flirting, exaggerating or discussing an attraction to someone else, and spending time with others. According to White (1980), individuals may try to make their partners jealous as a way of testing the relationship (e.g., see whether the partner still cares) and/or increasing specific rewards (e.g., get more attention or affection). White found that women, especially those who thought they were more involved in the relationship than their partners, were more likely to induce jealousy in a relationship than men.

RECENT RESEARCH

Peretti and Pudowski (1997) studied jealousy in 95 undergraduate men and 95 undergraduate women who were presently dating and going steady with their dating partner. The effects that the jealous partner had on the nonjealous partner included loss of affection, rejection, insecurity, anxiety, and low self-esteem.

Examples of internal causes of jealousy include being mistrustful, having low self-esteem, being highly involved in and dependent on the relationship, and having no perceived alternative partners available (Pines, 1992). These internal causes of jealousy are explained below.

1. Mistrust. If an individual has been deceived or "cheated on" in a previous relationship, that individual may learn to be mistrustful in subsequent relationships. Such mistrust may manifest itself in jealousy.

2. Low self-esteem. Individuals who have low self-esteem tend to be jealous because they lack a sense of self-worth and hence find it difficult to believe anyone can value and love them. Feelings of worthlessness may contribute to suspicions that someone else is valued more.

3. High degree of relative involvement or dependency. In general, individuals who are more involved in the relationship than their partners or who are more dependent on the relationship than their partners are prone to jealousy (Radecki et al., 1988).

4. Lack of perceived alternatives. Individuals who have no alternative person or who feel inadequate in attracting others may be particularly vulnerable to jealously. They feel that if they do not keep the person they have, they will be alone.

Consequences of Jealousy

Jealousy can have both desirable consequences (reinforce one's sense of value, confirm unacceptability of outside relationship, result in reevaluation of relationship, increase communication) and undesirable consequences (increase stress, ignite self-fulfilling prophecy, lead to homicide or suicide).

Desirable outcomes   Jealousy may be functional if it occurs at a "low level" and results in open and honest discussion about the relationship. Not only may jealousy keep the partner aware that he or she is cared for (the implied message is "I love you and don't want to lose you to someone else"), but also the partner may learn that the development of other romantic and sexual relationships is unacceptable. One wife said:

> When I started spending extra time with this guy at the office my husband got jealous and told me he thought I was getting in over my head and asked me to cut back on the relationship because it was "tearing him up" and he couldn't stay married to me with these feelings. I felt he really loved me when he told me this and I chose to stop having lunch with the guy at work (authors' files).

Jealousy may improve a couple's relationship in yet another way. When the partners begin to take each other for granted, involvement of one or both partners outside the relationship can encourage them to reevaluate how important the relationship is and can help to recharge it. Bringle and Buunk summarized the potential desirable outcomes of jealousy:

> Suspicious jealousy is not necessarily unhealthy jealousy. When there is a pattern of minor incidents suggesting that the partner might be involved

Earlier in this chapter we suggested that love may be viewed on a continuum from romanticism to realism. This chapter's scale allows you to identify your level of romanticism and realism and compare it with others'.

Love Attitudes Scale*

This scale is designed to assess the degree to which you are romantic or realistic in your attitudes toward love. There are no right or wrong answers.

DIRECTIONS: After reading each sentence carefully, circle the number that best represents the degree to which you agree or disagree with the sentence.

1 Strongly agree
2 Mildly agree
3 Undecided
4 Mildly disagree
5 Strongly disagree

	1 SA	2 MA	3 U	4 MD	5 SD
1. Love doesn't make sense. It just is.	1	2	3	4	**5**
2. When you fall "head over heels" in love, it's sure to be the real thing.	1	2	**3**	4	5
3. To be in love with someone you would like to marry but can't is a tragedy.	**1**	2	3	4	5
4. When love hits, you know it.	**1**	2	3	4	5
5. Common interests are really unimportant; as long as each of you is truly in love, you will adjust.	**1**	2	3	4	5
6. It doesn't matter if you marry after you have known your partner for only a short time as long as you know you are in love.	**1**	2	3	4	5
7. If you are going to love a person, you will "know" after a short time.	1	2	3	4	**5**
8. As long as two people love each other, the educational differences they have really do not matter.	**1**	2	3	4	5
9. You can love someone even though you do not like any of that person's friends.	**1**	2	3	4	5
10. When you are in love, you are usually in a daze.	**1**	2	3	4	**5**
11. Love "at first sight" is often the deepest and most enduring type of love.	**1**	2	3	4	**5**
12. When you are in love, it really does not matter what your partner does because you will love him or her anyway.	**1**	2	3	4	**5**

	SA	MA	U	MD	SD
13. As long as you really love a person, you will be able to solve the problems you have with the person.	1	2	3	4	5
14. Usually you can really love and be happy with only one or two people in the world.	1	2	3	4	5
15. Regardless of other factors, if you truly love another person, that is a good enough reason to marry that person.	1	2	3	4	5
16. It is necessary to be in love with the one you marry to be happy.	1	2	3	4	5
17. Love is more of a feeling than a relationship.	1	2	3	4	5
18. People should not get married unless they are in love.	1	2	3	4	5
19. Most people truly love only once during their lives.	1	2	3	4	5
20. Somewhere there is an ideal mate for most people.	1	2	3	4	5
21. In most cases, you will "know it" when you meet the right partner.	1	2	3	4	5
22. Jealousy usually varies directly with love; that is, the more you are in love, the greater your tendency to become jealous.	1	2	3	4	5
23. When you are in love, you are motivated by what you feel rather than by what you think.	1	2	3	4	5
24. Love is best described as an exciting rather than a calm thing.	1	2	3	4	5
25. Most divorces probably result from falling out of love rather than failing to adjust.	1	2	3	4	5
26. When you are in love, your judgment is usually not too clear.	1	2	3	4	5
27. Love often comes only once in a lifetime.	1	2	3	4	5
28. Love is often a violent and uncontrollable emotion.	1	2	3	4	5
29. When selecting a marriage partner, differences in social class and religion are of small importance compared with love.	1	2	3	4	5
30. No matter what anyone says, love cannot be understood.	1	2	3	4	5

SCORING: Add the numbers you circled. 1 (strongly agree) is the most romantic response and 5 (strongly disagree) is the most realistic response. The lower your total score (30 is the lowest possible score), the more romantic your attitudes toward love. The higher your total score (150 is the highest possible score) the more realistic your attitudes toward love. A score of 90 places you at the midpoint between being an extreme romantic and an extreme realist. Data from 883 undergraduates revealed that both women and men are slightly more realistic than romantic and similar in their romantic/realism score. (Knox & Schacht, 1995).

*From D. Knox, *The love attitudes inventory,* rev. ed. (Saluda, N.C.: Family Life Publications, 1983).

Sometimes the reaction to one's partner's jealousy encourages further jealous behavior. Suppose John accuses
Mary of being interested in someone
else, and Mary denies the accusation
and responds by saying "I love you"
and by being very affectionate. From a
behavioral or social learning perspective, if this pattern continues, Mary will
teach John to continue being jealous.
When John acts jealous, good things
happen to him—Mary showers him
with love and physical affection. Inadvertently, Mary may be reinforcing
John for exhibiting jealous behavior.

To break the cycle, Mary should
tell John of her love for him and be affectionate when he is *not* exhibiting
jealous behavior. When he does act
jealous, she might say she feels bad
when he accuses her of something she
isn't doing and ask him to stop. If he
does not stop, she might terminate the
interaction until John can be around
her and not act jealous.

with someone else, vigilance to determine what is happening may be a
prudent response that reflects reasonable concern and good strategies to
cope with the situation. Furthermore, emotional reactions to these events
may forewarn the partner of what will happen if there are serious trans-
gressions and thereby serve the role of *preventing* extradyadic involvements (1991, 137).

Undesirable outcomes   Shakespeare referred to jealousy as the "green-eyed
monster," suggesting that jealousy has undesirable outcomes for relationships.
Three researchers (Barnett et al., 1995) confirmed the link between jealousy
and marital dissatisfaction. Jealousy that stems from low self-esteem may also
cause the partner of the jealous person to leave the relationship. Walsh (1991)
explained:

> An individual with feelings of negative self-worth . . . is continually imag-
> ining that no one could really be faithful to such an undeserving soul. If a
> person feels this way about him or herself, that atmosphere of insecurity,
> possessiveness, and accusations . . . makes it more probable that the
> mate will eventually come to share the self-evaluation and go forth to
> seek someone more deserving of his or her love. If such an event does oc-
> cur, it merely seems to vindicate what we've known all along—we're no
> good (1991).

Jealousy may also result in the jealous partner's trying to control what the
partner does and whom the partner sees. Stets studied 509 people in dating relationships (95 percent had never been married) and observed that control was
more common among nonwhite males who had low trust and high conflict with
their partners. Stets also found that younger males were more prone to control—that older males had "more alternative identities that serve as the basis
of their self esteem" (1995, 498).

In its extreme form, jealousy may have devastating consequences. In the
name of love, people have stalked the beloved, shot the beloved, and killed
themselves in reaction to rejected love. Symbolic interactionists note that the
concept of romantic love has been so constructed as to be "used as an excuse
for irresponsible, unpredictable behavior" (Winton, 1995, 141).

Being jealous is not an easy emotion to cope with. Partners who cope best
have very positive self-concepts, an array of interpersonal options, and a range
of social skills to attract and maintain new partners.

Jealousy may also create a self-fulfilling prophecy whereby the partner who is accused of having an interest
in or being involved with someone else reacts by engaging in the behavior of which he or she is accused.
The self-fulfilling prophecy is illustrated by the partner who says, "If I'm going to get accused all the time for
something I'm not guilty of, I might as well go ahead and sleep with others. After all, I get accused of it
whether I do or I don't, so I might as well enjoy what I'm being accused of."

agape love style   Love style characterized by a focus on the well-being of the love object with little regard for reciprocation. The love of parents for their children is agape love.

codependent relationship   A relationship in which each partner requires constant affirmation from the other and may tolerate almost anything to maintain the relationship.

conjugal love   The type of love experienced by couples who have been married several years, which is characterized by companionship, calmness, comfort, and security.

eros love style   Love style characterized by passion and romance.

jealousy   An emotional response to a perceived or real threat to an important or valued relationship.

ludic love style   Love style in which love is viewed as a game whereby the love interest is kept at a distance.

mania love style   An out-of-control love whereby the person "must have" the love object. Obsessive jealousy and controlling behavior are symptoms of manic love.

pragma love style   Love style that is logical and rational. The love partner is evaluated in terms of pluses and minuses and regarded as a good or bad "deal."

romantic love   An intense love whereby the lover believes in love at first sight, only one true love, and love conquers all.

storge love style   A love consisting of friendship that is calm and nonsexual.

SUMMARY

Love remains one of the most powerful emotions in relationships.

Conceptions of Love

Sternberg conceptualized love as a triangular interplay of three basic elements—intimacy, passion, and commitment. The ways in which people regard love consist of varying amounts of these three elements. Liking, for example, consists of intimacy that is devoid of both passion and commitment.

John Lee identified various love styles that describe the way lovers relate to each other. Ludic lovers play games, storge lovers have a strong friendship, and pragma lovers are very businesslike in their decisions about love. Eros lovers, who are consumed with passion and romance, are more likely to maintain their relationship than ludic, game-playing, lovers.

Romantic love is characteristic of both men and women. Men are more romantic when romantic love is defined as falling in love very quickly. Women are more romantic when love is defined as the tendency to form intense and lasting bonds.

Theories on the Origins of Love

Theories of love include evolutionary (love provides the social glue needed to bond parents with their dependent children and spouses with each other to care for their dependent offspring), social learning (positive experiences create love feelings), psychosexual (love results from a blocked biological drive), ego-ideal (love springs from a dissatisfaction with one's self), ontological (love represents people's urge to be reunited with the other half), biochemical (love involves feelings produced by biochemical events), and attachment (a primary motivation in life is to be connected with other people).

Conditions for the Development of Love

Love occurs under certain conditions. Social pressures encourage some love relationships and discourage others. Psychological conditions involve a positive self-concept and a willingness to disclose one's self to others. Physiological and cognitive conditions imply that the individual experiences a stirred-up state and labels it "love."

Problems Associated with Love

Problems of love include being in love with two people at the same time, being in love with someone who is abusive or perhaps an alcoholic, and allowing "love" to create a context in which we make dangerous choices.

Jealousy

Jealousy is an emotional response to a perceived or real threat to a valued relationship. Jealous feelings may have both internal and external causes and may have both positive and negative consequences for a couple's relationship.

Jealousy may be functional for a couple's relationship in that it may help the partners reassess their value to each other. For example, if one partner is perceived as being interested in someone new, the other partner may experience jealousy and ask what is going on. Such a question need not be accusatory but may express a concern for keeping one's partner satisfied in the relationship.

REFERENCES

Ackerman, Diane. 1994. *A natural history of love.* New York: Random House.

Barnett, Ola W., T. E. Martinez, and B. W. Bluestein. 1995. Jealousy and romantic attachment in maritally violent and nonviolent men. *Journal of Interpersonal Violence* 10: 473–86.

Brehm, S. S. 1992. *Intimate relationships.* 2d ed. New York: McGraw-Hill.

Bringle, R. G., and B. P. Buunk. 1991. Extradyadic relationships and sexual jealousy. In *Sexuality in close relationships,* edited by K. McKinney and S. Sprecher. Hillsdale, N.J.: Lawrence Erlbaum As, 135–54.

Buunk, B., and R. B. Hupka. 1987. Cross-cultural differences in the elicitation of sexual jealousy. *Journal of Sex Research* 23: 12–22.

Carey, Anne R., and Marcy E. Mullins. 1996. Gender and savings goals. *USA Today,* 31 January, B1.

Cassidy, M. L., and G. Lee. 1989. The study of polyandry: A critique and synthesis. *Journal of Comparative Family Studies* 20: 1–11.

Christensen, C. D., and D. M. Busby. 1992. Homogamy of personality variables and relationship stability. In *Families and work.* Vol. 2 of *Proceedings.* 54th Annual Conference of the National Council on Family Relations, no. 1, 35.

Freud, Sigmund. 1905[1938]. Three contributions to the theory of sex. In *The basic writings of Sigmund Freud,* edited by A. A. Brill. New York: Random House.

Fromm, E. 1963. *The art of loving.* New York: Bantam Books.

Gallmeier, C. P., M. E. Zusman, D. Knox, and L. Gibson. 1997. Can we talk? Gender differences in disclosure patterns and expectations. *Free Inquiry in Creative Sociology* 25: 219–225.

Goode, Wi. J. 1959. The theoretical importance of love. *American Sociological Review* 24: 38–47.

Hall, Cindy, and Marcia Staimer. 1995. Love at first sight? *USA Today,* 10 July, 1d.

Hatfield, Elaine, and Richard L. Rapson. 1996. *Love and sex: Cross-cultural perspectives.* Boston: Allyn and Bacon.

Hendrick, C., and S. S. Hendrick. 1991. Dimensions of love: A sociobiological interpretation. *Journal of Social and Clinical Psychology* 10: 206–30.

Hendrick, S. S., C. Hendrick, and N. L. Adler. 1988. Romantic relationships: Love, satisfaction, and staying together. *Journal of Personality and Social Psychology* 54: 980–88.

Hill, C. A., J. E. O. Blakemore, and P. Drumm. 1997. Mutual and unrequited love in adolescence and young adulthood. *Personal Relationships* 4: 15–23.

Inman-Amos, Jill, Susan S. Hendrick, and Clyde Hendrick. 1994. Love attitudes: Similarities between parents and between parents and children. *Family Relations* 43: 456–61.

Jankowiak, W. R., and E. F. Fischer. 1992. A cross-cultural perspective on romantic love. *Ethnology* 31: 149–55.

Johnson, Susan, and Hara Estroff Marano. 1994. Love: The immutable longing for contact. *Psychology Today*, March/April, 32 passim.

Knee, C. R. 1998. Implicit theories of relationships: Assessment and prediction of romantic relationship initiation, coping, and longevity. *Journal of Personality and Social Psychology* 74: 360–370.

Knox, David, and Caroline Schacht. 1995. Love attitudes inventory, abridged version data. Unpublished data, East Carolina University, Greenville, N.C.

Knox, D., and M. E. Zusman. 1998. Unpublished data collected for this text.

Knox, D., M. Zusman, and W. Nieves. In press. What I did for love: Dangerous behaviors of college students in love. *College Student Journal*.

Knox, D., and M. Zusman. Unpublished. Love among college students. Submitted to *College Student Journal*.

Lamm, H., and U. Wiesmann. 1997. Subjective attributes of attraction: How people characterize liking, their love, and their being in love. *Personal Relationships* 4: 271–284.

Lee, J. A. 1973. *The colors of love: An exploration of the ways of loving.* Don Mills, Ontario: New Press.

————. 1988. Love-styles. In *The psychology of love*, edited by R. Sternberg and M. Barnes. New Haven, Conn.: Yale University Press, 38–67.

Levine, Robert, Suguru Sato, Tsukasa Hashimoto, and Jvoti Verma. 1995. Love and marriage in eleven cultures. *Journal of Cross-Cultural Psychology* 26: 554–71.

Lewis, David. 1985. *In and out of love: The mystery of personal attraction.* London: Methuen.

Loftus, Mary. 1995. Frisky business. *Psychology Today*, March/April, 35–41.

Michael, R. T., J. H. Gagnon, E. O. Laumann, and G. Kolata. 1994. *Sex in America.* Boston: Little, Brown.

Mitchell, Grace, and Patricia M. Fandt. 1995. Examining the relationship between role-defining characteristics and self-esteem of college students. *College Student Journal* 29: 96–102.

Montgomery, Marilyn J., and Gwendolyn T. Sorell. 1997. Differences in love attitudes across family life stages. *Family Relations* 46: 55–61.

Peele, S., and A. Brodsky. 1976. *Love and addiction.* New York: New American Library.

Peretti, P. O., and B. C. Pudowski. 1997. Influence of jealousy on male and female college daters. *Social Behavior and Personality* 25: 155–160.

Petrie, J., J. A. Giordano, and C. S. Roberts. 1992. Characteristics of women who love too much. *Affilia: Journal of Women and Social Work* 7: 7–20.

Pilkington, Constance J., Whitney Kern, and David Indest. 1994. Is safer sex necessary with a "safe" partner? Condom use and romantic feelings. *Journal of Sex Research* 31: 203–10.

Pines, A. M. 1992. *Romantic jealousy: Understanding and conquering the shadow of love.* New York: St. Martin's.

Radecki Bush, C. R., J. P. Bush, and J. Jennings. 1988. Effects of jealousy threats on relationship perceptions and emotions. *Journal of Social and Personal Relationships* 5: 285–303.

Reik, T. *Of love and lust.* 1949. New York: Farrar, Straus, and Cudahy.

Ripke, Marika N., and Ted L. Huston. 1997. The labor of love in marriage: A sentimental analysis of household work. Paper presented at the 59th Annual Conference of the National Council on Family Relations, Crystal City, Va., November.

Rodriquez, M., and E. L. Devall. 1997. Gender role attitudes, relational self-disclosure, and marital satisfaction. Paper presented at 59th Annual Conference of the National Council on Family Relations, Crystal City, Va., November.

Shepard, Janet G. 1993. One-time engaged individuals and multiple engaged individuals: A comparison of the degree of romanticism. Master's thesis, Southeast Missouri State University. Used by permission.

Silliman, Benjamin, and Walter R. Schumm. 1955. Client interests in premarital counseling: A further analysis. *Journal of Sex and Marital Therapy* 21: 43–56.

Statistical Abstract of the United States: 1997. 117th ed. Washington, D.C.: U.S. Bureau of the Census.

Stets, Jan E. 1995. Modeling control in relationships. *Journal of Marriage and the Family* 57: 489–501.

Sternberg, R. J. A triangular theory of love. 1986. *Psychological Review* 93: 119–35.

Sternberg, R. J., and A. E. Beall. 1991. How can we know what love is? An epistemological analysis. In *Cognition in close relationships,* edited by Garth J. O. Fletcher and F. D. Fincham. Hillsdale, N.J.: Lawrence Erlbaum Associates, 257–78.

Stone, C. R., and J. L. Philbrick. 1991. Attitudes toward love among members of a small fundamentalist community in South Africa. *Journal of Social Psychology* 131: 219–33.

Toufexis, Anastasia. 1993. The right chemistry. *Time.* 15 February, 49–51.

Umana, Adriana J., C. A. Surra, and S. E. Jacquet. 1997. Predicting commitment to wed among Hispanic and Anglo partners. Paper presented at the 59th Annual Conference of the National Council on Family Relations, Crystal City, Va., November. Used by permission.

Walsh, Anthony. 1991. *The science of love: Understanding love and its effects on mind and body.* Buffalo, N.Y.: Prometheus Books.

Walster, E., and G. W. Walster. 1978. *A new look at love.* Reading, Mass.: Addison-Wesley.

White, G. L. 1980. Inducing jealousy: A power perspective. *Personality and Social Psychology Bulletin* 6: 222–27.

Winton, C. A. *Frameworks for studying families.* Guilford, Conn.: Dushkin Publishing Group.

CHAPTER 4 · Dating and Mate Selection

Keep your eyes wide open before marriage, half shut afterwards.

Benjamin Franklin, Inventor

While *dating* is often solely for recreation it may also be a mechanism whereby men and women pair off and form exclusive, committed relationships for the reproduction, nurturing, and socialization of children. In this chapter, we look at the issues confronting dating couples, review theories of mate selection, and examine the consequences of living together. We begin with a look at how dating patterns have changed in the last fifty years.

Dating: Then and Now

The Industrial Revolution transformed dating norms. This transition and changes in the last fifty years are detailed below.

The Industrial Revolution

The transition from a courtship system controlled by parents to the relative freedom of mate selection experienced by today's youth occurred in response to a number of social changes. The most basic change was the Industrial Revolution, which began in England in the middle of the eighteenth century. No longer were women needed exclusively in the home to spin yarn, make clothes, and process food from garden to table. Commercial industries had developed to provide these services, and women transferred their activities in these areas from the home to the factory. The result was that women had more frequent contact with men.

Women's involvement in factory work decreased parental control: parents were unable to dictate the extent to which their offspring could interact with those they met at work. Hence, values in mate selection shifted from the parents to the children. In the past, the "good wife" was valued for her domestic aptitude—her ability to spin yarn, make clothes, cook meals, preserve food, and care for children. The "good husband" was evaluated primarily as an economic provider. Though these issues may still be important, contemporary mates are more likely to be selected on the basis of personal qualities, particularly for love and companionship, than for either utilitarian or economic reasons.

Changes in the Last Fifty Years

Though the Industrial Revolution had a profound effect on courtship patterns, these patterns have continued to change in the past fifty years. The changes include marrying later, which allows a person a longer period of time to become involved with more people. Marrying at age 29 provides a greater opportunity to date more people than marrying at age 24. The dating pool for persons in their late twenties also includes an increasing number of individuals in their thirties who have been married before. These individuals often have children, which changes the nature of a date from two adults going out alone to see a movie to renting a movie and baby-sitting in the apartment or home of one of the partners.

Today, young people in high school and early college sometimes date in the traditional sense of phoning a person on Wednesday for a Saturday night date. In addition, contemporary patterns of dating among youth include dating in groups and dating in nonexclusive or exclusive relationships. Some young people date by "hanging out" and "getting together" in groups of various sizes; others prefer one-on-one relationships. The latter may be "open" (each partner may date others) or "closed" (the partners date each other exclusively). Such exclusive dating may or may not be oriented toward marriage.

However, dating in the traditional sense still occurs for some older adults. Often these individuals are divorced or widowed and tend to revert to the dating patterns they exhibited when they were dating in high school and early college.

As we will see later in this chapter, cohabitation has become more normative. For some couples, the sequence of date, fall in love, and get married has been replaced by date, fall in love, and live together. Such a sequence results in marriages between couples who are more relationship-savvy than those who dated and married out of high school. In addition to individuals dating more partners and potentially living together, gender role relationships have become more egalitarian. Though the double standard still exists, women today are more likely than women in the 1950s to ask men to go out, to have sex with them without requiring a commitment, and to time marriage when their own educational and career goals have been met. Women no longer feel desperate to marry but consider marriage one of many goals they have for themselves.

Unlike during the fifties, both sexes today are aware of and somewhat cautious of the threat of HIV infection. Sex has become potentially deadly, and condoms have become accepted as a prelude to sexual activity. The embarrassment in the fifties over asking a pharmacist for a condom is being replaced by the confidence and mundaneness of buying condoms along with one's groceries.

Finally, couples of today are more aware of the impermanence of marriage. However, most couples continue to feel that divorce will not happen to them and remain committed to domestic goals. Seventy-two percent of the first-year students identified in the national data section reported "an essential or important goal" for them is "raising a family" (American Council on Education and University of California, 1997).

Dating Issues

Issues in dating include finding a partner, problems in dating, dating after divorce, and dating in the later years. This section explores each of these dating issues.

Finding a Partner

Finding a partner is a common problem. Country-and-western singer Pam Tillis pointed out the difficulty as one gets older. In her 1997 hit "All the Good Ones Are Gone" she sings of a 34-year-old single woman who goes to bars with her friends looking for a partner: "Back then there were so many but now there just aren't any . . . it seems like all the good ones are gone."

> If you want to catch more
> fish, use more hooks.
> George Allen
> Football coach

"On-line" is one of the most popular ways of finding a new partner.

> The odds are good—but the
> goods are odd.
> Anonymous

> I flee the one that chases me
> and chase the one that flees
> me.
> Ovid
> Roman poet

> I only like two kinds of men—
> domestic and foreign.
> Mae West
> American film actress

NATIONAL DATA

Of America's approximately 2,000 firms specializing in finding a partner, about 600 use video technology (Ahuvia & Adelman, 1992). Great Expectations is the largest videodating service in the United States. It has over twenty-one centers and 65,000 clients nationwide.

A national sample of people who were dating identified how they met their partners (Michael et al., 1994). Most (42%) reported that they introduced themselves; 36 percent said they had been introduced by mutual friends and 11 percent, by coworkers. Only 8 percent said that their parents provided the introduction (Michael, et al., 1994). With regard to self-introduction, Bill Clinton was asked how he and Hillary met. He described the occasion when he and Hillary were studying at opposite ends of a long table in the Yale law library. They had seen each other on campus and in classes but had never spoken. Then one evening Hillary walked the length of the table and said, "I'd like to meet you."

The place the respondents in the national sample met was also variable, with about equal percentages meeting at school (22%), party or gym (20%), work (17%) and "elsewhere" (23%). Sixteen percent said that they met through a personal ad, at a bar, or on vacation (Michael et al., 1994).

Personal ads include the use of magazines, newspapers, and videodating clubs. In a review of studies on advertising in magazines and newspapers, Ahuvia and Adelman (1992) reported that women received an average of fifteen replies, men eleven. Factors associated with men who received the most replies included (1) being older and taller, (2) having higher education/professional status, and (3) seeking an attractive woman but avoiding sexual references. For women, those factors associated with receiving the most replies included (1) physical attractiveness, (2) being younger, physically fit, and interested in sports, and (3) mentioning or alluding to sex. "For both men and women, writing an ad with originality or flair increased the number of respondents, as did seemingly trivial traits like possessing a certain hair color" (Ahuvia & Adelman, 1992, 457).

How successful are commercial dating services in matching people with others with whom they have something in common and whom they eventually marry? The available evidence suggests that such avenues do help to screen out partners with whom one would have little or nothing in common. The result is that partners who end up meeting are fairly similar in their interests and characteristics. Persons who are matched as to optimism-pessimism and abstract versus concrete thinking are the most likely to like the person with whom they are matched and to end up getting married (Ahuvia & Adelman, 1992).

I bet you that within ten years, the on-line community is going to change the way a majority of people meet and mate.

Deborah Baumbrucker
who met her husband on-line

Searching for a partner on the Internet is a version of the mail order matching that was prevalent in the 1830s–1860s. With the ratio of men to women as high as twenty to one in Colorado in 1868, men would advertise for a wife in Eastern newspapers and propose after a brief courtship by correspondence. A woman took a big chance agreeing to marry a man twice her age who lived in the wilderness in very harsh conditions. (But the stigma of "old maid" was worse).

Some magazines feature ads marketed to a particular group of singles. Sparrow (1991) noted, "Increasingly, blacks are using personal ads in black and other singles magazines to meet members of the other sex. Such magazines include *Black Gold* in Chicago, *Chocolate Single* in New York, and *Positively Black Professionals* in New Jersey" (105).

Another magazine, *Cherry Blossoms* (now in its twenty-seventh year), is designed for the individual seeking a partner from another country. Each issue features ads from more than 500 women from forty countries. A twelve-month subscription costs over $500.

Another method of finding a partner is to be interviewed on videotape and to let others watch your videotape in exchange for your watching tapes already on file. Once you have seen someone that you like, the person is contacted by the service and invited in to review your tape. If the interest is mutual, you and the other person will meet.

The newest method of finding a partner is using a computer. *Chat rooms, E-mail,* and *IMs* (instant messages) are computer terms that reflect the revolution in finding new partners on the Internet. Once an initial contact is made, individuals assess each other's interests and determine whether they would like to meet. Rush Limbaugh reported that he met his wife through an E-mail on the Internet. Meeting at the movies in the fifties has been replaced by meeting on-line in the twenty-first century.

One of the allures of an on-line meeting is that it shifts the focus from the visual to inner thoughts and feelings—a quick trip to intimacy. In the absence of the visual to distract or eliminate would-be partners, the focus of the interaction may become common interests, values, and goals. But there are drawbacks. Deception about weight, age, looks, and even one's sex (called gender bending) is not unusual. Beware, and take a friend with you to the first meeting if there is one.

Some Internet exchanges have disastrous consequences. Howard Eskin discovered Marlene Stumpf on-line (they were both married). He ended up sending her a dozen red roses, which were intercepted by Marlene's husband, Raymond, who allegedly knifed her to death over her "Internet affairs" (O'Neill et al., 1997). Such an outcome is unusual but does emphasize the need for caution.

Problems in Dating

After finding a dating partner, individuals are confronted with problems or issues they must manage. Table 4.1 describes dating problems reported by 527 undergraduate students. When "casual" and "involved" daters were compared, both reported that communication was the number one problem. However, the number two problem for "casual" daters was "lack of commitment," and this was the tenth most frequent problem reported by the "involved" (Zusman and Knox, 1998).

Dating After Divorce

Over two million Americans get divorced each year. As evidenced by the fact that over three-quarters of the divorced remarry, most of the divorced are

Table 4.1

Top Ten Problems Experienced by Casual and Involved Daters (N = 527) (of 620 respondents, 527 reported problems)			
Casual Daters (N = 240)		**Involved Daters (N = 287)**	
Communication	19.6%	Communication	22.3%
Lack of commitment	12.5	Other problems	15.3
Jealousy	12.1	Jealousy	13.9
Other problems	9.6	No problems	13.2
No problems	8.3	Time for relationship	9.1
Different values	7.9	Lack of money	5.2
Honesty	7.5	Places to go	4.9
Shyness	5.4	Honesty	4.2
Unwanted sex pressure	2.1	Different values	3.1
Acceptance	1.7	Lack of commitment	2.4

Source: Zusman and Knox, 1998.

"looking again" for a new partner. But there are differences between this "single again" population and those becoming involved for the first time.

1. Divorced individuals are, on the average, ten years older than those who are in the marriage market who have never been married before. Widows and widowers are usually forty and thirty years older, respectively, when they begin to date the second time around. When divorced men remarry, they are about age 37; divorced women, 34. When widowed men remarry, they are about age 63; widowed women, 54. In contrast, men and women marrying for the first time are 27 and 25, respectively (Saluter, 1996).

2. Most men and women who are dating the second time around find fewer partners from which to choose than when they were dating before their first marriage. The large pool of never-marrieds (23 percent of the population) and currently marrieds (60 percent of the total population) is not considered an option (*Statistical Abstract of the U.S. 1997,* Table 58). Most divorced and widowed persons date and marry others who have been married before.

3. The older an unmarried person, the greater the likelihood of having had multiple sexual partners (Michael et al., 1994), which has been associated with increased risk of contracting HIV and other STDs. Therefore, individuals entering the dating market for the second time are advised to be more selective in choosing their sexual partners because the likelihood of

RECENT RESEARCH

Among the "other dating problems" mentioned by the respondents in Table 4.1 was being physically separated from one's dating partner—being involved in a long-distance dating relationship. Van Horn et al., 1997, compared 80 college students' long-distance romantic relationships with 82 relationships that were not long distance and found that the former were more likely to be characterized by less self-disclosure, less companionship with the partner, less satisfaction, and less certainty that the relationship would endure. Relationship satisfaction was the greatest predictor of the stability of the relationship.

NATIONAL DATA

Men age fifty and older were one-sixth as likely to use condoms during sex and one-fifth as likely to have been tested for HIV as a comparison group of at-risk individuals in their twenties (Stall & Catania, 1994).

those partners' having had a higher number of sexual partners is greater. In addition, the divorced are much less likely to be monogamous than married adults (Greeley, Michael, & Smith, 1990), so dating an older divorced person may involve even greater risk of contracting an STD. An increasing number of individuals are becoming concerned about HIV and other infections and discuss being HIV tested with new partners before having sex with them. Such a discussion may be particularly important for those reentering the dating market. Although only 10 percent of all HIV cases diagnosed in the United States have been among Americans fifty years of age and older, older adults continue to engage in high-risk behaviors (Stall & Catania, 1994).

4. More than half of those dating again have children from a previous marriage. How these children feel about their parents' dating, how the partners feel about each other's children, and how the partners' children feel about one another are complex issues. Deciding whether to have intercourse when one's children are in the house, what the children should call the new partner, and how to deal with terminations of previous relationships are other issues familiar to many people dating for the second time.

A team of researchers (Darling et al., 1989) studied the effects of children on dating the second time around as experienced by 155 single parents. They concluded, "The presence of children appears to be a major obstacle in the development of new relationships by parents" (p. 241). Mothers reported more interference than fathers, since the former are more often the custodial parents.

Nothing grows again more easily than love.
Seneca
Philosopher

5. Previous ties to the ex-spouse in the form of child support or alimony, phone calls, and the psychological memory-experience of the partner's first marriage will have an influence on the new dating relationship. If the separation/divorce was bitter, the partner may be preoccupied or frustrated in his or her attempts to cope with the harassing ex-spouse.

So far as is known, no widow ever eloped.
E. W. Howe
County Town Sayings

6. Divorced people who are dating again tend to have a shorter courtship period than first marrieds. In a study of 248 individuals who remarried, the median length of courtship was nine months as opposed to seventeen months the first time around (O'Flaherty & Eells, 1988). A shorter courtship may mean that sexual decisions are confronted more quickly—timing of first intercourse, discussing the use of condoms/contraceptives, and discussing when and whether the relationship is to be monogamous.

NATIONAL DATA

In 1996, there were 1,718,000 unattached men (never-married, widowed, divorced), compared with 4,543,000 unattached women (never-married, widowed, divorced) between the ages of 65 and 74 (*Statistical Abstract of the United States: 1997,* Table 59).

Dating in the Later Years

As a result of divorce and late-life widowhood, the number of elderly (over age 65) people without partners is one of the fastest growing segments of our population.

Particularly vulnerable to being without a partner are the women over age 80 for whom there are only fifty-three men for every one hundred women. Patterns that are developing to adjust to this lopsided man-woman ratio include

While entering the dating market as an older woman is a decided liability, older men are also not immune to being screened out. Charles Simmons, a novelist, lamented that as an older man, he has become invisible to women. The grocery store clerks, he said, used to flirt with him. Now, he wrote, "There comes a time when, like the boy putting groceries in your cart, she doesn't see you. She sees the cold cuts and the beer, the half a loaf of bread, and that's all." He concludes: "You're being screened out." (Michael et al., 1994, 87).

Individuals who marry are influenced by endogamous and exogamous cultural factors.

Marriage has no natural relation to love. Marriage belongs to society; it is a social contract.
Samuel Rogers
Table-Talk

women dating younger men, romance without marriage, and "share-a-man" relationships.

Over 3 million women are married to men who are at least ten years younger than they. Although traditionally women were socialized to seek older, financially "established" men, the sheer shortage of men has encouraged many women to seek younger partners.

Women in their later years have also moved away from the idea that they must remarry and have become more accepting of the idea that they can enjoy the romance of a relationship without the obligations of a marriage. Ken Dychtwald (1990) in his book *Age Wave* observed that many elderly women were interested in romance with a man but were not interested in giving up their independence. "Many say they do not have the same family-building reasons for marriage that young people do. For women especially, divorce or widowhood may have marked the first time in their lives that they have been on their own, and many now enjoy their independence" (p. 222). In addition, many elders are reluctant to marry because of a new mate's deteriorating health. "They would not want to become the caretaker of an ill spouse, especially if they had been through an emotionally draining ordeal before" (p. 222).

Faced with a shortage of men but reluctant to marry those who are available, some elderly women are willing to share a man. In elderly retirement communities such as Palm Beach, Florida, women count themselves lucky to have a man who will come for lunch, take them to a movie, or be an escort to a dance. They accept the fact that the man may also have luncheon, movie, or dancing dates with other women. "The alternative may be no male companionship at all" (Dychtwald, 1990, 226).

Mate Selection

The meeting and marriage of the couples you know did not occur by chance. Rather, various cultural, sociological, psychological, and—some sociobiologists say—biological factors were influential in their union.

Cultural Aspects of Mate Selection

OTHER CULTURES As noted earlier, cultural norms for mate selection vary. The degree of freedom individuals have in choosing a marriage partner depends on the culture of the society in which they live. Although the tradition is becoming more rare, some parents in cultures such as China, India, and Palestine select the spouses for their children. They are more concerned about the religion and social class of the person's family than the love feelings between the partners. Love is expected to follow, not precede, marriage.

A sample of Indians (from India) (mostly between the ages of 17 and 21) identified the advantages of their arranged marriages as having support from their parents and having compatible backgrounds. They listed the disadvantages as not knowing the partner well, greedy in-laws demanding a huge dowry from the woman's father, and incompatibility (Sprecher & Chandak, 1992). ●

Endogamous-exogamous pressures   Whereas some societies exert specific pressure on individuals to marry predetermined mates, other societies apply more subtle pressure. The United States has a system of free choice that is not exactly free. Social approval and disapproval restrict your choices so that you do not marry just anybody. *Endogamous pressures* encourage you to marry those within your own social group (racial, religious, ethnic, educational, economic); *exogamous pressures*, in the form of legal restrictions, require you to marry outside your family group (to avoid sex with and marriage to a sibling or other close relative).

The pressure toward an endogamous mate choice is especially strong when race is concerned. One white woman said, "Some of my closest friends are black. But my parents would disown me if I were to openly date a black guy." In contrast, a black man said, "I would really like to date a girl in my introductory psychology class who's white. But my black brothers wouldn't like it, and while my parents wouldn't throw me out of the house, they would wonder why I wasn't dating a black girl."

These endogamous pressures are not operative on all people at the same level or may not work at all. Those who are older than thirty, who have been married before, or who live in large urban centers are more likely to be color-blind in their dating and marrying. In Hawaii, interracial dating and marriage are normative.

Endogamous dating and mating pressure are also evident among religious groups such as the Mormons. One researcher (Markstrom-Adams, 1991) compared thirty-six Mormon high school students with forty-seven non-Mormons and found only 22 percent of Mormons would "advise dating between Mormons and non-Mormons," in contrast to 45 percent of non-Mormons who would advise such dating.

In contrast to endogamous marriage pressures, exogamous pressures are mainly designed to ensure that individuals who are perceived to have a close biological relationship do not marry each other. Incest taboos are universal. In no society are children permitted to marry the parent of the other sex. In the United States, siblings and (in some states) first cousins are also prohibited from marrying each other. The reason for such restrictions is fear of genetic defects in children.

RECENT RESEARCH

Public reaction to interracial marriage is often negative. A study of twenty black-white couples who had been married for at least a year found that all had experienced negative public reactions—stares, disapproving expressions, harassment at the work site, etc. (Killian, 1997).

Sociological Aspects of Mate Selection

Sociological aspects of mate selection include homogamy and propinquity.

Homogamy   While endogamy is a concept that refers to cultural pressure, *homogamy* refers to individual initiative toward sameness. The homogamy theory of mate selection states that we tend to be attracted to and become involved with those who are similar to ourselves in such characteristics as age, race, religion, and social class. Selecting for sameness seems particularly true for older couples, as "younger couples are more attracted to somewhat opposite types" (Sherman & Jones, 1994). Considerable research suggests that homogamy, or "like attracts like," in the selection of a marriage partner is associated with more durable and satisfying relationships (Houts et al., 1996; Michael et al., 1994). The various homogamous factors follow:

If you would be married fitly, wed your equal.
Ovid
Philosopher

INSIGHT

One of the unique qualities of universities is that they provide an environment in which to meet hundreds or even thousands of possible partners of similar age, education, social class, and general goals. This opportunity will probably not be matched later in the workplace or where one lives following graduation.

Only equals should mate.
O. Henry
Novelist

RECENT RESEARCH

Waris (1997) studied 300 married couples living in Kansas City and found that the husbands tended to have a higher occupational status and to be older than their wives.

1. *Age.* When a friend gets you a date, you assume the person will be close to your age. Your peers are not likely to approve of your becoming involved with someone twice your age. A concern for age homogamy is particularly characteristic of individuals who have never married. Those who are divorced or widowed are much more likely to become involved with someone who is less close to their age.

2. *Education.* The level of education you attain will also influence your selection of a mate. A sophomore who worked in a large urban department store during the summer remarked, "The time Todd and I spent working in the store was great. But our relationship never gathered momentum. I was looking forward to my last two years of school, but Todd said college was a waste of time. I don't want to get tied to someone who thinks that way."

   This student's experience suggests that you are likely to marry someone who has also attended college. Not only does college provide an opportunity to meet, date, and marry another college student, but it also increases the chance that only a college-educated person will be acceptable to you. The very pursuit of education becomes a value to be shared and, when educational levels are similar, helps to insulate one's relationship against future divorce (Tzeng, 1992).

3. *Social Class.* You have been reared in a particular social class that reflects your parents' occupations, incomes, and educations as well as your residence, language, and values. If you were brought up in the home of a physician, you probably lived in a large house in a nice residential area. You were in a higher social class than you would have been if your parents were less educated and had lower incomes.

   The social class in which you were reared will influence how comfortable you feel with a partner. "I never knew what a finger bowl was," recalled one man, "until I ate dinner with my girlfriend in her parents' Manhattan apartment. I knew then that while her lifestyle was exciting, I was more comfortable with paper napkins and potato chips. We stopped dating." Indeed, women who have high incomes may either be less attractive as partners or have less need for a husband. Greenstein (1992) noted in a national sample of 2,375 never-married women that high-income women were less likely to marry.

   The *mating gradient* refers to the tendency for husbands to be more advanced than their wives with regard to age, education, and occupational success. Two researchers assessed the expectations of 131 single female and 103 male college students and found that most women expected their husbands to be "superior in intelligence, ability, success, income, and education. Less than 10 percent of the women in this sample expected to exceed their marriage partner on any of the variables measured" (Ganong & Coleman, 1992, 61).

   As a result of the mating gradient, some high-status women remain single. Upper-class women typically receive approval from their parents and peers only if they marry someone of equal status. Educated women, rather than drop their standards, may go unmarried. Conversely, those who improve their own educational and economic opportunities "improve the likelihood of marrying a man of means." This finding is based on a national sample of 1,711 women in their first marriages (Lichter et al., 1995, 429).

4. *Race.* While 95 percent of people marry someone of the same race (*Statistical Abstract of the United States: 1997*, Table 62), there is an increasing willingness to date interracially.

The greatest tolerance seems to be for whites dating Hispanics. Black-white pairings are still infrequent. Only 17 percent of white teens who date report having dated someone who is black (33% have dated a Hispanic). Nevertheless, the trend is toward more interracial dating of all races. Only 13 percent of the respondents say that they would not consider dating interracially. Reasons for dating interracially include "curiosity" (75%), "trying to be different" (54%), and "to rebel against parents" (47%) (Peterson, 1997, 2A). As for parental reactions, 69 percent of the parents in the teen study reported that "it's all right for blacks and whites to date each other" (Peterson, 1997, 10A).

5. *Physical Attractiveness.* In general, people tend to become involved with those who are similar in physical attractiveness. However, a partner's attractiveness is usually more important to men than women.

6. *Body-Clock Compatibility.* Some of us are morning people, and some of us are night people. Morning people arise early and feel most energetic in the morning. Night people feel most energy late at night. Night people like to go to sleep at dawn—just the time when morning people are getting up.

A team of researchers (Larson et al., 1991) studied the marital relationships of 150 couples who were mismatched or matched in terms of body clocks. The mismatched couples reported significantly less marital adjustment, more marital conflict, less time spent in serious conversation, less time spent in shared activities, and less frequent sexual intercourse than matched couples. Loneliness was also a problem. For mismatched couples, both spouses commonly complained of feeling lonely. The morning person would wake up and have coffee alone, since the partner was still asleep; similarly, the night person would watch late-night TV alone, since the partner was already asleep. Couples on different sleep patterns may work on this aspect of their relationship through compromise or accommodation or may develop positive ways of viewing their differences.

7. *Other Factors.* Marital status and religion are other factors involved in homogamous mate selection. The never-married tend to select the never-married, the divorced tend to select the divorced, and the widowed tend to select the widowed as partners to marry. In addition, although religious homogamy is decreasing because we are becoming increasingly pluralistic and secularized as a society, this factor is still operative. Two researchers analyzed the couple formation of a sample of college students and observed that spirituality, Christianity, and a view that marriage is a lifetime commitment were important considerations in the selection of a partner. Some of the respondents also noted that God played a vital role in their formation as a couple (Young & Schvaneveldt, 1992). Booth et al.

It is possible that anxious persons and persons from troubled backgrounds come together by necessity because they were not seen as attractive mates by others. Anxious persons and persons from troubled backgrounds may also share common interpersonal styles that serve as the basis for their attraction. Or, persons from troubled backgrounds may be disproportionately likely to choose partners who have behavioral and psychological problems because they have low levels of self-confidence (McLeod, 1995, 211).

(1995) found that religiosity is also related to marital happiness in that it may be associated with a common activity, such as attending church.

Similarity of perceptions is another homogamous factor that researchers have found to be related to marital happiness. Family therapists have found that spouses who have similar perceptions of themselves, their relationship, their children, and family life in general tend to report more marital satisfaction and more harmonious family functioning (Deal, Wampler, & Halverson, 1992).

Another homogamous variable in mate selection is similar interpersonal values. Kilby (1993) identified the following list of interpersonal values:

- being treated with respect
- granting the other autonomy or independence
- giving the other support
- doing things for others
- respecting the privacy of others
- having complete candor and no secrets

You might consider the degree to which each of these interpersonal values is important to you and the degree to which you would like your partner to feel the same.

Propinquity   *Propinquity* means nearness and, in mate selection, suggests that you are most likely to marry someone that you live near, work with, or go to school with. Propinquity is best illustrated by the marriage of servicemen overseas. During World War II over 70,000 American GIs in England married British women and upwards of 100,000 servicemen stationed in the Far East married Asian women. Servicepeople continue to marry persons they are stationed near.

Personnel directors of large corporations who hire individuals to work in the corporation are, in effect, matchmakers. Selecting people with a certain level of education as well as job skills for the corporation produces a pool of similarly educated and skilled people who will be in daily contact. The theory of propinquity says that they are likely to end up dating and marrying one another.

Psychological Aspects of Mate Selection

Psychologists have focused on complementary needs, exchanges, parental characteristics, and personality qualities with regard to mate selection.

Complementary-needs theory   "In spite of the women's movement and a lot of assertive friends, I am a shy and dependent person," remarked a transfer student. "My need for dependency is met by Warren, who is the dominant, protective type." The tendency for a submissive person to become involved with a dominant person (one who likes to control the behavior of others) is an example of attraction based on complementary needs. *Complementary-needs theory* states that we tend to select mates whose needs are opposite and complementary to our own needs. Partners can also be drawn to each other on the basis

A difference of taste in jokes is a great strain on the affections.

George Eliot (Mary Ann Evans)
English novelist

Getting along in marriage is not a function of compatibility but adaptability.

Joe Hancock

Married 24 years

of nurturance versus receptivity. These complementary needs suggest that one person likes to give and take care of another, while the other likes to be the benefactor of such care. Other examples of complementary needs may involve responsibility versus irresponsibility and peacemaker versus troublemaker. The idea that mate selection is based on complementary needs was suggested by Winch (1955), who noted that needs can be complementary if they are different (for example, dominant and submissive) or if the partners have the same need at different levels of intensity. As an example of the latter, two individuals may have a complementary relationship when they both want to do advanced graduate study but do not both have a need to get a Ph.D. The partners will complement each other if one is comfortable with his or her level of aspiration as represented by a master's degree but still approves of the other's commitment to earn a Ph.D.

Winch's theory of complementary needs, commonly referred to as "opposites attract," is based on the observation of twenty-five undergraduate married couples at Northwestern University. The findings have been criticized by other researchers who have not been able to replicate Winch's study. Two researchers said, "It would now appear that Winch's findings may have been an artifact of either his methodology or his sample of married people" (Meyer & Pepper, 1977).

Three questions can be raised about the theory of complementary needs:

1. Couldn't personality needs be met just as easily outside the couple's relationship as through mate selection? For example, couldn't a person who has the need to be dominant find such fulfillment in a job that involved an authoritative role, such as head of a corporation or an academic department?

2. What is a complementary need as opposed to a similar value? For example, is desire to achieve at different levels a complementary need or a shared value?

3. Don't people change as they age? Could a dependent person grow and develop self-confidence so that he or she might no longer need to be involved with a dominant person? Indeed, the person might no longer enjoy interacting with a dominant person.

Exchange theory   Exchange theory emphasizes that mate selection is based on assessing who offers the greatest rewards at the lowest cost. Five concepts help to explain the exchange process in mate selection.

1. Rewards are the behaviors (your partner looking at you with the "eyes of love"), words (saying "I love you"), resources (being beautiful or handsome, having money), and services (driving you home, typing for you) your partner provides for you that you value and that influence you to continue the relationship.

2. Costs are the unpleasant aspects of a relationship. One man said, "I have to drive across town to pick her up, listen to her nagging mother before we can leave, and be back at her house by midnight."

3. Profit occurs when the rewards exceed the costs.

4. Loss occurs when the costs exceed the rewards.

5. No other person is currently available who offers a higher profit.

Most people have definite ideas about what they are looking for in a mate. The currency used in the marriage market consists of the socially valued characteristics of the persons involved, such as age, physical characteristics, and economic status. In our free-choice system of mate selection, we typically seek as much in return for our social attributes as we can.

Once you identify a person who offers you a good exchange for what you have to offer, other bargains are made about the conditions of your continued relationship. Forty years ago, two researchers (Waller & Hill, 1951) observed that the person who has the least interest in continuing the relationship can control the relationship. This *principle of least interest* is illustrated by the woman who said, "He wants to date me more than I want to date him, so we end up going where I want to go and doing what I want to do." In this case, the woman trades her company for the man's acquiescence to her choices.

Parental characteristics Whereas the complementary-needs and exchange theories of mate selection are relatively recent, Freud suggested that the choice of a love object in adulthood represents a shift in libidinal energy from the first love objects—the parents. Role theory and modeling theory emphasize that a son or daughter models after the parent of the same sex by selecting a partner similar to the one the parent selected. This means that a man looks for a wife that has similar characteristics to those of his mother and that a woman looks for a husband who is very similar to her father.

OTHER CULTURES *Desired Personality Characteristics* In an impressive cross-cultural study, Buss asked over 10,000 men and women from thirty-seven countries, located on six continents and five islands, to identify the personality characteristics they most desired in a potential mate. The preferences for the top four characteristics identified by both men and women were identical—mutual attraction (love), dependable character, emotional stability/maturity, and pleasing disposition (Buss et al., 1990). One of the differences between men and women was men's greater emphasis on good looks and the greater importance women placed on ambition/economic potential. However, Derochers asked ninety-six American female college women what types of men they were most attracted to and found that economic considerations became important only after personality criteria had been met. "Feminine males were preferred as friends and romantic partners over masculine males" (1995, 375).

In traditional China, a future bride's social and economic standings were the most important criteria the man's parents considered in selecting a wife for their son and a future daughter-in-law for themselves. Personal attraction and love between the bride and the groom were regarded as unnecessary, even harmful.

> Once social and economic eligibility was determined, genealogical and horoscopic data had to be obtained and analyzed. Traditional Chinese believed that "marriage is made in Heaven," and people are "prematched." Horoscopes were examined to ascertain whether the given couple were meant for each other (Engel, 1982, 6, 7). ●

RECENT RESEARCH

Subramanian (1997) reported data from over 3,000 interviews in twenty-one cities and found that very attractive women believed that they were entitled to and expected an economically successful partner.

RECENT RESEARCH

Subramanian (1997) analyzed data from interviews with over 3,000 U.S. residents in twenty-one cities and found that single women ranked economic potential as the most important factor in mate selection; men ranked physical attractiveness.

Sociobiological Aspects of Mate Selection

Sociobiology suggests a biological basis for all social behavior—including mate selection. Based on Charles Darwin's theory of natural selection, which states that the strongest of the species survive, sociobiologists contend that men and women select each other as mates on the basis of their concern for producing offspring who are most capable of surviving.

According to sociobiologists, men look for a young, healthy, attractive, sexually conservative woman who will produce healthy children and who will invest in taking care of the children. Women, in contrast, look for an ambitious man who has good economic capacity who will invest in her children. This pattern of men seeking young healthy women and women seeking economically ambitious men was observed in thirty-seven countries throughout the world (Buss et al., 1990).

The sociobiological explanation for mate selection is extremely controversial. Critics argue that women may show concern for the earning capacity of men because women have been systematically denied access to similar economic resources and selecting a mate with these resources is one of their remaining options. In addition, it is argued that both women and men, when selecting a mate, think more about their partners as companions than as future parents of their offspring.

Figure 4.1 summarizes the cultural, sociological, and psychological filters involved in mate selection.

Living Together (Cohabitation)

After finding each other and pair bonding, about 40 percent of individuals in the United States live together before they get married (Rogers & Amato, 1997, 1099). Also called *cohabitation*, this arrangement is defined as two unrelated adults involved in an emotional and sexual relationship who sleep in the same residence on a regular basis. POSSLQ is the acronym used by the U.S. Census Bureau for People of the Opposite Sex Sharing Living Quarters.

In the 1920s, Judge B. B. Lindsey suggested that couples live together before getting married. His suggestion grew out of his concern for the parade of divorcing couples he saw in his courtroom. He reasoned that if couples lived together before marriage, they might be better able to assess their compatibility. Similarly, Margaret Mead suggested a two-stage marriage. The first stage would involve living together without having children. If the partners felt that their relationship was happy and durable, they would get married and have children.

Although Judge Lindsey's suggestion was made in the 1920s, cultural support for cohabitation has not flourished until today.

Reasons for the increase in cohabitation include fear of marriage; career or educational commitments; increased tolerance from society, parents, and peers; improved birth-control technology; and the desire for a stable emotional and sexual relationship without legal ties. Cohabitants also regard living together as a vaccination against divorce. Later, we will review studies which emphasize that this hope is more often an illusion.

The rise in cohabitation shows that people want intimate relations that are more flexible and less binding than legal marriage.
Charles Jones
Lorne Tepperman
Susannah Wilson
The Futures of the Family

NATIONAL DATA

Almost 4 million (3,958,000) unmarried U.S. couples are living together (*Statistical Abstract of the U.S.: 1997,* Table 61). In a national random sample of 947 individuals, 60 percent of couples who had been married five years or less reported that they cohabited before they married (*Stanley & Markman, 1997*).

Figure 4.1

Cultural, Sociological, and Psychological Filters in Mate Selection

Cultural Filters

. . . .

For two people to consider marriage to each other,

Endogamous factors   and   Exogamous factors

(same race)   (not blood related)

must be met.

After the cultural prerequisites have been satisfied, sociological and psychological filters become operative.

Sociological Filters

. . . .

Propinquity = the tendency to select a mate from among those who live, work, or go to school nearby.

Homogamy = the tendency to select a mate similar to one's self with regard to the following:

Age	Physical appearance
Education	Body clock compatibility
Social class	Religion
Race	Marital status
Intelligence	Interpersonal values

Psychological Filters

. . . .

Complementary needs

Reward-cost ratio for profit

Parental characteristics

Desired personality characteristics

OTHER CULTURES   Iceland is a homogeneous country of 250,000 descendants of the Vikings. Their sexual norms include early (age 14) protected intercourse, nonmarital parenthood, and living together before marriage. Indeed, a wedding photo often includes not only the couple but the children they have already had. One American woman who was involved with an Icelander noted, "My parents were upset with me because Ollie and I were thinking about living together but his parents were upset that we were not already living together" (authors' files). ●

RECENT RESEARCH

Sixty-seven percent of 620 never-married undergraduates reported that they would live with a partner without being married (Knox and Zusman, 1998).

Types of Cohabiting Couples

Six major types of cohabiting couples include the:

1. Here and now—partners who have an emotional/sexual relationship and want to live together because they enjoy each other. They are focused on the here and now, not the future of the relationship.

Elderly couples often live together unmarried to keep their pensions from previous spouses from whom they are widowed or divorced.

RECENT RESEARCH

Axinn and Barber (1997) compared individuals who had cohabited with those who had not and found that cohabitants had more negative attitudes toward marriage.

INSIGHT

It is sometimes suggested that "living together is the same as being married." However, Rindfuss and VandenHeuvel (1992) observed that living together is closer to being single than to being married. They noted that "cohabitors' fertility expectations, nonfamilial activities, and home ownership rates resemble those of the singles" (136).

2. Testers—emotionally/sexually involved partners who want to assess whether staying together and getting married would be right for them.

3. Engaged—partners who are committed to marry each other and want to live together until they get married.

4. Money savers—partners who live together primarily out of economic necessity. They are open to the possibility of a future together, but saving money motivated their initial moving in together.

5. Pension partners—partners who live together because marriage would result in the loss for one or both partners of benefits from a previous marriage. A widow or widower receiving the Social Security and/or pension benefits of her or his deceased spouse loses these benefits upon remarriage.

6. Cohabitants forever—couples who view living together as a permanent alternative to marriage. They plan to stay together but never to marry. They regard marriage as a relationship defined by the couple rather than as an institution defined by society/culture. They may also have experienced a bad marriage or see no need to change a good living-together relationship. Actors Goldie Hawn and Kurt Russell contend that they will continue to live together and never marry.

Consequences of Cohabitation

Although living together before marriage does not ensure a happy, stable marriage, it has some potential advantages. Cohabitation also has potential disadvantages.

Potential advantages of cohabitation Many unmarried couples who live together report that it is an enjoyable, maturing experience. Other potential benefits of living together include the following:

1. *Sense of Well-Being.* When cohabiting persons (in a national sample) were compared with individuals who lived alone, the cohabitants reported a greater sense of well-being. However, when they were compared with marrieds, they were less happy (Kurdek, 1991). One explanation suggests that living together affords the companionship that living alone does not but that cohabitation does not afford the security often associated with marriage.

2. *Delaying Marriage.* Individuals who marry in their middle and late twenties are more likely to stay married and to report higher levels of marital satisfaction than those who marry earlier. To the degree that living together functions to delay the age at which a person marries, it may be considered beneficial. "I married when I was twenty," remarked one woman, "because you just didn't live together in those days. I wish I had waited to get married and had had the option of living together in the meantime." Lana Turner, the Hollywood star of the forties, said one of the reasons she had had seven marriages was that when two people became involved, they were not expected to live together—they were expected to marry.

3. *Gaining Information about Self and Partner.* Living with an intimate partner provides an opportunity for individuals to learn more about themselves and their partners. For example, individuals in living-together relationships may find that their role expectations are more (or less) traditional than they had previously thought.

   Learning more about one's partner is a major advantage of living together. A person's values, habits, reactions, behavior patterns, and relationship expectations are sometimes more fully revealed in a living-together context than in a traditional dating context.

4. *Easier Adjustment to Stepfamily for Children.* Children who have mothers who choose to cohabit with their future husbands before getting married may have an easier adjustment to stepfamily living than children who have mothers who do not cohabit. A team of researchers studying remarriages of mothers with children who lived with their partners before marrying them found that cohabitation served as a way of introducing the children to the new stepdads (and the sooner after the divorce the better). A quick transition from divorce to cohabitation minimizes the time that children must adjust to routines in a single-parent home (Montgomery et al., 1992).

5. *Terminating Unsatisfactory Relationship before Marriage.* Even though ending a living-together relationship may be a difficult and stressful experience, it is probably more traumatic to end a relationship after a couple has married. Since living-together relationships usually involve fewer legal ties, it may be easier to disengage from such a relationship than from a marriage. According to Rindfuss and VandenHeuvel (1992), discord is more likely to lead to the termination of a cohabiting relationship than to termination of a marital relationship.

It brings comfort and encouragement to have companions in whatever happens.
Dio Chrysostom
Third Discourse on Kingship

The easiest person to deceive is one's self.
Lord Lytton
The Disowned

Living together is a sure way to avoid a possible disaster
Abigail Van Buren
(Dear Abby)

Potential disadvantages of cohabitation "Never again," said a man who had formerly been involved in a living-together relationship. "I invested myself completely and felt we would eventually get married. But she never had that in mind and just took me for a ride. The next time, I'll be married before moving in with someone." Living together can have negative consequences for some people, who may feel used or tricked. In addition, as noted earlier, cohabitants may experience problems with their parents. Finally, living-together couples are often not granted the same economic benefits bestowed by a marriage license.

1. *Feeling Used or Tricked.* When levels of commitment are uneven in a relationship, the partner who is more committed feels used. "I always felt I was giving more than I was getting," said one partner. "It's not a good feeling." In a one-sided convenience relationship, one partner manipulates the other to fulfill sexual, domestic, or other needs while withholding any semblance of commitment. There is little reciprocity, and the relationship becomes exploitative.

   Living-together partners may also feel used or tricked if their relationship does not lead to marriage. "I always felt we would be getting married, but it turns out that she was seeing someone else the whole time we were living together and had no intention of marrying me," recalled one partner.

For male respondents in particular, the negative part of having a live-in partner was the loss of personal freedom.

Maureen Lynch,
Leslie Richards
Family Life Researchers

2. **Problems with Parents.** Some cohabiting couples must contend with parents who disapprove of or do not fully accept their living arrangement. For example, cohabitants commonly report that when visiting their parents' homes, they are required to sleep in separate beds in separate rooms. Some cohabitants who have parents with traditional values respect their parents' values, and sleeping in separate rooms is not a problem. Other cohabitants feel resentful of parents who require them to sleep separately.

   Some parents express their disapproval of their child's cohabiting by cutting off communication with, as well as economic support for, their child. Other parents display lack of acceptance of cohabitation in more subtle ways. One woman who had lived with her partner for two years said that her partner's parents would not include her in the family's annual photo portrait. Emotionally, she felt very much a part of her partner's family and was deeply hurt by the exclusion (authors' files).

3. *Economic Disadvantages.* Some economic liabilities exist for those who live together instead of getting married. People who live together typically do not benefit from their partner's health insurance, Social Security, or retirement benefits. Only spouses qualify for such payments.

Cohabitation has been shown to attract a different type of couple than marriage, and to foster attitudes that contribute to divorce.

Steven Nock
Sociologist

Cohabitation as Preparation for Marriage?

One of the motivations for living together is the opportunity to screen out a partner with whom marriage might not work. But do couples who live together before they get married have a greater chance of staying married than couples who do not live together before they get married? The answer is no. According to a comparison of national samples of couples who did and did not live together before they got married, "the proportion separating or divorcing within ten years is a third higher among those who lived together before marriage than among those who did not—36 percent versus 27 percent" (Bumpass & Sweet, 1989, 10). These results are similar to those reported by DeMaris and Vaninadha Rao (1992) for U.S. couples and by Balakrishnan and colleagues (1987) for Canadian couples.

Other researchers have found lower commitment to the institution of marriage (belief that marriage is not a lifetime commitment) and greater perceived likelihood of divorce among couples who had cohabited (Thomson & Colella, 1992; Stets, 1993).

Nock compared a national sample of cohabitants and marrieds and found that the former were characterized by lower-quality relationships. He noted that

> lower quality existed because cohabitation is an incomplete institution. No matter how widespread the practice, nonmarital unions are not yet governed by strong consensual norms or formal laws. What is the legitimate role of a parent in his or her offspring's cohabiting union? What is the nonmarital equivalent of an in-law? Answers to such questions will emerge if cohabitation persists as a popular form of intimate relationship. For the time being, however, the absence of such institutional norms is a plausible explanation for much of the poorer quality of cohabiting relationships (1995, 74).

Another explanation for why cohabitants may have lower-quality relationships, lower institutional commitment, and less hope for the future of their relationships is that cohabitation may draw people who are not ready to commit to each other. Cohabitants may also have developed "bad habits with respect to the development and maintenance of a relationship, and these problems get imported into subsequent relationships" (Stets, 1993, 255). For example, they may have a greater tendency to withdraw from a relationship and separate than to negotiate disagreements. Finally, cohabitants tend to be people who are willing to violate social norms and live together before marriage. Once they marry, they may be more willing to break another social norm and divorce if they are unhappy than unhappily married persons who tend to conform to social norms and have no history of unconventional behavior.

Schoen (1992) also found higher divorce risks among first-time marrieds who had previously cohabitated but noticed that this phenomenon was largely true of persons who were born before the late forties. Those who were born between 1948 and 1952 and between 1953 and 1957 who lived together and married did *not* show a significantly higher divorce risk. Schoen argued that this might be because in earlier decades, cohabitation was practiced primarily by persons who were less conventional and stable—characteristics that may be related to relationship instability. In more recent decades, cohabitation has been practiced by a wider range of individuals—not only those who are less conventional and stable. However, the respondents in the DeMaris and Vaninadha Rao (1992) study of cohabitation were from a 1972 high school cohort. This cohort did evidence a greater divorce rate, which is contrary to what Schoen (1992) would have predicted.

An ill marriage is a spring of ill fortune.
Thomas Drake
Bibliotheca

Before you run in double harness, look well to the other horse.
Ovid
Roman poet

Engagement

Engagement moves the relationship of a couple from a private to a public experience. Family and friends are invited to enjoy the happiness and commitment of the individuals to a future marriage.

Using the Engagement Productively

The engagement period is your last opportunity to systematically examine your relationship, ask each other specific questions, recognize dangerous relationship patterns, visit future in-laws, and participate in premarital counseling.

Examine your relationship   In a commercial for an oil filter, a mechanic says he has just completed a ring job on a car engine that will cost the owner over $600. He goes on to say that a $5.98 oil filter would have made the job unnecessary and ends his soliloquy with, "Pay me now, or pay me later." The same idea applies to the consequences of using or not using your engagement period to examine your relationship. At some point, you will take a very close look at your partner and your relationship; but will you do it now—or later? Doing so now may be less costly than doing so after the wedding.

Recognize dangerous relationship patterns   As you examine your relationship, you should be sensitive to patterns that suggest you may be on a collision course. Three such patterns include breaking the relationship frequently, constant arguing, and inequality resulting from such differences as education and social class. A roller-coaster engagement is predictive of a marital relationship that will follow the same pattern. The same is true of frequent arguments in a relationship. If arguing is a pattern, the engaged couple may want to consider learning more productive communication techniques (Chapter 6, Communication and Conflict Resolution).

Recognize potentially problematic personality characteristics   Just as relationships may have qualities that predict future difficulty, individuals may have characteristics that should be viewed with concern. Snyder and Regts (1990) identified three such personality characteristics that predispose individuals toward impaired functioning in marriage: poor impulse control, hypersensitivity to perceived criticism, and exaggerated self-appraisal.

Persons who have poor impulse control have little self-restraint and may be prone to aggression and violence. Lack of impulse control is also problematic in marriage because the person is less likely to consider the consequences of his or her actions. For example, having an affair to some people might sound like a good idea at the time but may have devastating consequences for the marriage.

Hypersensitivity to perceived criticism involves getting hurt easily. Any negative statement or criticism is received with a greater impact than intended by the partner. The disadvantage of such hypersensitivity is that the partner may learn not to give feedback for fear of hurting the hypersensitive partner. Such lack of feedback to the hypersensitive partner blocks information about what the person does that upsets the other and could do to make things better.

A potential groom is likely to be a husband much like his father. A potential bride is likely to be a wife much like her mother.

RECENT RESEARCH

Holman and Olsen (1997) studied 467 couples and found that those reporting positive childhood relationships with their fathers and mothers were more likely to report high-quality marriages. This was especially true for females.

NATIONAL DATA

In a national random sample, 36.3 percent of the respondents who had been married five years or less reported that they had participated in premarital counseling through a religious organization *(Stanley & Markman, 1997)*

Hence, the hypersensitive one has no way of learning that something is wrong, and the partner has no way of alerting the hypersensitive partner. The result is a relationship in which the partners can't talk about what is wrong.

An exaggerated sense of one's self is another way of saying the person has a big ego and always wants things to be his or her way. A person with an inflated sense of self may be less likely to consider the other person's opinion in negotiating a conflict and prefer to dictate an outcome. Such disrespect for the partner can be damaging to the relationship.

Observe your future in-laws Engagements often mean more frequent interaction with each partner's parents, so you might seize the opportunity to assess the type of family your partner was reared in and the implications for your marriage. When visiting your in-laws-to-be, observe their standard of living, the way they relate to each other, and the degree to which your partner is similar to his or her same-sex parent. How does their standard of living compare with that of your own family? How does the emotional closeness (or distance) of your partner's family compare with that of your family? Such comparisons are significant because both you and your partner will reflect your respective home environments to some degree. If you want to know what your partner may be like in twenty years, look at his or her parent of the same sex. There is a tendency for a man to become like his father and a woman to become like her mother.

Consider premarital education A survey of over 150 college students revealed their interest in brief, high-quality, low-cost, voluntary premarital programs led by well-trained clergy or mental health professionals (Silliman & Schumm, 1995). Blacks also evidence more interest in marriage preparation programs than whites (Duncan et al., 1996). This latter finding suggests that black college students are more concerned about their chances for a successful marriage. Perhaps this concern arises from awareness of the high divorce rate among Blacks or from respect for traditional family values (p. 88).

Although the Catholic Church has a mandatory requirement for marriage preparation, most clergy suggest three premarital sessions before the wedding. These sessions consist of information about marriage, an assessment of the couple's relationship, and/or resolving conflicts that have surfaced in the couple's relationship.

Those who do not plan to be married in a church or synagogue or who do not choose to see a member of the clergy for premarital counseling sometimes see a marriage counselor. A professional can be helpful in assisting a couple to assess their relationship. Although the couple might deny the existence of a problem for fear that looking at it will break up the relationship, the counselor can help them examine the problem and work toward solving it.

Some premarital counselors use inventories to help identify couples who are likely to get divorced. Larsen and Olson (1989) developed an inventory, Premarital Personal and Relationship Evaluation (PREPARE), which assesses expectations, communication, conflict resolution skills, and background origins. They gave the inventory to 164 premarital couples and conducted a follow-up three years later. They found that couples who had unrealistic expectations,

Every couple is not a pair.
Cheales
Proverbial Folk-Lore

When you break up before marriage, they call it a broken engagement. When you break up after marriage, they call it a divorce and it goes on your record.
Bob Sammons
Psychiatrist

Marry in haste and repent at leisure.
John Ray
English Proverbs

RECENT RESEARCH

Swann and Gill (1997) studied 103 couples to assess how confident and accurate the respective partners were about each other's sexual histories and activities. Results revealed that length of the relationship and intensity of involvement increased a partner's confidence in the perceptions of the other. However, this confidence was unrelated to being accurate although it sometimes contributed to relationship quality.

poor communication patterns, absence of conflict resolution skills, etc., were more likely to have separated than those scoring high in these areas.

Alternatively, RELATE is an instrument that may be taken by anyone interested in either how he or she might do in a relationship or how he or she might improve a current relationship. Hence, it may be taken by an individual not currently involved or by partners in an ongoing relationship. You and/or your partner can obtain RELATE from the following address, complete the 250-item questionnaire in about an hour and a half, mail it back for analysis, and receive a written report (within ten days) with graphs and tables for easy comprehension of results. The cost is under five dollars per person. Persons who take RELATE will have more information about their strengths and weaknesses in a relationship. RELATE can be obtained from The Marriage Study Consortium, P.O. Box 25391, Brigham Young University, Provo, UT 84602-5391. E-mail is RELATE@byu.edu, and phone is 801-378-4359.

Finally, the Relationship Dynamics Scale in the Personal Application section at the end of this chapter features a way for you to evaluate your relationship yourself in terms of a "green light" (good/great shape) or a "red light" (trouble ahead).

Because divorce is often an agonizing personal experience for both spouses and their children and has negative societal consequences (e.g., greater poverty for women and children), to what degree might a social policy requiring premarital education be valuable? We examine this issue in this chapter's social policy section.

When to Consider Calling Off the Wedding

Earlier we mentioned some signs that your relationship may be in danger and some personality characteristics of one's future spouse that might predict future trouble. If your relationship is characterized by all of these factors and those that follow, you might consider calling off the wedding at least until the most distressing issues are resolved. Stanley and Markman (1997) noted that "it is now possible to look at a variety of factors and predict marital stability vs divorce with up to 90% (or even greater) accuracy" (p. 9).

Age 18 or younger   The strongest predictor of getting divorced is getting married as an adolescent. Individuals who marry in their teens have three times the divorce rate of those marrying in their twenties. Adolescents who marry in their teens may evidence a tendency toward

rash decision making, antisocial behavior, or juvenile irresponsibility, that could threaten the stability of a marriage. Similarly, people who marry young may have been driven out of an abusive or combative family of origin whose values are inimical to marital stability (South, 1995, 446).

Short courtship   Impulsive marriages where the partners knew each other for less than a month (e.g., Julia Roberts and Lyle Lovett) are associated with a higher-than-average divorce rate. Indeed, partners who date each other for at

Mandatory Premarital Education?

Becoming a licensed driver requires passing a written examination and taking a road test. Before applying for a driver's license, many individuals take a driver's education course. But there is no required course that couples must take before becoming licensed spouses. In most states, filling out a few forms and paying a nominal fee is the extent of the requirement to pull out on the road of marital commitment.

But should marriage licenses be obtained so easily? Should couples be required, or at least encouraged, to participate in premarital education before saying "I do"? Should the public schools offer teenagers courses to help them prepare for marriage? Given the high rate of divorce today, policymakers and family scholars are considering these questions.

Several states have proposed legislation requiring or encouraging premarital education. For example, Michigan proposed a provision requiring engaged couples to participate in a premarital educational program or pay a higher fee for their marriage license. In Lenawee County, Michigan, local civil servants and clergy have made a pact—they will not marry a couple unless that couple has attended marriage education classes. Other states that are considering policies to require or encourage premarital education include Arizona, Illinois, Iowa, Maryland, Minnesota, Mississippi, Missouri, Oregon, and Washington.

Proposed policies include not only mandating premarital education and lowering marriage license fees for those who attend courses but imposing delays on issuing marriage licenses for those who refuse premarital education. Michigan State Representative Jessie Dalman introduced a premarital education bill that would require couples who refuse premarital counseling to wait sixty days for a license instead of three.

Michael J. McManus, a columnist on religion and ethics and author of the book *Marriage Savers*, has traveled to over forty cities campaigning to convince clergy not to marry couples unless they first participate in premarital education (Clark, 1996). Traditionally, most Protestant pastors and Catholic priests require premarital counseling before they will perform marriage ceremonies. Couples who do not want to participate in premarital education can simply get married in a secular ceremony.

Advocates of mandatory premarital education refer to numerous studies that point to the value of such education in creating strong marriages. For example, researchers have found that couples who participated in a widely used couples' education program called PREP (the Prevention and Relationship Enhancement Program), had a lower divorce/separation rate five years after completing the program than did couples who did not participate (Stanley et al., 1995). PREP couples also showed significant improvement in conflict management skills, maintained higher levels of satisfaction, and reported significantly lower levels of aggression than did the controls.

But despite such evidence of the promise of premarital education in strengthening marriages, mandatory premarital education raises some concerns and unanswered questions. Who will decide who may teach these programs? What qualifications will be required to teach couples? Should the government mandate premarital education, or should it be a couple's choice? Diane Sollee, Director of the Coalition for Marriage, Family and Couples Education suggests an alternative to policies that require couples to participate in marriage education courses. Instead, she suggests making marriage preparation courses inexpensive and widely available (Peterson, 1997). But whether couples in love (and denial) will take such courses is questionable.

REFERENCES

Clark, Charles S. 1996. Marriage and divorce. *CO Researcher* 6, no. 18: 409–32.

Peterson, Karen S. 1997. States flirt with ways to reduce divorce rate. *USA Today,* 10 April, 1D-2D.

Stanley, Scott M., Howard J. Markman, Michelle St. Peters, and B. Douglas Leber. 1995. Strengthening marriages and preventing divorce: New directions in prevention research. *Family Relations* 44: 392–401.

First thrive, then wive.
English proverb

RECENT RESEARCH

Holman and Olsen (1997) studied 467 couples and found that premarital approval by significant others was positively related to marital quality one year after the marriage (but not six years after the marriage).

If you want to get rid of a man, I suggest saying, "I love you . . . I want to marry you . . . I want have your children." Sometimes they leave skid marks.
Rita Rudner
Comedian

least two years before getting married report the highest level of marital satisfaction (Grover et al., 1985). A short courtship does not allow a partner to observe and scrutinize the behavior of the other in a variety of settings.

Financial stress Two researchers (Johnson & Booth, 1990) observed that spouses who were economically stressed reported that the situation had a negative effect on their marital happiness and that they thought about divorce more often. Being economically stressed does not necessarily mean having a low income. Many couples with high incomes are under financial stress. Two other researchers (Berry & Williams, 1987) suggested that marital satisfaction depends not on the income itself but on the couple's feelings that their income is adequate. Some research suggests that African-American women are more likely than Caucasian women to insist on economic resources' being in place before the marriage and resist a marital commitment to someone who has limited resources (Bulcroft & Bulcroft, 1993).

Parental disapproval A parent recalled, "I knew when I met the guy that it wouldn't work out. I told my daughter and pleaded that she not marry him. She did, and they are divorced." Such parental predictions (whether positive or negative) often come true. If the predictions are negative, they may contribute to stress and conflict once the couple marries.

Even though parents who reject the commitment choice of their offspring are often regarded as unfair, their opinions should not be taken lightly. The parents' own experience in marriage and their intimate knowledge of their offspring combine to help them assess how their child might get along with a particular mate. If the parents of either partner disapprove of the marital choice, the partners should try to evaluate these concerns objectively. Parental insights might prove valuable.

Premarital childbearing Having a child (rather than just being pregnant) before marriage is also associated with divorce. However, "the effect of premarital childbearing on divorce may be stronger for Caucasians than for African Americans and stronger in the early years" (Larson & Holman, 1994, 233).

Conflictual relationship A team of researchers (Fowers et al. 1996) identified ninety-five premarital couples who reported considerable conflict in their relationships. They were dissatisfied with their partner's personality and habits and reported problems in their ability to communicate and discuss problems in their relationship. They also reported disagreements over leisure activities, parents, friends, and their sexual relationship. When these couples were studied two to three years later, 40 percent had called off the wedding and almost half (48%) of those who married were separated or divorced.

Dysfunctional relationship dynamics Stanley and Markman (1997) surveyed 947 adults and identified communication and conflict management patterns predictive of relationship trouble. Partners who belittle each other, hold back what they really feel, withdraw from an argument, and think seriously about what it would be like to be with someone else are less likely to stay together.

Should you call off the wedding? If you are having second thoughts about getting married to the person to whom you have made a commitment, you are not alone. It is not uncommon for engagements to be broken. Ninety-two percent of the men and 73 percent of the women in a survey of 334 university respondents reported that they had been previously engaged (Laner, 1995, 151). Although some anxiety about getting married is normal (you are entering a new role), constant questions to yourself such as "Am I doing the right thing?" or thoughts such as "This doesn't feel right" are definite caution signals that suggest it might be best to call off the wedding. "When in serious doubt, don't go through with the wedding," says one marriage counselor.

Their Relationship Dynamics Scale presented in the Personal Application section at the end of this chapter is helpful in identifying the degree to which you and your partner may be at risk for a difficult relationship.

Why College Students End Relationships

Table 4.2 shows the percentage of 185 college students who gave various reasons for ending their last relationship.

The most frequently reported reason for ending a relationship involved another person. Respondents mentioned this in various ways: "cheating" (18%), "I met someone new" (15%), "My partner met someone new" (13%), "I went back to a previous lover" (6%), and "My partner went back to a previous lover" (5%). When these percentages were totaled, the major reason relationships terminated in almost 60 percent of the cases (57%) was "someone else." If we include the 18% who reported "dishonesty" as why their relationship ended, and assume that such dishonesty was in reference to another person, the percentage goes almost to 75 percent. And if we include the 15 percent who reported their relationships ended because of "separation," which provides the context for meeting someone else, the percentage rises to almost 90 percent.

That infidelity is a factor in divorce was identified by Charney and Parnass (1995), who studied sixty-two couples in which at least one spouse had had an affair. They found that 34 percent of these marriages ended in divorce that was attributable to the affair. Since dating partners are less committed than

RECENT RESEARCH

Miller (1997) studied attentiveness to desirable alternatives to one's partner in a sample of 99 men and 147 women undergraduates and found that there was no better predictor of relationship failure than high attentiveness to alternatives. The opposite was also true. Happy committed lovers are inattentive to their relationship alternatives.

RECENT RESEARCH

Lydon et al. (1997) studied 69 undergraduates who were involved in long-distance dating relationships and found that "moral commitment" to the relationship predicted the survival of the relationship.

Table 4.2

Reasons 185 College Students Identified for Why Last Relationship Ended (respondents could give more than one reason)	
Reason	**% Percentage**
Cheating	18
I met someone new	15
My partner met someone new	13
I went back to a previous lover	6
My partner went back to a previous lover	5
Dishonesty	18
Separation	15
Too many differences/different values	43
Got tired of each other	27
Parental disapproval	13
Violence/abuse	9
Alcohol/drugs	7

Source: Knox, D., L. Gibson, M. Zusman, and C. Gallmeier. Why College Student Relationships End. *College Student Journal* 1997. Reprinted by permission.

Individuals in relationships are often interested in assessing the strength of their relationship. The Relationship Dynamics Scale allows you not only to assess your own relationship but to compare your score with a national random sample of others who have taken the scale.

Relationship Dynamics Scale

Please answer each of the following questions in terms of your relationship with your "mate" if married, or your "partner" if dating or engaged. We recommend that you answer these questions by yourself (not with your partner), using the ranges following for your own reflection.

Use the following 3 point scale to rate how often you and your mate or partner experience the following:

1 = almost never
2 = once in awhile
3 = frequently

1 **2** 3   Little arguments escalate into ugly fights with accusations, criticisms, name calling, or bringing up past hurts.

1 2 3   My partner criticizes or belittles my opinions, feelings, or desires.

1 2 **3**   My partner seems to view my words or actions more negatively than I mean them to be.

1 **2** 3   When we have a problem to solve, it is like we are on opposite teams.

1 2 3   I hold back from telling my partner what I really think and feel.

1 2 3   I think seriously about what it would be like to date or marry someone else.

1 2 3   I feel lonely in this relationship.

1 2 **3**   When we argue, one of us withdraws . . . that is, doesn't want to talk about it anymore; or leaves the scene.

Who tends to withdraw more when there is an argument?

Male
Female
Both Equally
Neither Tend to Withdraw

Stanley and Markman, 1997, Copyright PREP, Inc. (303) 759-9931

Where Are You At in Your Marriage?

We devised these questions based on seventeen years of research at the University of Denver on the kinds of communication and conflict management patterns that predict if a relationship is headed for trouble. We have recently completed a nationwide, random phone survey using these questions. The average score was 11 on this scale. While you should not take a higher score to mean that your relationship is somehow destined to fail, higher scores can mean that

(Continued on following page)

your relationship may be in greater danger unless changes are made. (These ranges are based only on your individual ratings—not a couple total.)

8 to 12   "Green Light"

If you scored in the 8–12 range, your relationship is probably in good or even great shape at THIS TIME, but we emphasize "at THIS TIME" because relationships don't stand still. In the next twelve months, you'll either have a stronger, happier relationship, or you could head in the other direction.

To think about it another way, it's like you are traveling along and have come to a green light. There is no need to stop, but it is probably a great time to work on making your relationship all it can be.

13 to 17   "Yellow Light"

If you scored in the 13–17 range, it's like you are coming to a "yellow light." You need to be cautious. While you may be happy now in your relationship, your score reveals warning signs of patterns you don't want to let get worse. You'll want to be taking action to protect and improve what you have. Spending time to strengthen your relationship now could be the best thing you could do for your future together.

18 to 24   "Red Light"

Finally, if you scored in the 18–24 range, it's like approaching a red light. Stop, and think about where the two of you are headed. Your score indicates the presence of patterns that could put your relationship at significant risk. You may be heading for trouble—or already be there. But there is GOOD NEWS. You can stop and learn ways to improve your relationship now!

For more information on danger signs and constructive tools for strong marriages, see: Markman, H. J., Stanley, S. M., & Blumberg, S. L. (1994). *Fighting* for *Your Marriage: Positive Steps for a Loving and Lasting Relationship*. San Francisco: Jossey Bass, Inc. (PREP 1-800-366-0166)

To: Those interested in using the Relationship Dynamics Scale
From: PREP, Inc.

1. We wrote these items based on an understanding of many key studies in the field. The content or themes behind the questions are based on numerous in-depth studies on how people think and act in their marriages. These kinds of dynamics have been compared with patterns on many other key variables, such as satisfaction, commitment, problem intensity, etc. Because the kinds of methods researchers can use in their laboratories are quite complex, this actual measure is far simpler than many of the methods we and others use to study marriages over time. But the themes are based on many solid studies. Caution is warranted in interpreting scores.

2. The discussion of the Relationship Dynamics Scale gives rough guidelines for interpreting the meaning of the scores. The ranges we suggest for the measure are based on results from a nationwide, random phone survey of 947 people (85% married) in January 1996. These ranges are meant as a rough guideline for helping couples assess the degree to which

they are experiencing key danger signs in their marriages. The measure as you have it here powerfully discriminated between those doing well in their marriages/relationships and those who were not doing well on a host of other dimensions (thoughts of divorce, low satisfaction, low sense of friendship in the relationship, lower dedication, etc.). Couples scoring more highly on these items are truly more likely to be experiencing problems (or, based on other research, are more likely to experience problems in the future).

3. This measure in and of itself should not be taken as a predictor of couples who are going to fail in their marriages. No couple should be told they will not "make it" based on a higher score. That would not be in keeping with our intention in developing this scale or with the meaning one could take from it for any one couple. While the items are based on studies that assess such things as the likelihood of a marriage working out, we would hate for any one person to take this and assume the worst about their future based on a high score. Rather, we believe that the measure can be used to motivate high and moderately high scoring people to take a serious look at where their marriages are heading—and take steps to turn such negative patterns around for the better.

For more information on constructive tools for strong marriages, see: Markman, H. J., Stanley, S. M., & Blumberg, S. L. (1994). *Fighting* for *Your Marriage: Positive Steps for a Loving and Lasting Relationship.* If you have questions about the measure and the meaning of it, please write to us at:

PREP, Inc.
P. O. Box 102530
Denver, Colorado 80250-2530
e mail: PREPinc@AOL.com

Scott M. Stanley, Ph.D.
Howard J. Markman, Ph.D.

RECENT RESEARCH

Amato and Rogers (1997) interviewed a national sample of spouses and found that infidelity was one of the prime predictors of subsequent divorce.

marital partners to work out their differences, a higher percentage would be expected to terminate their relationship when the strain of an extradyadic involvement became evident.

Other reasons (unrelated to another person) for ending their last relationship included "too many differences/different values" (43%), "got tired of each other" (27%), "parental disapproval" (13%), "violence/abuse" (9%), and "alcohol/ drugs" (7%). The most frequent ways college students in one sample ended their relationships were either face-to-face or over the phone. Men were more likely than women to just say nothing and disappear (Knox et al., 1998).

GLOSSARY

cohabitation   A living arrangement in which two unrelated adults involved in an emotional and sexual relationship sleep in the same residence on a regular basis.

complementary-needs theory   The tendency to select mates whose needs are opposite and complementary to one's own needs—for example, a dominant person selecting a passive person and vice versa.

dating A mechanism whereby some men and women pair off for recreational purposes and/or to form exclusive, committed relationships for the reproduction, nurturing, and socialization of children.

endogamous pressure Social pressure to marry within one's own race, religion, and social class.

exogamous pressure Social pressure to marry outside your family group—to avoid sex and marriage with a sibling or other close relative.

homogamy Mate selection based on similar characteristics of the partners, such as values, age, race, religion, and education.

mating gradient The tendency for husbands to be more advanced than their wives with regard to age, education, and occupational success.

POSSLQ An acronym used by the U.S. Census Bureau that stands for People of the Opposite Sex Sharing Living Quarters.

propinquity The tendency for persons who live, work, or attend school in the same place to select each other as partners.

SUMMARY

While dating is often for recreation, it is also a mechanism whereby some men and women pair off and form exclusive, committed relationships for the reproduction, nurturing, and socialization of children.

Dating: Then and Now

Courtship patterns in seventeenth century England were very restrictive, with parents controlling the courtship process. The Industrial Revolution increased the economic independence of women who worked in factories, thereby shifting courtship control away from parents. Changes in courtship in the past fifty years have included marrying at a later age, an increase in cohabitation, greater gender role equality, greater fear of contracting sexually transmitted diseases, and an increased awareness of the impermanence of marriage.

Dating Issues

Individuals tend to meet one another though self-introductions or through friends. School, work, and parties are typical social contexts for meeting others. An increasing number of people are meeting on-line through their computers but newspaper and videomatching services are still being used.

Typical problems on dates include communication and jealousy. Dating after divorce may involve issues such as children and ex-spouses and increased risk of HIV and other STDs (due to increased age and number of sexual partners). In the later years, a shortage of men makes finding a dating partner difficult for divorced and widowed women.

Mate Selection

As dating moves from recreation to mate selection, the partners are influenced by various cultural, sociological, psychological, and perhaps biological factors. Although marriages in some cultures are arranged by parents or other relatives, our culture relies mainly on endogamous and exogamous pressures to guide mate choice.

Sociological aspects of mate selection include homogamy (people prefer someone like themselves) and propinquity (people are more likely to find a mate who lives nearby, attends the same school, or works for the same company).

Psychological aspects of mate selection include complementary needs, exchange theory, and parental image. Complementary-needs theory suggests that people select others who have opposite characteristics to their own. People may also seek each other out if they both have the same need at different levels of intensity. Most researchers find little evidence for the complementary-needs theory.

Exchange theory suggests that one individual selects another on the basis of rewards and costs. As long as an individual derives more profit from a relationship with one partner than with another, the relationship will continue. Exchange concepts influence who dates whom, the conditions of the dating relationship, and the decision to marry.

The sociobiological view of mate selection suggests that men and women select each other on the basis of their biological capacity to produce and support healthy offspring. Men seek young women with healthy bodies, and women seek ambitious men who will provide economic support.

Living Together (Cohabitation)

Types of living-together relationships include the here and now (live together for fun), testers (test the relationship), engaged (plan to marry), and cohabitants forever (intend never to marry). Most people who live together eventually get married but not necessarily to each other.

Potential benefits of living together include a sense of well-being, delaying marriage until one is older, ending unsatisfactory relationships before marriage, gaining information about one's self and one's partner, and providing an easier adjustment to stepfamily living. Potential disadvantages include feeling used or tricked, problems with parents, and economic disadvantages.

When compared with marrieds, cohabitants are less committed to each other. When cohabitants marry, they are more likely to divorce than individuals who did not live together before marriage. Cohabitants with children are more likely to stay together.

Engagement

Couples can use the engagement period productively by systematically examining their relationship, observing their future in-laws for clues about their partner's background and character, and getting premarital counseling. Conditions under which a couple might want to prolong their engagement include having known each other for less than two years, having an inadequate or unstable source of income, having parents who disapprove of the marriage, and having a child.

A sample of college students reported that the most frequent reason their last relationship ended involved another person. When the categories of "cheating," "met someone new," and "went back to previous lover" were totaled, almost 60 percent of relationships were ended over another person. Other reasons cited for ending their last relationship included different values, getting tired of each other, parental disapproval, violence/abuse, and alcohol/drugs. Most relationships ended with a face-to-face or telephone discussion.

REFERENCES

Ahuvia, A. C., and M. B. Adelman. 1992. Formal intermediaries in the marriage market: A typology and review. *Journal of Marriage and the Family* 54: 452–63.

Amato, P. R., and S. J. Rogers. 1997. A longitudinal study of marital problems and subsequent divorce. *Journal of Marriage and the Family* 59: 612–24.

American Council on Education and University of California. 1997. *The American freshman: National norms for fall, 1997.* Los Angeles: Los Angeles Higher Education Research Institute.

Axinn, William G., and Jennifer S. Barber. 1997. Living arrangements and family formation attitudes in early adulthood. *Journal of Marriage and the Family* 59: 595–611.

Balakrishman, T. R., K. V. Rao, E. Lapierre-Adamyck, and K. J. Krotki. 1987. A hazard model analysis of the covariates of marriage dissolution in Canada. *Demography* 24: 395–406.

Berry, R. E., and F. L. Williams. 1987. Assessing the relationship between quality of life and marital income satisfaction: A path analytic approach. *Journal of Marriage and the Family* 49: 107–16.

Booth, Alan, David R. Johnson, Ann Branaman, and Alan Sica. 1995. Belief and behavior: Does religion matter in today's marriage? *Journal of Marriage and the Family* 57: 661–71.

Bulcroft, R. A., and K. A. Bulcroft. 1993. Race differences in attitudinal and motivational factors in the decision to marry. *Journal of Marriage and the Family* 55: 338–55.

Bumpass, L. L., T. Castro Martin, and J. A. Sweet. 1991. The impact of family background and early marital disruption. *Journal of Family Issues* 12: 22–42.

Buss, D. M. et al. 1990. International preferences in selecting mates: A study of 37 cultures. *Journal of Cross-Cultural Psychology* 21, no. 4: 5–47.

Charny, I. W., and Parnass, S. 1995. The impact of extramarital relationships on the continuation of marriages. *Journal of Sex and Marital Therapy* 21: 100–15.

Daniel, H. J., Jr., and R. B. McCabe. 1992. Gender differences in the perception of vocal sexiness. In *The nature of the sexes: The sociobiology of sex differences*, edited by J. N. G. van der Dennen. Groningen, the Netherlands: Origin Press, 55–62.

Darling, C. A., J. K. Davidson, and W. E. Prish, Jr. 1989. Single parents: Interaction of parenting and sexual issues. *Journal of Sex and Marital Therapy* 15: 227–45.

Deal, J. E., K. S. Wampler, and C. F. Halverson, Jr. 1992. The importance of similarity in the marital relationship. *Family Process* 31: 369–82.

DeMaris, A., and K. Vaninadha Rao. 1992. Premarital cohabitation and subsequent marital stability in the United States: A reassessment. *Journal of Marriage and the Family* 54: 178–90.

Derochers, Stephan. 1995. What types of men are most attractive and most repulsive to women? *Sex Roles* 32: 375–91.

Duncan, S. T., G. Box, and B. Silliman. 1996. Racial and gender differences in perceptions of marriage preparation programs among college-educated young adults. *Family Relations* 45: 80–90.

Dychtwald, K. 1990. *Age wave*. New York: Bantam Books.

Engel, John W. 1982. Changes in male-female relationships and family life in People's Republic of China. Research Series 014. College of Tropical Agriculture and Human Resources, University of Hawaii.

Fowers, Blane J., Kelly H. Montel, and David H. Olson. 1996. Predicting marital success for premarital couple types based on PREPARE. *Journal of Marital and Family Therapy* 22: 103–19.

Gallmeier, C. P., M. E. Zusman, D. Knox, and L. Gibson. 1997. Can we talk?: Gender differences in disclosure patterns and expectations. *Free Inquiry in Creative Sociology*. 25: 219–25.

Ganong, L. W., and M. Coleman. 1992. Gender differences in expectations of self and future partner. *Journal of Family Issues* 13: 55–64.

Greeley, Andrew M., Robert T. Michael, and Tom W. Smith. 1990. Americans and their sexual partners. *Society*, July/August, 36–42.

Greenstein, T. N. 1992. Delaying marriage: Women's work experience and marital timing. Paper presented at the 54th Annual Conference of the National Council on Family Relations, Orlando, Florida.

Grover, K. J., C. S. Russell, W. E. Schumm, and L. A. Paff-Bergen. 1985. Mate selection processes and marital satisfaction. *Family Relations* 34: 383–86.

Holman, Thomas B., and Susanne F. Olsen. 1997. Family background, emotional health, social network approval, and couple interactional processes: Premarital factors predicting early marital quality with implications for education. Paper presented at the Annual Conference of the National Council on Family Relations, Crystal City, Va., November.

Houts, R. M., E. Robins, and T. L. Huston. 1996. Compatibility and the development of premarital relationships. *Journal of Marriage and the Family* 58: 7–20.

Johnson, D. R., and A. Booth. 1990. Rural economic decline and marital quality: A panel study on farm marriages. *Family Relations* 39: 159–65.

Kiecolt, K. J., and M. A. Fossett. 1997. Mate availability and marriage among African-Americans. *African American Perspectives* 3: 12–18.

Kilby, Richard W. 1993. *The study of human values*. Lanham, Md.: University Press of America.

Killian, Kyle D. 1997. What's the difference: Negotiating race, class and gender in interracial relationships. Paper presented at the Annual Conference of the National Council on Family Relations, Crystal City, Va., November.

Knox, D., and M. Zusman. 1998. Unpublished data collected for this text.

Knox, D., L. Gibson, M. Zusman, and C. Gallmeier. In press. Why college student relationships end. *College Student Journal* 31: 449–452.

Knox, D., M. E. Zusman, and W. Nieves. 1998. Breaking away: How college students end relationships. *College Student Journal* 32: in press.

Kurdek, L. A. 1991. The relations between reported well-being and divorce history, availability of a proximate adult, and gender. *Journal of Marriage and the Family* 253: 956–57.

Laner, Mary Riege. 1995. *Dating: Delights, discontents and dilemmas.* 2d ed. Salem, Mass.: Sheffield Publishing Co.

Larsen, Andrea S., and David H. Olson. 1989. Predicting marital satisfaction using PREPARE: A replication study. *Journal of Marital and Family Therapy* 15: 311–22.

Larson, J. H., D. R. Crane, and C. W. Smith. 1991. Morning and night couples: The effect of wake and sleep patterns on marital adjustment. *Journal of Marital and Family Therapy* 17: 53–65.

Larson, J. H., and T. B. Holman. 1994. Premarital predictors of marital quality and stability. *Family Relations* 43: 228–37.

Larson, J. H., T. B. Holman, D. M. Klein, D. M. Busby, R. F. Stahmann, and D. Peterson. 1995. A review of comprehensive questionnaires used in premarital education and counseling. *Family Relations* 44: 245–52.

Lichter, Daniel T., Robert N. Anderson, and Mark D. Hayward. 1995. Marriage markets and marital choice. *Journal of Family Issues* 16: 412–31.

Lydon, J., T. Pierce, and S. O'Regan. 1997. Coping with moral commitment to long distance dating relationships. *Journal of Personality and Social Psychology* 73: 104–13.

Markstrom-Adams, C. 1991. Attitudes on dating, courtship, and marriage: Perspectives on in-group versus out-group relationships by religious minority and majority adolescents. *Family Relations* 40: 91–96.

McLeod, Jane D. 1995. Social and psychological bases of homogamy for common psychiatric disorders. *Journal of Marriage and the Family* 57: 201–14.

McRae, S., 1997. Cohabitation: A trail run for marriage? 1997. *Sexual and Marital Therapy* 12: 259–73.

Meyer, J. P., and S. Pepper. 1977. Need compatibility and marital adjustment in young married couples. *Journal of Personality and Social Psychology* 35: 331–42.

Michael, Robert T., John H. Gagnon, Edward O. Laumann, and Gina Kolata. 1994. *Sex in America: A definitive survey.* Boston: Little, Brown. 1994.

Miller, Leslie. 1995. Looking for love in cyberspace. *USA Today,* 9 February, 6d.

Miller, R. S. 1997. Inattentive and contented: Relationship commitment and attention to alternatives. *Journal of Personality and Social Psychology* 73:758–66.

Montgomery, M. J., E. R. Anderson, E. M. Hetherington, and W. G. Clingempeel. Patterns of courtship for remarriage: Implications for child adjustment and parent-child relationships. 1992. *Journal of Marriage and the Family* 54: 686–98.

Nock, Steven L. 1995. A comparison of marriages and cohabiting relationships. *Journal of Family Issues* 16: 53–76.

O'Flaherty, Kathleen M., and L. W. Eells. 1988. Courtship behavior of the remarried. *Journal of Marriage and the Family* 50: 499–506.

O'Neill, A., A. Duigann-Cabrera, C. Wang, and L. McNeil. 1997. Webstruck. *People Magazine,* 2 June, 85–86.

Peterson, Karen S. 1997. Interracial dating: For today's teens, race 'not an issue anymore'. *USA Today.* 3 November, 1A.

Regan, P. C., and E. Berscheid. 1997. Gender differences in characteristics desired in a potential sexual and marriage partner. *Journal of Psychology and Human Sexuality* 9: 25–37.

Ridley, C. A., and I. E. Sladeczek. 1992. Premarital relationship enhancement: Its effect on needs to relate to others. *Family Relations* 41: 148–53.

Rindfuss, R. R., and A. VandenHeuvel. 1992. Cohabitation: A precursor to marriage or an alternative to being single? In *The changing American family,* edited by S. J. South and S. E. Tolnay. Boulder, Colo: Westview Press, 118–42.

Rogers, S. J., and P. R. Amato. 1997. Is marital quality declining? The evidence from two generations. *Social Forces* 75, no. 3: 1089–1100.

Ross, L. F. 1997. Mate selection preferences among African American college students. *Journal of Black Studies* 27: 554–69.

Schoen, R. 1992. First unions and the stability of first marriages. *Journal of Marriage and the Family* 54: 281–84.

Sherman, Ruth G., and Jane Hardy Jones. 1994. Exchange on the Myers-Briggs type indicator: A response to the article on "the validity of the Myers-Briggs type of indicator for predicting expressed marital problems." *Family Relations* 43: 94–95.

Silliman, Benjamin, and Walter R. Schumm. 1995. Client interests in premarital counseling: A further analysis. *Journal of Sex and Marital Therapy* 21: 43–56.

Snyder, D. K., and J. M. Regts. 1990. Personality correlates of marital dissatisfaction: A comparison of psychiatric, maritally distressed, and nonclinic samples. *Journal of Sex and Marital Therapy* 90: 34–43.

South, Scott J. 1995. Do you need to shop around? *Journal of Family Issues* 16: 432–49.

Sparrow, K. H. 1991. Factors in mate selection for single black professional women. *Free Inquiry in Creative Sociology* 19: 103–9.

Sprecher, S., and R. Chandak. 1992. Attitudes about arranged marriages and dating among men and women from India. *Free Inquiry in Creative Sociology* 20: 59–69.

Stall, Ron, and Joe Catania. 1994. AIDS risk behaviors among late middle-aged and elderly Americans. *Archives of Internal Medicine* 154: 57–63.

Stanley, S. M., and H. J. Markman. 1997. *Marriage in the 90s: A nationwide random phone survey.* Denver: PREP, Inc. (303-759-9931).

Statistical Abstract of the United States: 1997. 117th ed. Washington, D.C.: U.S. Bureau of the Census.

Stets, J. E. 1993. The link between past and present intimate relationships. *Journal of Family Issues* 14: 236–60.

Subramanian, Saskia. 1997. Economic considerations in mate selection criteria. Paper presented at the Annual Convention of the American Psychological Association, Chicago, 18 August.

Swann, W. B., and M. J. Gill. 1997. Confidence and accuracy in person perception: Do we know what we think we know about our relationship partners? *Journal of Personality and Social Psychology* 73: 747–57.

Thompson, E., and U. Colella. 1992. Cohabitation and marital stability: Quality or commitment? *Journal of Marriage and the Family* 54: 259–67.

Tzeng, Meei-Shenn. 1992. The effects of socioeconomic heterogamy and changes on marital dissolution for first marriages. *Journal of Marriage and the Family* 54: 609–19.

Van Horn, K. R., A. Arnone, K. Nesbitt, L. Desilets, T. Sears, M. Giffin, and R. Brudi. 1997. Physical distance and interpersonal characteristics in college students' romantic relationships. *Personal Relationships* 4: 25-34.

Waller, W., and R. Hill. 1951. *The family: A dynamic interpretation.* New York: Holt, Rinehart and Winston.

Waris, R. G. 1997. Age and occupation in selection of human mates. *Psychological Reports* 80: 1223–26.

Williams, Lee, and Joan Jurich. 1995. Predicting marital success after five years: Assessing the predictive validity of focus. *Journal of Marital and Family Therapy* 21: 141–53.

Winch, R. F. 1955. The theory of complementary needs in mate selection. Final results on the test of the general hypothesis. *American Sociological Review* 20: 552–55.

Young, M. H., and J. D. Schvaneveldt. 1992. The effects of religious orientation on couple formation among college students. Paper presented at the 54th Annual Conference of the National Council on Family Relations, Orlando, Florida. Used by permission.

Zusman, M., and D. Knox. 1998 . Relationship problems of casual and involved university students. In press, *College Student Journal.*

CHAPTER 5 Marriage Relationships

A genuinely happy and fulfilling marriage, resting on indestructible love and requiring nothing more, is rare. But it is the foundation of a unique happiness and, possibly, of the only perfect fulfillment of which we are capable.

Richard Taylor, *Love Affairs*

recently married individual was asked, "What's the difference between being single and being married?" "About twenty pounds" was the answer. But extra weight is not the only difference between singlehood and marriage. Moreover, there are many types of marriage relationships.

In this chapter we look at the transition/adjustment experiences of individuals as they move from lovers to spouses, at various types of marital relationships, such as African-American, Mexican-American, and interracial marriages, and at the characteristics of spouses who stay together and report that they are happy. We begin by examining what motivates people to marry and the meaning of commitment.

Motivations toward and Commitment in Marriage

In this section we look at why people marry and the three levels of commitment when they do so.

Motivations to Marry

The following are some of the reasons that motivate people to marry.

Personal fulfillment   We marry because we feel a sense of personal fulfillment in doing so. And we remain optimistic that our marriage will be a good one. Even young adults whose parents divorced believe that their own marriage will work out and are committed to ensuring that it does (Landis-Kleine et al., 1995).

Companionship   Talk show host Oprah Winfrey once said that lots of people want to ride in her limo, but that what she wants is someone who will take the bus when the limo breaks down. One of the motivations for marriage is to enter a structured relationship whereby you live with a genuine companion, a person who will take the bus with you when the limo breaks down. Sociologists refer to spouses as being in a primary group relationship with each other. *Primary groups* are characterized by relationships in which the partners are intimate, personal, emotional, and informal (and, one would hope, happy). The family in which you grew up is a primary group.

Although marriage does not ensure it, companionship is the greatest expected benefit of marriage in the United States. Companionship is talking about and doing things with someone you love; it is creating a history with someone. "Only my partner and I know the things we've shared," said one spouse. "The shrimp dinner at the ocean, the walk down Bourbon Street in New Orleans, and the robins that built the nest in our backyard are part of our joint memory bank."

Parenthood   Having children is an important lifetime goal. Although some people are willing to have children outside of marriage (in a cohabiting relationship or in no relationship at all), most desire to have children in a marital context. A particularly strong norm exists in our society (particularly for whites) that individuals should be married before they have children.

Security People also marry for the emotional and financial security that marriage can provide. However, since "half of all marriages entered into in recent years in this country will end in divorce or separation if recent marital dissolution rates continue" (Glenn, 1997, 5), marriage is not protection for the future.

Although individuals may be drawn to marriage for the preceding reasons on a conscious level, unconscious motivations may also be operative. Individuals reared in a happy family of origin may seek to duplicate this perceived state of warmth, affection, and sharing. Alternatively, individuals reared in unhappy, abusive, drug-dependent families may inadvertently seek to recreate a similar family because that is what they are familiar with.

Commitment to Marry

Though the reasons people marry may be private, the fact that they do is public. Marriage implies three levels of commitment—person to person, family to family, and couple to state.

Person-to-person commitment Commitment in American marriages can be defined as an intent to maintain the relationship. Saying "I do" in a marriage ceremony implies that you and your partner are making a personal commitment to love, support, and negotiate differences with each other. You are establishing a primary relationship with your partner. Although other existing relationships with parents and friends may continue to be important, they may become secondary.

OTHER CULTURES The primacy of the marriage relationship over family relationships is not shared by couples in all societies. Traditional Chinese, Japanese, and Korean marriages are regarded less as a commitment of the spouses to each other and more as a commitment of the eldest male offspring to take care of his parents. In these families, which emphasize patrilocal residence, the wife

This couple steals a kiss during their wedding reception.

(now daughter-in-law) moves into her husband's parents' home, where she lives with her husband. She is expected to cook, clean, and be obedient to the wishes of her husband's parents. Since her new husband expects this of her, the couple's marital relationship is subsumed under their obligations to the husband's parents. For this reason, some Asian women are reluctant to marry the eldest son because doing so requires living with his parents. Also, according to Confucian ethics, a son is expected to divorce his wife if she does not please his parents, even if he loves her (Engel, 1982, 1984). ●

Commitment may be defined not only by one's intention to maintain a relationship but by perceived relationship satisfaction and alternative quality. The latter refers to perceptions of how easy it would be to do better in another relationship than in the present one (Sacher & Fine, 1996). Commitment may also be measured as the imagined consequences should the marriage end. Nock (1995) hypothesized that the more dependent spouses were on each other and on their relationship, the more they were committed to staying together. He identified sources of dependency as

> including, but not limited to, various socially valued resources (income, educational attainment), negotiated arrangements for the solution of mundane household management (cooking, cleaning, paying bills, etc.), and numerous corollaries of the duration of the marriage and the ordinary events of the life course (i.e., marital-specific capital—children) (Nock, 1995, 507).

Hence, partners married to spouses with high incomes who do or share a large part of the domestic labor and with whom children and parenting are shared have more to gain by staying with their spouses than those with spouses producing no income or no help and with whom there are no children. Nock (1995) analyzed a national sample of 2,331 adults with regard to dependency and commitment and found that the greater the dependency, the greater the commitment. Lawler and Yoon (1996) also observed that being equals in terms of power in the relationship was associated with high rates of exchange and commitment to each other.

Family-to-family commitment   Marriage also involves commitments by each of the marriage partners to the family members of the spouse. Married couples are often expected to divide their holiday visits between both sets of parents. In addition, each spouse becomes committed to help his or her in-laws when appropriate and to regard family ties as part of marital ties. For some older couples, this means caring for disabled parents who may live in their home. "We always said that no parent was ever going to live with us," said one spouse. "But my wife's father died, and her mother had no place to go. Her living here was an initial strain, but we've learned to cope with the situation quite well."

Not all couples accept the family-to-family commitment. Some spouses have limited contact with their respective parents. "I haven't seen my folks in years and don't want to," said one spouse.

The Hmong of Laos live in extended families (over 60,000 have immigrated to the United States) where marriage commitments are formally negotiated between the respective families through mediators. Once the man and the woman decide that they would like to marry, the parents of the man send two mediators to the parents of the woman with a formal marriage proposal. "Upon the mediators' arrival at the girl's parents' residence, they will announce their mission and request permission to enter the home. The girl's parents will also find two able mediators who come to the negotiation table with the other mediators" (Thao, 1992, 56). One of the issues they discuss is the bride price (an amount of money) the man is to pay the family of the girl for their hardship of raising a child they are now giving up. The agreements are written down and signed, not necessarily by the man and the woman, but by the parents. "The proposal will become a binding contract between the two parties, the young man and his family on one side and the girl and her family on the other, witnessed by the mediators themselves" (Thao, 1992, 56). ●

Couple-to-state commitment   In addition to making person-to-person and family-to-family commitments, spouses become legally committed to each other according to the laws of the state in which they reside. This means they cannot arbitrarily decide to terminate their own marital agreement.

Just as the state says who can marry (not close relatives, the insane, or the mentally deficient) and when (usually at age 18 or older), legal procedures must be instituted if the spouses want to divorce. The state's interest is that a couple stay married, have children, and take care of them.

Changes after Marriage

Personal Changes

New spouses experience an array of changes in their lives. One initial consequence of getting married may be an enhanced self-concept. Parents and close friends usually arrange their schedules to participate in your wedding and give gifts to express their approval. In addition, the strong evidence that your spouse approves of you and is willing to spend a lifetime with you also tells you that you are a desirable person.

The married person also begins adopting new values and behaviors consistent with the married role. Although new spouses often vow that "marriage won't change me," it does. For example, rather than stay out all night at a party, which is not uncommon for singles who may be looking for a partner, spouses (who are already paired off) tend to go home early. Their roles of spouse, employee, and parent force them to adopt more regular hours. The role of married person implies a different set of behaviors than the role of single person. Although there is an initial resistance to "becoming like old married folks," the resistance soon gives way to the realities of the role.

Another common result of getting married is disenchantment. It may not happen in the first few weeks or months of marriage, but disenchantment is inevitable. Whereas courtship is the anticipation of a life together, marriage is the

---

NATIONAL DATA
About 2.4 million marriage licenses are issued to couples in the United States every year. These represent a legal bond, not only between individuals but also between the couple and the state (*National Center for Health Statistics*, 1998).

All women are married to two men—the one they married and the one they think they married.

Jay Leno, Comedian

The social function of the honeymoon (whether in the first or subsequent marriage) is to provide a time for the couple to solidify their identity as a married couple.

RECENT RESEARCH

When the spouses of 153 African-American couples and 160 white couples compared their ideals for their respective partners according to twelve characteristics with the reality of how their partners measured up, the greater the discrepancy, the lower the marital satisfaction a year later (Ruvolo & Veroff, 1997).

They dream in courtship, but wake in marriage
Pope
Wife of Bath's Prologue

It does not much signify whom one marries, as one is sure to find the next morning that it is someone else.
Samuel Rogers
Table-Talk

day-to-day reality of that life—and the reality does not always fit the dream. Daily marital interaction reveals both partners as they really are.

Disenchantment after marriage may result from each partner's shifting his or her focus of interest away from the other and toward work. When children come, the focus shifts again to the children. In any case, each partner usually gives and gets less attention in marriage than in courtship. If the partners do not discuss these changes, they may define each other's interests that are external to the relationship as betrayal. "The task is then to start down the rocky road of accepting differentness as enhancing the relationship" (Marano, 1992, 51).

Five Myths about Marriage

One of the reasons we are surprised by the actual experience of marriage is that we have a poor idea of what day-to-day living together in marriage is really like. Our assumptions are often different from reality. Some of the more unrealistic beliefs our society perpetuates about marriage are discussed here.

Myth 1: Our Marriage Will Be Different   All of us know married people who are bored, unhappy, and in conflict. Despite this, we assume that our marriage will be different. The feeling before marriage that "it won't happen to us" reflects the deceptive nature of courtship. If we are determined that our marriage will be different, what steps are we taking to ensure that it is? This question is relevant because many of us who enter marriage believing that it will be different for us blindly imitate the marriage patterns of others instead of making a conscious effort to manage our own relationship to make it as fulfilling as we expect it to be.

Myth 2: We Will Make Each Other Happy   We also tend to believe that we are responsible for each other's happiness. One woman recalls:

When my husband tried to commit suicide, I couldn't help but think that if I had been the right kind of wife he wouldn't have done such a thing. But I've come to accept that there was more to his depression than just me. He wasn't happy with his work, he drank heavily, and he never got over his twin brother's death.

Although you and your partner will be a tremendous influence on each other's happiness, each of you has roles (employee, student, sibling, friend, son or daughter, parent, and so on) other than the role of spouse. These role relationships will color the interaction with your mate. If you have lost your job or flunked out of school, your father has cancer, your closest friend moves away, or your mother can no longer care for herself but is resisting going to a retirement home, it will be difficult for your spouse to "make you happy." Similarly, although you may make every effort to ensure your spouse's happiness, circumstances can defeat you. Waiting for someone else to make you happy is quite likely to be a lifelong wait.

Myth 3: Our Disagreements Will Not Be Serious   Many couples acknowledge that they will have disagreements, but they assume theirs will be minor and "just part of being married." However, "insignificant" conflict that is not resolved can threaten any marriage. "All I wanted was for him to spend more time with me," recalls a divorced woman. "But he said he had to run the business because he couldn't trust anyone else. I got tired of spending my evenings alone and got involved with someone else."

Myth 4: My Spouse Is All I Need   All of us have needs that require the support of others. These needs range from wanting to see a movie with someone to needing someone to talk to about personal problems to needing the physical expression of a partner's love. Although it is encouraging to believe that our partner can satisfy all our intellectual, physical, and emotional needs, such a belief is not realistic.

Myth 5: The More the Love, the Less the Conflict   Partners sometime assume that the more they love each other, the fewer problems they will have. To the contrary, two researchers (Sprecher & Felmlee, 1993) found that higher levels of love were associated with increases in conflict over time. Individuals in a relationship who love each other are often frustrated when their partner does not act as expected or desired. Out of concern that the relationship be the best it can be, they will engage the other in conflict in the hope of improving the relationship. Partners who have emotionally withdrawn from each other are not motivated to discuss anything with each other, and so their relationship deteriorates.

A more optimistic way to think of these five beliefs about marriage is to recognize that our marriage *may* be different, that we will be *one* important influence on our partner's happiness, that our disagreements *may* not be serious (or if they are serious, we may develop skills to resolve them), that we will be able to satisfy many of our partner's needs, and that love and conflict go hand in hand.

Changes Involving Parents, In-laws, and Friendships

Marriage affects relationships with parents, in-laws, and the friends of both partners. Parents are likely to be more accepting of the partner following the wedding. "I encouraged her not to marry him," said the father of a recent bride, "but once they were married, he was her husband and my son-in-law, so I did my best to get along with him."

Just as acceptance of the mate by the partner's parents is likely to increase, interaction with the partner's parents is likely to decrease. This is particularly true when the newly married couple moves to a distant town. "I still love my parents a great deal," said a new husband, "but I just don't get to see them very often." Parents whose lives have revolved around their children may feel particularly saddened at the marriage of their last child and may be reluctant to accept the reduced contacts. Frequent phone calls, visits, invitations, and gifts may be their way of trying to ensure a meaningful place in the life of their married son or daughter (Goetting, 1990). Such insistence by the parents and in-laws may be the basis of the first major conflict between the spouses. There is no problem if both spouses agree on which set of in-laws or parents they enjoy visiting and the

RECENT RESEARCH

Malia and Blackwell (1997) helped to dispel the myth of the overbearing and interfering mother-in-law. Only four of twenty-two daughters-in-law identified their mothers-in-law as negative forces in their lives.

frequency of such get-togethers. But when one spouse wants his or her parents around more often than the partner does, frustration will be felt by everyone.

Two researchers (Serovich & Price, 1992) examined in-law relationships of 309 spouses. They found that most reported high relationship quality with in-laws, that wives reported equally satisfying relationships with both mothers-in-law and fathers-in-law, and that how close one lived to one's in-laws was not a significant factor in satisfaction with in-law relationships.

Marriage also affects relationships with friends of the same and other sex. Less time will be spent with friends because of the new role demands as a spouse. In addition, friends will assume that the newly married person now has a built-in companion and is not interested in (or would be punished by the spouse for) going barhopping, to movies, or whatever. More time will be spent with other married couples, who will become powerful influences on the new couple's relationship.

Legal Changes

Unless the partners have signed a prenuptial agreement specifying that their earnings and property will remain separate, the wedding ceremony is associated with an exchange of property. Once two individuals become husband and wife, each spouse automatically becomes part owner of what the other earns in income and accumulates in property. Although the laws on domestic relations differ from state to state, courts typically award to each spouse half of the assets accumulated during the marriage (even though one of the partners may have contributed a smaller proportion). For example, if a couple buy a house together, even though one spouse invested more money in the initial purchase, the other will likely be awarded half of the value of the house if they divorce. (Having children complicates the distribution of assets, since the house is often awarded to the custodial parent.) In the event of death, the spouses are legally entitled to inherit between one-third and one-half of the partner's estate, unless a will specifies otherwise.

Should a couple divorce after having children, both are legally responsible to provide for the economic support of their children. In a typical case, the mother is awarded primary physical custody of the children and the father is required by court order to pay about one-half of his gross income if there are two children (until the children graduate from high school). However, child support payments vary by state and, even when ordered, are often not paid. In spite of governmental attempts to get deadbeat parents to pay, many escape by changing jobs or moving to another state.

Sexual Changes

Sex will also undergo some changes during the first year of marriage. The frequency declines for most married couples, but the quality may improve. According to one wife:

> The urgency to have sex disappears after you're married. After a while you discover that your husband isn't going to vanish back to his apartment at midnight. He's going to be with you all night, every night. You

don't have to have sex every minute because you know you've got plenty of time. Also, you've got work and children and other responsibilities, so sex takes a lower priority than before you were married.

Division-of-Labor Changes

One result of the feminist and women's movement is that an increasing number of couples share the domestic work in their relationship. However, research consistently shows that household work is still largely performed by women. Hochschild's study of professional couples noted that after women get home from their regular jobs, they begin what is known as the second shift (Hochschild & Machung, 1989). During courtship, couples may share equally in household work. After marriage, however, there is a tendency for women to drift into more traditional roles. One woman remarked:

> When we were dating, Joe cooked elaborate meals on the weekends, did his own laundry, and we went grocery shopping together. Now that we are married, he hardly ever cooks, and I'm the one who ends up doing his laundry and going to the grocery store.

Traditional versus Egalitarian Expectations

While women end up doing more child care and house care than men in most relationships, they do so with greater acceptance in traditional marriages. Such acceptance is often associated with several other characteristics in traditional marriages. Table 5.1 examines traditional versus egalitarian marriage relationships. Though the relationship a couple has changes after marriage, the nature and extent of the change will depend on the spouses who bring traditional or egalitarian expectations into the marriage. Table 5.1 shows several of these differences.

Table 5.1

Traditional versus Egalitarian Expectations

Traditional Marriage	Egalitarian Marriage
Emphasis on ritual and roles.	Emphasis on companionship.
Couples do not live together before marriage.	Couples may live together before marriage.
Wife takes husband's last name.	Wife may keep her maiden name.
Man dominant; woman submissive.	Neither spouse dominant.
Rigid roles for husband and wife.	Flexible roles for spouses.
One income (the husband's).	Two incomes.
Husband initiates sex; wife complies.	Sex initiated by either spouse.
Wife takes care of children.	Parents share childrearing.
Education important for husband, not for wife.	Education equally important for both spouses.
Husband's career decides family residence.	Family residence decided by career of either spouse.

Interactional Changes

The way wives and husbands perceive and interact with each other continues to change throughout the course of the marriage. Two researchers studied 238 spouses who had been married over thirty years and observed that (across time) men changed from being patriarchal to collaborating with their wives and that women changed from deferring to their husbands' authority to challenging that authority (Huyck & Gutmann, 1992).

Marriage is a battlefield and not a bed of roses.
P. G. Wodehouse
Uncle Fred in the Springtime

Racial Variations in Marriage Relationships

Racial background affects marital relationships because of the cultural heritage of the spouses. Whereas Anglo-American (Euro-American) marriages are characterized by the values of independence, equality, materialism, and competition, less is known about other racial variations. We begin with a discussion of African-American marriages.

African-American Marriages

African-American families have often been described in negative terms, such as being low-income families, having high birthrates among unmarried mothers, being one-parent families, having absentee fathers, and being spouses with limited educations. Such a pathological view of African-American family life is a result of researchers' approaching the African-American family as a deviation from the white norm. Alternatively, the African-American family should be regarded as a unique form with its own history and culture rather than as a deficient white family.

While some Africans migrated to the United States as explorers (McAdoo, 1998), others were brought as slaves without choice.

> Instead of bettering their circumstances, their forced departure from the West African coast resulted in pervasive losses . . . loss of community, loss of original languages, and the loss of status as human beings. . . Biological families were split up so slave families came to place less emphasis on the role of biological parents because most children were separated from and not raised by them. Rather, children were informally "adopted" and raised by other people in their immediate community in extended rather than nuclear family arrangements (Greene, 1995, 29).

In view of the difficulties African-American families encountered, negative labels are inappropriate. Rather, these families may be described in terms of their strengths, including strong kinship bonds, favorable attitudes toward their elderly, adaptable roles, strong achievement orientations, strong religious values, and a love of children.

The context of racism   African-American marriages occur in the context of continued racism, discrimination, and economic insufficiency. *Racism* is the belief that some groups are, as a result of heredity, inferior to other groups. African-

Never has it been more obvious that the African-American experience is not one reality.
Harriet Pipes McAdoo
Black Families

We're not where we want to be. And we're not where we're going to be. But we sure are a long way from where we were.
Martin Luther King, Jr.
Civil rights leader

Where can one better be than in the bosom of one's family?
Unknown
French song

If we are to achieve a richer culture, rich in contrasting values, we must recognize the whole gamut of human potentialities, and so weave a less arbitrary social fabric, one in which each diverse human gift will find a fitting place.
Margaret Mead
Anthropologist

Americans live in a white society in which many whites are prejudiced against African-Americans. This prejudice manifests itself in discrimination against African-Americans. The Mark Fuhrman tapes that surfaced during the O. J. Simpson trial provided a shocking insight into the racist mind. Johnson and Farrell studied almost 2,000 able-bodied men from several ethnic groups in Los Angeles and concluded:

> Despite what opponents of affirmative action contend, the dividing line in American society is still based on race, not class. In employment, race still matters a great deal, and so does skin tone. . . . We found that only 10.3 percent of light skinned African-American men with 13 or more years of schooling were unemployed, compared with 19.4 percent of their dark-skinned counterparts with similar education (1995, A-48).

In the face of continued racism, "all African Americans must make psychological sense out of their disparaged condition, deflect hostility from the dominant group, and negotiate racial barriers under a wide range of circumstances" (Greene, 1995, 30).

Another clear reflection of racism is economic inequality in terms of more limited earning capacity and employment opportunities. "The history of the Black family in the United States includes consistent and persistent economic strain" (Heaton & Jacobson, 1994, 304).

Kinship ties   As previously noted, biological families were often separated during slavery, necessitating a redefinition of kin to include not just nuclear but also extended ties. Children looked not just to their biological mothers for nurturing and support but also to their grandmothers, aunts, cousins, close friends, or people considered kin to a child's mother. "These arrangements also emphasized the important role for elder members of the family" (Greene, 1995, 31).

Even today, African-American spouses continue to maintain close ties with their parents and other kin after they are married. About half of the 423 African-American mothers in a national sample reported that they had received baby-sitting help from family members (most often their own mothers) in the past month. Other relatives are also available to help with care for sick or out-of-school children (Benin & Keith, 1995).

Marital roles   The strong emphasis on ties to one's parents and the larger kinship system seem to affect the African-American marriage relationship. In many cases, the mother-child relationship seems to take precedence over the wife-husband relationship. "The consanguineal bond is often stronger than the marriage relationship. Marriages may end, but blood ties last forever" (McAdoo, 1998, p. 370). It is uncertain whether this tie is maintained because the African-American wife feels that her economically disadvantaged husband will not be able to support her or that he will not stay around to do so (desertion rates among African-American men are higher than among white men.)

African-American women have always been strong, independent women. Forced to take charge of the family because of the father's absence as a result of separation during slavery or his lack of job opportunities in a white racist society, the African-American woman has always worked outside the home. "The dominant cultural norm of women remaining in the home while men worked outside the home was never a practical reality for African American families" (Greene, 1995, 31). Chapter 9, Work, Leisure, and the Family, emphasizes that employed women who bring in more money than their husbands have increased power in the marriage, which may be associated with less stable marriages.

African-American men labor under negative stereotypes. Some of these stereotypes become self-fulfilling prophecies because the dominant society is structured in a way that prevents many African-American men from achieving socially approved goals. In spite of their disadvantaged socioeconomic status, most African-American men function in a way that gains the respect of their mates, children, and community (Staples, 1988). As to the parenting role, Mirande reviewed the literature and observed that in spite of the traditional view that the father in the African-American family is absent or insignificant, many fathers play an integral role in the family and "typically assume an authoritative role" (1991, 58).

Marital satisfaction   While married African-Americans report being happier than never-married, separated, and divorced African-Americans (Taylor et al., 1991), African-American spouses tend to be less happy than white spouses (Staples, 1988). Reasons for their lower sense of marital satisfaction are primarily in reference to economic and social discrimination. African-American wives may be particularly unhappy because, with fewer partners to select from, they may be forced to settle for husbands who have less education than they do. The sense of inadequacy the African-American husband may feel, coupled with the African-American wife's feeling that she has selected someone who is less than her ideal mate, may have a negative impact on both partners (Ball & Robbins, 1986). African-American husbands are also influenced by economic concerns: the higher the family income, the greater the satisfaction with family life (Staples, 1988).

Mexican-American Marriages

It is estimated that by the year 2000 there will be 31 million Mexican-Americans (about 11 percent of the population) living in the United States. Most of them will live in five southwestern states—California, Texas, New Mexico, Arizona, and Colorado (*Statistical Abstract of the United States: 1997*, Table 25). The term *Mexican-American* refers to people of Mexican origin or descent living in America. The breakdown of the Hispanic population in the United States is 64 percent Mexican, 11 percent Puerto Rican, 5 percent Cuban, and 14 percent from various Spanish-speaking Central and South American nations. The remaining 6 percent are categorized as other Hispanics (Ortiz, 1995).

The term *Mexican-American* is sometimes used synonymously with Chicano, Spanish American, Hispanic, Mexican, Californian, and Latin American (Latino). The term *Mexican-American* is sometimes regarded as derogatory.

> The term Mexican American is what I call a government label; it is used to identify Americans of Mexican descent. It is a term that attempts to represent Mexican Americans as full-fledged members of American society.

The divorce level for African-American women is twice as high as for white women.
Harriet Pipes McAdoo,
Black Families

INSIGHT

Stewart (1994) analyzed data on African-American husbands from the National Survey of Black Americans and found only one variable—length of marriage—related to marital satisfaction. Employment status and education were not related. Stewart concluded that African-American married men who stay married do so because they are happy. Hence, they are not likely to stay in unhappy marital relationships.

By the year 2050, one in four Americans will be Hispanic.
George Spencer
Census Bureau

However, by adding the ethnic/national marker Mexican, it relegates them to a subtle form of second-class citizenship (Garcia, 1995, 161).

The treatment of Mexican-Americans as second-class citizens has historical roots. When America annexed Texas in 1845, Mexico became outraged, and the Mexican War followed (1846–48). In the Treaty of Gaudalupe Hidalgo, Mexico recognized the loss of Texas and accepted the Rio Grande as the boundary between Mexico and the United States. Although the war was over, hostilities continued, and the negative stereotyping of Mexican-Americans as a conquered and subsequently inferior people became entrenched. Such stereotyping and discrimination have increased the stress to which Mexican-American spouses have been exposed. Compared with Anglos, Mexican-Americans have less education, earn lower incomes, and work in lower-status occupations (Becerra, 1998).

The husband-wife relationship   Great variability exists among Mexican-American marriages. What is true in one relationship may not be true in another, and the same relationship may change over the passage of time. Nevertheless, some "typical" characteristics of Mexican-American relationships are detailed here.

Male Dominance   Although role relationships between women and men are changing in all segments of society, traditional role relationships between the Mexican-American sexes are characterized by male domination.

> Male dominance means the designation of the father as the head of the household, the major decision maker, and the absolute power holder in the Mexican American family. In his absence, this power position reverts to the oldest son. All members of the household are expected to carry out the orders of the male head (Becerra, 1998, 159).

Mexican-American history is full of examples of women who have deviated from the submissive role.
Rosina M. Becerra
UCLA School of Public Policy and Social Research

Female Submissiveness   The complement to the male authority figure in the Mexican-American marriage is the submissive female partner. Traditionally, the Latina is subservient to her husband and devotes her time totally to the roles of homemaker and mother. As more wives begin to work outside the home, the nature of the Mexican-American husband-wife relationship is becoming more egalitarian in terms of joint decision making and joint childrearing (Vega, 1991).

INSIGHT

Strong familistic values   Mexican-Americans, regardless of their national origin, value close relationships with both nuclear and extended family members (familism). Mexican-Americans report receiving a high level of family support and desiring geographical closeness. Although familism may decrease with increasing levels of acculturation, "even among the most acculturated individuals, Latinos' attitudes and behaviors are still more familistic than Anglo's" (Huratado, 1995, 49).

The relationship between Mexican-American children and their parents has traditionally been one of respect. This respect results not from fear but from parents (including fathers) being warm, nurturing, and companionable. In addition, it is common for the younger generation to pay great deference to the older generation. When children speak to their elders, they do so in a formal way.

Native Americans continue to be the most rural of any ethnic group in the United States.
Robert John
Minority Aging Research Institute

Intergenerational support Dietz (1995) analyzed data from 2,299 interviews with a random sample of Mexican-Americans over the age of 65 and found frequent contact with their adult children. Almost half reported seeing their children on a daily basis. However, despite such contact, only 31 percent reported help with heavy household chores, and only 6 percent received any money. This latter percentage is important, since almost 60 percent of these elderly had incomes below the poverty line. Researcher Dietz concluded that the needs of elderly Mexican-Americans are in large part not being met and called on "policymakers and service providers . . . to provide assistance to them as they have done for other older populations in the United States" (1995, 344). Amey and colleagues (1995) also observed that Mexican-American families are the least likely to have health insurance coverage, making the elderly even more vulnerable.

Native American Marriages

About 2 million individuals define themselves as Native Americans (*Statistical Abstract of the United States: 1997*, Table 52). The term refers not only to American Indians but also to Inuit (Eskimos) and Aleuts (native people of the Aleutian Islands). American Indians comprise over 95 percent of all Native Americans and are the group to which we will refer.

Native Americans, the original Indian inhabitants, are an incredibly diverse group. They comprise 510 federally recognized tribes and 278 reservations speaking 187 different languages (Coburn et al., 1995, 226). Family patterns of Native Americans are also diverse. When viewed across time, Native American families have been patrilineal (heritage traced through males), matrilineal (heritage traced through females), monogamous, polygynous, and polyandrous. Tribal identity has consistently taken precedence over family identity, and the values of a family reflect those of the particular tribe.

Given these caveats, Coburn and colleagues (1995), Yee (1992), and John (1998) made the following observations about Native Americans:

1. Mate selection is based on romantic love.

2. Little stigma is attached to having a child without being married. Children are highly valued in the Native American community.

3. Intermarriage rates are the highest of any racial group. The most frequent intermarriage involving a Native American is between a white husband and a Native American wife.

4. Divorce among Native Americans is not regarded as a traumatic event and is usually not associated with guilt, recriminations, or adverse effects.

5. Elders are viewed as important and are looked up to. They are given meaningful economic, political, religious, and familial roles within the tribe.

6. Extended families are the norm.

7. Role relationships between husbands and wives are becoming less traditional as Native American women become increasingly involved in working outside the home.

Riches take away more
pleasure than they give.
S. G. Champion
Chinese proverb

8. Native American family values include concern for the group, generosity, and disdain for material possessions. These values are counter to the individualism and materialism characteristic of mainstream American culture.

Asian and Pacific Islander American Marriages

It is estimated that by the year 2000 there will be about 11 million Asian and Pacific Islander Americans representing about 4 percent of the U.S. population (*Statistical Abstract of the United States: 1997*, Table 25). These two groups comprise Chinese, Filipino, Japanese, Korean, Vietnamese, Cambodian, Thai, Lao Hmong, Burmese, Samoan, and Guamanian. Each group is different, depending on its cultural heritage, immigration history, American response to its arrival or presence, and its resulting socioeconomic and social adaptation (Yee, 1992). Immigration history is relevant to family patterns in that, historically, immigration laws restricted whole families from immigrating. Rather, only men were permitted entry, because they were a source of inexpensive labor, which resulted in splintered families (wives and children later joined the immigrated men).

The cultural heritage brought to the states included Confucian philosophy and the importance of familism. Confucian principles emphasized superiority of husbands and elders over wives and children. These values were in conflict with egalitarian values that Asian children socialized in America were taught. Familism as a value also emphasized the importance of the family group over the individual and led to a much lower divorce rate among Asians and Pacific Islanders.

Japanese-American marriages differ depending on how long the spouses have lived in the United States. Issei, or first-generation, Japanese-Americans were born in Japan and immigrated to the United States in the early 1900s. They are now in their 80s and live with their children, in nursing homes, or in senior citizen housing projects such as the Little Tokyo Towers in Los Angeles. Their values, beliefs, and patterns reflect those of traditional Japanese families: (1) offspring not allowed to select their own spouse, (2) a stronger parent-child bond than the bond between husband and wife, (3) male dominance, (4) rigid division of labor by sex, and (5) precedence of family values over individual values.

The Japanese are the oldest of
all Asian groups with a
median age of 36, compared
to the total Asian median age
of 30 years.
Kerrily J. Kitano
Harry H. L. Kitano
Social workers

The Nisei, or second-generation, children "were U.S. citizens by birth and not 'aliens ineligible for citizenship'" (Kitano & Kitano, 1998, 315). The younger Nisei believe in romantic love, select their own mates, regard the husband-wife relationship as more important than the parent-child relationship, and have egalitarian sex roles. However, family gatherings of extended kin are common.

The Sansei are the third-generation children and reflect even greater Americanization than the Nisei. The Sansei marry for love (Mirande, 1991) and do not hesitate to marry someone who is not Japanese (over 60 percent of marriages are out-group marriages) (Kitano & Kitano, 1998). The reasons for high outgroup marriages include integrated housing, loss of family, control over marital choices, changes in the law, and a more open generation (p. 328). Native-born Japanese women are also willing to delay marriage and, increasingly, not to marry at all. In a national sample of native-born Japanese-American women, the average age of those who married was 25.6, in contrast to native-

NATIONAL DATA

Of the more than 55.6 million married couples in the United States, only 1,260,000 (2.3%) are interracial (*Statistical Abstract of the United States: 1997*, Table 62).

NATIONAL DATA

Of the over 55.6 million married couples in the United States, 337,000 (about one-half of 1%) are black-white couples (*Statistical Abstract of the United States: 1997*, Table 62).

RECENT RESEARCH

Lewis et al. (1997) analyzed data on 292 individuals who married black-white. Nonracial factors were more important than racial factors for spouse selection among individuals in this sample. For example, more than 70 percent of the respondents reported that common interests and attractiveness of the person, irrespective of racial group membership, were very important factors in selecting a mate.

Children who have interracial friends are less influenced by the prejudice against interracial dating and marriage as adults.

born whites, whose average age at marriage was 22.6. Also, 34 percent of the Japanese-Americans were never-married, in contrast to 14 percent of the white Americans (Ferguson, 1995). Similar changes were observed by Kibria (1995) with regard to the demise of the traditional family hierarchical organization of the Vietnamese family as a result of immigration and Americanization.

Other Variations in Marriage Relationships

Partners in interracial, cross-national, interreligious, commuter, and age-discrepant marriages are also unique.

Interracial Marriages

Interracial marriages may involve many combinations, including American white, American black, Indian, Chinese, Japanese, Korean, Mexican, Malaysian, and Hindu mates. In Hawaii, interracial marriages are not unusual. In 1990, 46 percent of the marriages were interracial (*State of Hawaii Data Book, 1992*). In general, however, interracial marriages are rare in the United States. Black-white marriages are even more rare.

Kalmijn suggested the reason for the low frequency of such pairings:

> The exceptionally strong black/white color line in marriage summarizes the three main features of racial differentiation in American society. The color line is linked to high degrees of racial prejudice, it is the natural outcome of strong patterns of residential and school segregation, and it is in part the heritage of a long history of racial inequality in the economic sphere (1993, 119).

Examples of African-American men who are married to Caucasian women are Quincy Jones, Gregory Hines, and Charles Barkley. Indeed, most black-white pairings involve black men married to white women. One explanation for this phenomenon emphasizes that the African-American community gives African-American men more freedom of choice than it gives African-American women—the latter being under tighter social control. One exception is the interracial marriage of Roger Ebert (of Siskel and Ebert *Sneak Previews*), whose wife is African-American.

Black-white spouses are more likely to have been married before, to be age-discrepant, to live far away from their families of origin, to have been reared in racially tolerant homes, and to have educations beyond high school. Such spouses also tend to seek contexts of diversity. One of the respondents in a study of twenty-one black-white couples observed:

> It's a conscious choice on my part to seek out people of color and to seek out situations that are different from what my experience was beforehand (Rosenblatt et al., 1995, 50).

Another said:

> The type of young people that we were growing up, we were . . . very inquisitive and interested in differences, and to that extent, I would say we

Black-white couples see themselves
not as different but as ordinary people
and spouses.

Everyone else is looking
at us as a black and white
couple, which I think is
really stupid because we
are just a married couple.
We are no different than
a Chinese couple that
were married or a white
couple that were married
or a black couple that
were married. We're just
a couple who decided we
wanted to be with one
another (Rosenblatt et al.,
1995, 25).

RECENT RESEARCH

Airne and Baugher (1997) found that
biracial individuals tend to feel most
comfortable with the racial group corresponding with that of their primary
caregiver or caregivers. They also are
likely to be characterized as warm and
cooperative on the basis of their responses to the 300-item Adjective
Check List.

RECENT RESEARCH

It is generally assumed that interracial
couples have lower marital quality than
same-race couples. However, Chan
and Wethington (1997) compared reports from both husbands and wives in
Asian-white marriages and found no
significant differences from those of
same-race Asian and same-race white
couples.

were the personality types that would not at all shy away from this type
of relationship (Rosenblatt et al., 1995, 50).

Partners in interracial relationships also report that they are self-confident,
autonomous, and somewhat immune to what others think. One respondent
echoed this feeling:

As I matured and grew more self-confident and cared less about what
other people thought about what I was doing, I dated a couple of black
women (Rosenblatt et al., 1995, 51).

Finally, partners who crossed racial lines to marry noted that they felt very
similar to the person they married, and the similarity was an attraction.

If you're really looking at who out of all the people I ever dated is most
like me in terms of values and what we want to do with life, Patricia is
more like me than anybody I've ever known (Rosenblatt et al., 1995, 58).

The parents of the African-American spouse are typically more accepting
of the white partner than the parents of the white spouse are accepting of the
African-American partner. Rosenblatt et al. (1995) emphasized that this difference can be understood in reference to the culture of the African-American
family, in which the mother tends to exude a lot of influence:

In many black families, mothers play the key role in accepting or not
accepting an interracial relationship, much more than in white families. In
white families, fathers were more often major players, as were siblings,
grandparents, and other kin. . . . If women respond more often with openness and efforts to relate and less often with prejudice to the relationship
choices of sons and daughters, then the fact that the crucial person in
black families is most often a woman means there may be more acceptance of a family member's entry into an interracial couple (118).

Children of interracial parents may experience confusion—their own and
the confusion of others—regarding their racial identity. Some parents teach
their biracial children to view themselves as and to tell others that they are biracial. One mother observed:

These kids kept coming to my mother's house saying to our son, "Are you
black or white?" . . . They weren't tormenting him with it. They felt like
they really needed to put him in a category, and finally he says, "I'm
neither" or "I'm both" or something like that, and they went away. They
never bothered him again or said "hello" to him again or wanted to play
with him again. They had their answer (Rosenblatt et al., 1995, 197).

In effect, biracial children "risk encounters with racism on a daily basis. Being categorized as 'half white' or 'mixed' or something else biracial will not
erase the problems caused by racism. In fact, being categorized as biracial
brings an additional load of stereotypes" (Rosenblatt et al., 1995, 197). Some
interracial parents feel helpless to protect their biracial children from the racism
they will confront.

Black-white interracial marriages are likely to increase. In spite of the publicity given to the racism of Mark Fuhrman during the O. J. Simpson trial, few

NATIONAL DATA

Approximately 454,000 foreign students are enrolled at more than 2,500 colleges and universities in the United States. Almost 65 percent (290,000) are from Asia (*Statistical Abstract of the United States: 1997*, Table 285).

RECENT RESEARCH

Jones et al. (1997) studied the effects of different forms of homogamy (religious, generational, and educational) on group closure as reflected by attitudes toward ethnic inmarriage and found that while tendencies toward inmarriage remain strong, they have weakened due to modernization and assimilation.

 INSIGHT

Cultural differences do not necessarily cause stress in cross-national marriage, and degree of cultural difference is not necessarily related to degree of stress. Much of the stress is related to society's intolerance of cross-national marriages which is manifested in attitudes of friends and family. Japan and Korea place an extraordinarily high value on racial purity. At the other extreme is the racial tolerance evident in Hawaii, where a high level of out-group marriage is normative (Cottrell, 1993).

people have anything but contempt for his views. Not only has white prejudice against African-Americans in general declined, but segregation in school, at work, and in housing has decreased, permitting greater contact between the races. The changes are slow, but they are there (Kalmijn, 1993).

Cross-National Marriages

The number of international students studying at American colleges and universities is considerable.

Since American students take classes with foreign students, it sometimes happens that dating and romance lead to marriage. When the international student is male, more likely than not his cultural mores will prevail and will clash strongly with his American bride's expectations, especially if the couple should return to his country.

One female American student described her experience of marriage to a Pakistani, who violated his parents' wishes by not marrying the bride they had chosen for him in childhood. The marriage produced two children before the four of them returned to Pakistan. The woman felt that her in-laws did not accept her and were hostile toward her. The in-laws also imposed their religious beliefs on her children and took control of their upbringing. When this situation became intolerable, the woman wanted to return to the United States. Because the children were viewed as being "owned" by their father, she was not allowed to take them with her and was banned from even seeing them. Like many international students, the husband was from a wealthy, high-status family, and the woman was powerless to fight the family. The woman has not seen her children in six years.

Some cross-national couples solve their cultural differences by making as complete a break as possible from their cultural past. In the preceding case, the American woman could have accepted the traditions of her husband. However, Cottrell (1993) emphasized that for her to do so would have been quite difficult:

> The Western wife does not know how to play her primary role in a traditional society, that of mother and wife. She does not know the heritage which is her responsibility to transmit to her children, and she does not understand the institutional arrangements for cultural transmission (1993, 97).

Alternatively, the couple could have stayed in America and reared their children. Doing so would have been difficult for the Pakistani husband, since he would have had to give up his family ties. He would not have been likely to do this, since he was reared with familistic values. Because of the potential problems faced by cross-national couples, it might be advisable for the partners to spend considerable time in each other's cultural homes before making a marital commitment.

Interreligious Marriages

As Americans have become more secular, religion has become less influential as a criterion for selecting a partner. However, cultural pressure toward reli-

INSIGHT

The impact of a mixed religious marriage may depend more on the devoutness of the partners than on the fact that the partners are of different religions. If both spouses are devout in their religious beliefs, problems in the relationship are more likely. Less problematic is the relationship in which one spouse is devout but the partner is not. If neither spouse in an interfaith marriage is devout, problems regarding religious differences may be minimal or nonexistent.

gious endogamy and individual religious homogamous mate selection continues to exist.

Are people in interreligious marriages less satisfied with their marriages than those who marry someone of the same faith? The answer depends on a number of factors. First, people in marriages in which one or both spouses profess "no religion" tend to report lower levels of marital satisfaction than those in which at least one spouse has a religious tie. People with "no religion" are often more liberal and less bound by traditional societal norms and values—they feel less constrained to stay married for reasons of social propriety.

Second, men in interreligious marriages tend to report less marital satisfaction than men in marriages in which the partners have the same religion. This may be because children of interreligious marriages are typically reared in the faith of the mother, so that the father's influence is negligible. Third, wives who marry outside their faith do not seem any less happy than wives who marry inside their faith. Catholics who marry someone of a different faith are just as likely to report being happily married as Catholics who marry Catholics (Shehan et al., 1990).

Commuter Marriages*

Because more women now pursue careers (and many men are supportive of wives who do), couples are more often confronted with the decision to adopt a commuter relationship and lifestyle in which the spouses maintain separate residences in different cities and reunite regularly. Although some of these commuter marriages are those of high visibility celebrities such as Phil Donahue in Chicago and Marlo Thomas in New York (he later moved to New York), not all such marriages involve professionals. When a factory closes, one worker, usually the husband, moves to a new job in another town while the rest of the family stays in the old house that doesn't sell because the large company for whom the husband worked has gone out of business.

Characteristics of commuter marriages Three characteristics help to define commuter marriages: equal career commitment, distance, and a preference for living together (Gerstel & Gross, 1984).

Equal Career Commitment In commuter marriages, both spouses are equally dedicated to the advancement and success of their respective careers. Her career is as important as his career. Like Brutus, who said of Caesar in Shakespeare's *Julius Caesar* (act III, scene two), it's "not that I loved Caesar less, but that I loved Rome more," spouses in commuter marriages might say it's not that they love each other less but that they love their careers more.

The degree to which work represents a meaningful part of a commuter spouse's life is illustrated by a wife who said:

> I go to pieces when I don't work. I get bored when I am not working. We
> probably work too hard and occasionally feel guilty about it. But we're not

*Information from Gerstel and Gross (1984) based on and reprinted by permission of the publisher of *Commuter Marriage* by N. Gerstel and H. Gross. Copyright © 1984 by The Guilford Press, New York.

the kind of people who can just relax. We think we have to do something (Gerstel & Gross, 1984, 33).

Distance In commuter marriages, the distance between the spouses is great enough to require the establishment of two separate households. Commuter spouses cannot live in the same place and commute to their separate work-places. They must live near their work and commute to see each other. The distance can range from 200 miles for domestic marriages to 5,500 miles for bi-coastal marriages.

Preference for Living Together Although separated in reference to their careers, spouses in commuter marriages wish they could be together. They are not sep-arated because they are having marital problems or are drifting toward a di-vorce. They look forward to an undefined time in the future when they can have their careers and live together, too. In the meantime, they spend a lot of time, energy, and money traveling across the country so that they can be together. It is not unusual for commuter partners to spend about $6,000 per year traveling to be with each other (Gerstel & Gross, 1984).

Unique problems in commuter marriages When dual-career couples in com-muter marriages are compared with dual-career couples who have the same residence, the former report less marital and family happiness (Bunker et al., 1992). Some of the problems experienced by commuter couples are examined in the following subsections.

Fragmented Conversations Because commuters don't return to the same house each evening, their spouse is not there to share the intimate details of life and work. Most miss the presence of their partner and use the telephone as a sub-stitute for face-to-face interaction. In one study, 42 percent of commuter spouses phoned each other every day, and 30 percent called every other day (Gerstel & Gross, 1984).

But just as spouses who live together don't always view their communica-tion positively, neither do commuters. One husband recalls:

> Sometimes she will call me, and I'll be really tired. I just won't have any life in me. And she'll want something more from the call. There's a clashing. Or it happens the other way around. I'll feel good, and she'll be focused on something she's doing. It's hard to shift gears to get into someone else's mood when there is no forewarning and the phone call will soon be over (Gerstel & Gross, 1984, 58).

Lack of Shared Leisure Each partner in a commuter marriage can talk with the other during the week, but going out to dinner, seeing a movie, or attending a concert or play with the spouse is not an option. Each spouse often misses be-ing able to spend leisure time with the other partner. As a result of this high companionship need, most commuter spouses get together on weekends.

Marital Sex Commuter spouses obviously are not sexually available to each other every evening. But even spouses who live together rarely have intercourse every night. In commuter marriages, however, the partners' options of when they can have sex are compressed into smaller time periods. Even when the partners are not in the mood, they may feel that they should have sex because the weekend will soon be over. This places the unrealistic burden on the relationship that the limited time the couple does spend together should be perfect.

Some commuting women and men also experience the "stranger effect," reporting that they need a period of time to reacquaint themselves with their partners before they feel comfortable about having sex. "It takes me at least a day to feel close to him again," said one woman.

Children Children may be an additional problem for the commuter couple. In most cases, young children will stay in the home of one parent, in effect making a single parent out of one spouse. Although some spouses enjoy the role of primary caregiver, others feel resentful of the spouse who is unhampered by the responsibilities of childrearing.

Benefits of commuter marriage In spite of the problems, a commuter marriage has its benefits.

Higher Highs "I'd rather have two terrific evenings a week with my spouse than five average ones," illustrates the view that the time commuter spouses have for each other is, in some ways, like courtship time—more limited but definitely enjoyable. Each spouse makes a special effort to make the time they have together good time. Some commuter partners feel that the periods of separation enhance the love feelings in their relationship. One woman said:

> It's added some romance. There are a lot of comings and goings. We give each other presents. When I come home, there's a huge welcome. And there are tears at parting. I usually arrive looking exhausted. Show up completely collapsed. And my husband has a bottle of wine, no kidding, with a bow around it and flowers or a bottle of Chanel. And he makes a bath for me (Gerstel & Gross, 1984, 76).

Limited Bickering To ensure that the limited time is positive, commuters often make a point of avoiding petty bickering that sometimes creeps into the relationships of spouses who see each other every day. "We just don't want to argue over the laundry when we're together. We don't want to spoil the time we have together," said one commuter.

Work is the only capital that never missed dividends.
Unknown
Poor Richard Jr.'s Almanac

Greater Satisfaction with Work In one study, 90 commuters (defined as apart overnight for more than two nights a week) and 133 single-resident dual-career respondents were asked how satisfied they were with their work situation, their job, the time they have available for their work, the ability to be the worker they want to be, and the degree to which they meet their supervisor's work expectations (Bunker et al., 1992). Commuters gave more positive answers to each of the questions, suggesting much greater job satisfaction than two-earner couples in the same residence.

More Personal Time In the same study (Bunker et al., 1992), spouses in commuter marriages also reported that having more personal time to do what they wanted was a benefit of their lifestyle. The presence of a spouse in the evening necessarily involves negotiation with the spouse about how time is to be spent. With no one else there, one's personal time is free of family relationships or obligations.

Several factors influence the degree to which commuter couples are satisfied with their lifestyle. In a study of thirty-nine commuter couples, Anderson (1992) found that couples who are satisfied with the commuter lifestyle tend to respond well to spending time by themselves and are able to afford the financial costs of commuting. Further, "families who are at later stages of the family life cycle when children are older and less dependent, simply have more flexibility and fewer day-to-day burdens when implementing a commuter lifestyle" (p. 19). Finally, couples who use a systematic or planned decision-making style tend to be more satisfied with their decision to commute. In using a systematic decision-making style, couples consider the alternatives to commuting, collect and evaluate information regarding their decision, mutually choose a plan of action, and make specific plans to reevaluate their decision at a later date.

Age-Discrepant Marriages

Also referred to as ADMs (age-dissimilar marriages), age-discrepant marriages received nationwide visibility when Anna Nicole Smith, a Guess! jeans model, at age 26 married the late 89-year-old J. Howard Marshall II. The sixty-three years between them is unusual—the fact that the man is older than the woman is not. Other celebrities and the number of years they are older than their spouses include Tony Randall, fifty; Tony Bennett, forty; Woody Allen, thirty-five; Johnny Carson, twenty-six. Though the situation is less common, some women are significantly older than their husbands. Martha Raye was thirty-three years older than her husband. Mary Tyler Moore and Kim Basinger are sixteen and five years, respectively, older than their husbands. Whereas most ADMs that come to public attention are those with very wide differences in age, most ADMs fall within a much narrower range (Berardo et al., 1993). And though there is considerable publicity about ADMs, most people marry someone close to their own age. This pattern has continued and shows no abatement (Vera et al., 1990).

Researchers (Shehan et al., 1991; Berardo et al., 1993) have identified the background characteristics of those most likely to be involved in age-discrepant marriages:

1. Race. African-Americans are 1.5 times more likely than Caucasians to be in marriages in which the husband is ten or more years older than the wife. Similarly, wives of Spanish origin are 1.5 times more likely than non-Spanish women to be in marriages in which the husband is ten or more years older than the wife. One explanation for the higher incidence of age-discrepant relationships among African-Americans and Hispanics is the greater likelihood that older minority men will have higher incomes than younger minority men. Such higher incomes of these older males may make them more attractive mates.

2. Interracial and interethnic marriages. Interracial and interethnic marriages are more likely than nonmixed unions to involve relationships in which the husband is ten years older than the wife. Since both age-discrepant and interracial/interethnic relationships are not normative,

1. Age-discrepant relationships are happy. Eighty percent reported that they were happy in their relationships. Forty percent agreed with the statement, "I am happy in my current relationship" and 40 percent reported "strong agreement." Only 4 percent disagreed with the statement. Over 60 percent said that they would become involved in another age-discrepant relationship if their current relationship ended.

2. Age-discrepant relationships lack social approval and support. Only a quarter of the respondents reported that their friends, mothers, and fathers provided clear support for their relationship. Fathers were least approving, with over 40 percent not approving.

3. Age-discrepant relationships are not without problems. In addition to lack of support, the respondents in this study reported a range of problems they attributed to the age difference with their partners including money, in-laws, and recreation.

4. Women perceive benefits of involvement with older partners. Respondents noted financial security (58%), maturity (58%), and dependability (51%) as the primary advantages of involvement with an older man. Higher status was regarded as less important as 28 percent of the respondents identified this as a benefit.

5. Friends of the couple are joint friends. Over 70 percent (71%) of respondents reported that when they did something recreationally with another couple, the friends were likely to be both of theirs. However, if the friends were friends of only one of them, it was more likely to be the man (22%) than the woman (5%).

individuals willing to break one set of normative barriers may be more likely to break another.

3. Remarriage. Anna Nicole Smith (referred to at the opening of this section) was in her second marriage; her husband was in his third marriage. Wives in second or subsequent marriages are 2.7 times more likely to be younger than their husbands. The higher income of older males is one explanation for younger women marrying older men.

4. Older individuals. ADMs have higher proportions of individuals ten years older than those in first marriages.

Though age-discrepant couples must cope with having different interests, whether to have additional children, sexual compatibility, and early widowhood for the younger partner, only the last issue is unique to age-discrepant relationships. Whether spouses in age-discrepant relationships have a higher rate of divorce than couples in homogamous-age marriages remains an unanswered question (Berardo et al., 1993).

Successful Marriages

Judith Wallerstein, a wife for fifty years, a clinical psychologist, and coauthor with Sandra Blakeslee of *The Good Marriage* (1995), identified herself as having a successful marriage. But what is a successful marriage?

Definition of Marital Success

Marital success is measured in terms of marital stability and marital happiness. Stability refers to how long the spouses are married and how permanent they view their relationship, whereas marital happiness refers to more subjective aspects of the relationship. In describing marital success, researchers have used the terms *satisfaction, quality, adjustment, lack of distress,* and *integration.*

Marital success is often measured by asking spouses how happy they are, how often they spend their free time together, how often they agree about various issues, how easily they resolve conflict, how sexually satisfied they are, and how often they have considered separation or divorce.

Wallerstein and Blakeslee (1995) studied fifty financially secure couples in stable (from ten to forty years) and happy marriages with at least one child. These couples defined marital happiness as feeling respected and cherished. They also regarded their marriage as a work in progress that needed continued attention lest it become stale. No couple said that they were happy all the time. Rather, a good marriage is a process. However, how a couple evaluate their own marriage and how others evaluate it may differ. Two researchers (Porter & Wampler, 1997) found that a couple's parents are more likely to evaluate the marriage of their children as more adjusted and cohesive than the children themselves.

Billingsley and colleagues (1995) interviewed thirty happily married couples who had been wed an average of thirty-two years and had an average of 2.5 children. Six themes permeated the interviews with these couples.

- Commitment. Divorce was not considered an option. The couples were committed to each other for personal reasons rather than societal pressure. They also did not leave themselves open to affairs. One respondent said:

  > on any number of occasions attractive people had given clear signals of availability, however, whenever I sense something like this, I'm thrilled to talk about my wife. Talking about the beauty of our relationship helps to build a hedge in what could be a troublesome situation (Billingsley et al., 1995, 288).

- Religiosity. A strong religious orientation provided the couples with social support from church members, spiritual support, and moral guidance in working out problems. It was also a shared value.

- Role models. The couples spoke of positive role models in their parents or negative ones they were intent on improving on.

- Finances and work. Being nonmaterialistic, being disciplined, and being flexible with each other's work schedules and commitments were characteristic of these happily married couples.

- Common interests. The spouses talked of sharing values, children, work, travel, goals, dependability, and the desire to be together.

- Communication. Openness about likes and dislikes, feelings, preferences, etc., was a theme for these couples. The couples also discussed problems with each other and listened to each other's point of view.

What can we learn from these successful marriages? Commitment, common interests, communication skills, and a nonmaterialistic view of life all seem related to being involved in and maintaining a successful marriage. Couples might strive to include these as part of their relationship.

As we have noted, durability is only one criterion for a successful marriage. Satisfaction is another. It is important to note that marital satisfaction is not consistent throughout marriage but varies with the stage of the family life cycle. Figure 5.1 provides the retrospective marital enjoyment evaluation of fifty-two white college-educated husbands and wives over a period of thirty-five years together. The couples reported the most enjoyment with their relationship in the beginning, followed by less enjoyment during the childbearing stages, and a return to feeling more satisfied after the children left home. Husbands also reported being more satisfied throughout the marriage than the wives (Vaillant & Vaillant, 1993).

Marriages that Last Fifty-Plus Years

What is marriage like for those who still have partners in late, late life? Field and Weishaus (1992) reported interview data on seventeen couples who had been married an average of fifty-nine years and found that the husbands and wives viewed their marriage very differently. Men tended to report more marital satisfaction, more pleasure in the way their relationships had been across time, more pleasure in shared activities, and closer affectional ties. While club activities, including church attendance, were related to marital satisfaction, financial stability, amount of education, health, and intelligence were not. Sex

Walters et al. (1997) provided cross-cultural data on the perceived marital quality by both husbands and wives in 1,718 families in the United States, Georgia (south of Moscow), Poland, and Samara (south-central Russia). Both spouses had been married seven years or less and had at least one child. Approximately 20 to 30 percent of all the couples in all the countries perceived the quality of their relationship to be low.

Figure 5.1

Retrospective Marital Enjoyment

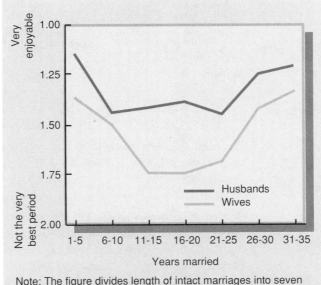

Note: The figure divides length of intact marriages into seven consecutive 5-year periods and compares the retrospectively assessed marital satisfaction of 52 husbands and their wives.

Source: Caroline O. Vaillant and George E. Vaillant. Is the U-curve of marital satisfaction an illusion? A 40-year study of marriage. *Journal of Marriage and the Family,* 1993, 55, 237, Figure 6. Copyrighted 1993 by the National Council on Family Relations, 3989 Central Avenue NE, Suite 550, Minneapolis, MN 55421. Reprinted by permission.

Traveling in the company of those we love is home in motion.

Leigh Hunt

The Indicator

This couple reports that "Our kids are gone and we're off to see the world."

was also more important to the husbands. While every man in the study reported that sex was always an important part of the relationship with his wife, only four of the seventeen wives had the same report.

The wives, according to the researchers, presented a much more realistic view of their marriage. For this generation of women, "it was never as important for them to put the best face on things" (Field & Weishaus, 1992, 273). Hence, these wives were not unhappy; they were just more willing to report disagreements and changes in their marriages across time.

Obstacles to Predicting Marital Success

Although everyone wants to have a successful marriage, predicting that one's own marriage will be successful is difficult, if not impossible. There are several reasons why.

Illusion of the perfect mate   The illusion that you have found the perfect partner—one who will meet all your needs—is often operative in courtship. The reality that surfaces after marriage can be very different. You have not found the perfect mate—there isn't one. Indeed, anyone that you marry will probably have at least one characteristic that you do not like. And this one trait may take on enormous proportions after the wedding. Some people who discover that their partner has a characteristic that they dislike (unfaithful, substance-abusing, not ambitious) consider divorce.

Deception during courtship   Your illusion of the perfect mate is helped along by some deception during courtship on the part of your partner. At the same time, you are presenting only favorable aspects of yourself to the other person. These deceptions are merely attempts to withhold the undesirable aspects of yourself for fear that your partner will not like them. One male student said he knew he drank too much and if his date found out, she would be disappointed and might drop him. He kept his drinking hidden throughout their courtship. They married and are now divorced. She said of him, "I never knew he drank whiskey until our honeymoon. He never drank like this before we were married." Courtship and marriage are different contexts and they elicit different behaviors in the partners.

Confinement of marriage   Although premarital norms permit relative freedom to move in and out of relationships, marriage involves a binding legal contract. One recently married person described marriage as an iron gate that clangs shut behind you, and getting it open is emotionally and financially expensive. One's freedom to leave a relationship is transformed by the wedding ceremony. Thereafter, there is tremendous individual and social pressure (particularly if there are children) to work things out and a feeling of obligation to do so that was not previously present. The new sense of confinement to one person and a routine set of activities may bring out the worst in partners who seemed very cooperative in the context of courtship.

Chains do not hold a marriage together. It is threads, hundreds of tiny threads which sew people together through the years. That is what makes a marriage last—more than passion or even sex.
Simone Signoret
Actress

Maidens! Why should you worry in choosing whom you should marry? Choose whom you may, you will find you have got someone else.
John Hay
Distiches

There are more lies told in dating relationships than any other time in life.
Jack Turner
Psychologist

Matrimony is not a word but a sentence.
H. W. Thompson
Body, Boots, and Britches

Stress of career and children   Predicting that one's own marriage will be successful is also difficult because the partners must necessarily shift their focus from each other to the business of life. Careers and children emerge as concerns that often take precedence over spending time with each other, going to parties, and seeing movies. Time and energy spent on jobs and childrearing often leave marriage partners too tired to interact with each other. "I never see my partner" is a statement often heard by marriage counselors. Also, when partners do get together they may be very stressed and yell at each other. "Whenever we do get together, we fight," said one partner. (Work and leisure are discussed in greater detail in Chapter 9, Work, Leisure, and the Family.)

Inevitability of change   Just as you are not the same person you were ten years ago, you will be different ten years from now. The direction and intensity of these changes are not predictable for you or for your partner. You, your partner, and your relationship will not be the same two years (or two days) in a row. Reflecting on change in her marriage, one woman recalled:

> When we were married we were very active in politics. Now I have my law degree and am enjoying my practice. But Jerry is totally immersed in meditating and taking health food nutrients. He also spends four nights a week playing racquetball. I never imagined that we'd have nothing to say to each other after only three years of marriage.

Other spouses may maintain similar interests across the years, but either or both may undergo a dramatic change in mental functioning. In recent years, Alzheimer's disease and its attendant problems have been featured in the media. One woman reflected on the effect of her marriage:

> After saving for retirement and planning to spend our later years traveling, my spouse developed Alzheimer's disease. I look at her as she stares into space wondering who I am. We have been robbed of our intimacy as a couple. It is sad.

Other changes that can dramatically affect a couple are physical (e.g., Christopher Reeve's paralysis below the neck) or philosophical (e.g., a change from being religiously devout to being an atheist—or vice versa). Still other changes may be event-related. The death of one's parents or child or the loss of one's job can have enormous effects on the spouse that require substantial adjustment on the part of the partner. How change will affect spouses and their relationship cannot be predicted.

Inadequate parental models   In spite of our desire to have a happy and fulfilling marriage, few of us have come from homes in which our parents served as good role models. Only 5 out of 100 spouses in a study on happy marriages reported that they wanted to have a marriage like that of their parents (Wallerstein & Blakeslee, 1995).

Denial of trouble ahead   Most lovers getting married think of divorce as something that happens to other couples. Barbara Whitehead (1997), author of *The Divorce Culture*, noted that courtship is not a teachable moment and that lovers

We marry someone who is ideally suited to us NOW, in our present "headspace," present "needs," and present environment. All are likely to change and it is expected that we accommodate to the change and remain devoted to our partners.
Tim Britton
Sociologist

Parents are patterns.
Thomas Fuller
Gnomologia

Strengthening Marriage through Divorce Law Reform

Some family scholars and policymakers advocate strengthening marriage by reforming divorce laws to make divorce harder to obtain. Since California became the first state to implement "no-fault" divorce laws in 1969, every state has passed similar laws allowing couples to divorce without proving in court that one spouse was "at fault" for the marital breakup. The intent of no-fault divorce legislation was to minimize the acrimony and legal costs involved in divorce, making it easier for unhappy spouses to get out of their marriage. Under the system of no-fault divorce, a partner who wanted a divorce could get one, usually by citing irreconcilable differences, even if his or her spouse did not want a divorce.

In recent years, about a dozen states have considered, and in some cases implemented, measures to revise divorce laws (Brienza, 1996). In most cases, these measures are designed to make breaking up harder to do by requiring proof of fault (such as infidelity, physical or mental abuse, drug or alcohol abuse, and desertion), or extending the waiting period required before a divorce is granted. In most divorce law reform proposals, no-fault divorces would still be available to couples who mutually agree to end their marriages. Only two states (New York and Mississippi) require mutual consent in granting divorces.

The vice president of the conservative Family Research Council and supporter of divorce law reform says,

> We don't want to keep people in a marriage that neither wants. But we do want to encourage them to go the extra mile to try to preserve it. . . . People [who divorce] have this mistaken notion they're gaining freedom. But what they actually take on is a whole new set of problems, and many think that maybe they abandoned the marriage too quickly (cited in Clark, 1996, 415).

Opponents argue that divorce law reform measures would increase acrimony between divorcing spouses (which harms the children as well as the adults involved), increase the legal costs of getting a divorce (which leaves less money to support any children), and delay court decisions on child support and custody and distribution of assets. In addition, critics point out that ending no-fault divorce would add countless court cases to the dockets of an already overloaded court system. Some opponents of reviving the old fault system of divorce also object to the concept of fault in a marriage. Lynn Gold-Bikin, an attorney and former chairperson of the American Bar Association's section on family law points out the problematic aspect of determining fault in a marital breakup:

> Some women who are desperately lonely are married to a guy who won't sleep with them because he's too busy with business, so she goes out and has an affair. Now who's at fault? (Cited in Clark, 1996, 416.)

Efforts in many state legislatures to repeal no-fault divorce laws have largely failed.

However, in June 1997, the Louisiana legislature became the first in the nation to pass a law creating a new kind of marriage contract that would permit divorce only in narrow circumstances. Under the new Louisiana law, couples can voluntarily choose between two types of marriage contracts: the standard contract that allows a no-fault divorce or a "covenant marriage" that permits divorce only under conditions of fault (such as abuse, adultery, or imprisonment on a felony) or after a marital separation of more than two years. Couples who choose a covenant marriage are also required to get premarital counseling from a clergy member or another counselor. Couples who choose the standard marriage contract can obtain a divorce after six months' separation, even if one spouse doesn't want the divorce. The Christian Coalition says Louisiana's new law will keep families together. Representative Tony Perkins, who sponsored the bill, believes the new law will prevent potentially weak marriages. Perkins explains:

> When a man says he wants a no-fault marriage and a woman says she wants a covenant marriage, that's going to raise some red flags. She's going to say "What? You're not willing to have a lifelong commitment to me?" (Louisiana Divorce, 1997, p. 1.)

How will Louisiana's new law affect the divorce rate in that state? Will it strengthen marriages by making divorce harder and weeding out potentially weak marriages? It will be years before family scholars may be able to answer that question. In the meantime, some states are moving to pass similar legislation.

REFERENCES

Brienza, Julie. 1996. At the fault line: Divorce laws divide re-formers. *Trial* 32, no.9: 12–14.

Clark, Charles S. 1996. Marriage and divorce. *CQ Researcher* 6, no.18: 409–32.

Louisiana Divorce. 1997. Women's Connection Online, Inc. http://www.womenconnect.com. 21 July.

If we share our troubles we halve them.

Patricia Wentworth

Moss Silver Deals with Death

refuse to contemplate troubles and adversity ahead. She cites the absence of prenuptial agreements implying that something can go wrong.

In an effort to help couples take their marriages seriously and to work on their marriages rather than abandon them when problems arise, American society is moving toward strengthening marriage by means of divorce law reform. We examine this issue in this chapter's social policy.

GLOSSARY

commitment   An intent to maintain a relationship.

marital success   Relationship in which the partners have a durable and self-reported happy relationship.

Mexican-American   Referring to people of Mexican origin or descent living in America.

primary group   Small, intimate, emotional, and informal group. Marriages and families are primary groups.

racism   The belief that some groups are, as a result of heredity, inferior to other groups.

SUMMARY

About 95 percent of individuals in U.S. society eventually marry. The relationships they have are very diverse.

Motivations toward and Commitment in Marriage

Persons marry for reasons that include personal fulfillment, companionship, legitimacy of parenthood, and emotional and financial security from marriage. Regardless of the reason, marriage is a commitment between individuals, between families, and between the couple and the state issuing the marriage license.

Changes after Marriage

Marriage involves an array of changes. Personal changes include an enhanced self-concept, a satisfying sex life (when compared with that of singles), improved acceptance of

Since companionship is a major reason for and benefit of marriage, this chapter's personal application scale focuses on this aspect of a couple's relationship.

Companionship Scale

This scale is designed to assess the degree to which partners experience companionship in their relationship. After reading each item, circle the number that best approximates your answer.

0 = strongly disagree
1 = disagree
2 = undecided
3 = agree
4 = strongly agree

	SD	D	UN	A	SA
1. We enjoy the same recreational activities.	0	1	2	3	4
2. I share in a few of my partner's interests. (reverse scored)	0	1	2	3	4
3. We like playing together.	0	1	2	3	4
4. We enjoy the out-of-doors together.	0	1	2	3	4
5. We seldom find time to do things together. (reverse scored)	0	1	2	3	4
6. I feel we share some of the same interests.	0	1	2	3	4

82
2.7

SCORING: Look at the numbers you circled above. Reverse score the numbers for items 2 and 5. For example, if you circled a 0, give yourself a 4; if you circled a 3, give yourself a 1, etc. Add the numbers and divide by 6, the total number of items. The lowest possible score would be 0, reflecting the complete absence of a companionship relationship; the highest score would be 4, reflecting a relationship characterized by great companionship. The average score of ninety-four male partners who took the scale was 3.11; the average score of ninety-four female partners was 3.29. Thirty-nine percent of the couples were married, 38 percent were unmarried, 23 percent were living together. The average age was just over 24.

Source: Susan Sprecher, Sandra Metts, Brant Burleson, Elaine Hatfield, and Alicia Thompson. 1995. Domains of expressive interaction in intimate relationships: Associations with satisfaction and commitment. *Family Relations,* 44: 203–10. Copyrighted © 1995 by the National Council on Family Relations, 3989 Central Avenue, NE, Suite 550, Minneapolis, MN 55421. Reprinted by permission.

the mate by one's parents, less time with one's parents, and less time with one's same-sex friends. Marriage also results in division-of-labor changes and in legal changes, such as the exchange of property and certain economic responsibilities in the event the couple divorces.

The degree to which the couple have traditional role relationships will depend on their expectations that their marriage will be traditional or egalitarian. The latter is characterized by flexible roles for the spouses, two incomes, coparenting of children, and family residence decided by either spouse.

Racial Variations in Marriage Relationships

African-American marriages are characterized by strong kinship ties and strong mother-child relationships. A national study of African-American marriages revealed that African-American men who stay married do so because they are happy.

Mexican-American marriages have traditionally been characterized by male dominance and female submissiveness. As more wives work outside the home, the power of the wife will increase and the balance of power will continue to shift.

Native American marriages are not so easily categorized. Since tribal identity supersedes family identity and the values and beliefs vary widely among the 300 federally recognized tribes, there are few fixed characteristics of Native American marriages.

There are over 10 million Asian and Pacific Islander Americans. Japanese-American marriages differ, depending on the degree of socialization of the spouses in the United States. Issei (first-generation) marriages are very traditional, in contrast to Sansei (third-generation) marriages, in which the spouses are very Americanized.

Other Variations in Marriage Relationships

Interracial marriages are also increasing. Spouses in such marriages have usually been married before, are older, were reared in racially tolerant homes, have parents who live far away, and seek contexts of diversity. They are also self-confident and autonomous and do not see their marriage as different from any other marriage.

An increasing number of marriages are interreligious. Although mixed religious marriages do not necessarily imply a greater risk to marital happiness, marriages in which one or both spouses profess no religion are in the greatest jeopardy. Also, husbands in interreligious marriages seem less satisfied because children are usually reared in the faith (or nonfaith) of the wife.

Commuter marriages reflect the desire of spouses to pursue their respective careers and maintain their marriage. Doing so is difficult and may provide more benefit to the career than to the marriage. Most couples view this lifestyle as temporary until they can arrange their work lives so that they can be together.

Age-discrepant marriages lack social approval, with fathers of daughters in such relationships being the least approving. However, wives in such relationships report that they are happy and over half would become involved in another such relationship. Benefits include higher financial security and a more dependable mate. Data on a higher divorce rate among spouses in age-discrepant marriages are mixed.

Successful Marriages

Successful marriages are those that are both durable and happy. Characteristics of spouses in such relationships include commitment, common interests, open communication, religiosity, and parents in stable/happy relationships. The early and later years of marriage are reported as being the most happy. Elderly husbands and wives view their marriages differently, with the husbands being happier.

Predicting that one's own marriage will be successful is difficult, since courtship is deceptive, marriage may be defined as confining, partners inevitably change, and lovers refuse to look closely at potential problems in their relationships before they marry. Despite the fact that almost half of couples get divorced, almost all newlyweds deny that they will have serious problems.

REFERENCES

Airne, Michelle, and Shirley L. Baughe. 1997. Self-description and factors contributing to racial identification of biracial individuals. Paper presented at the Annual Conference of the National Council on Family Relations, Crystal City, Va., November.

American Council on Education and University of California. 1996. *The American freshman: National norms for fall, 1996.* Los Angeles: Los Angeles Higher Education Research Institute.

American Council on Education and University of California. 1997. *The American freshman: National norms for fall, 1997.* Los Angeles: Los Angeles Higher Education Research Institute.

Amey, Cheryl, Karen Seccombe, and R. Paul Duncan. 1995. Health coverage of Mexican American families in the U.S. *Journal of Family Issues* 16: 488–510.

Anderson, Elaine A. 1992. Decision-making style: Impact on satisfaction of the commuter couples lifestyle. *Journal of Family and Economic Issues* 13: 5–21.

Ball, R. E., and L. Robbins. 1986. Marital status and life satisfaction among black Americans. *Journal of Marriage and the Family* 48: 389–94.

Becerra, R. M. 1998. The Mexican-American family. In *Ethnic families in America: Patterns and variations,* 4th ed. Edited by Charles H. Mindel, Robert W. Habenstein, and Roosevelt Wright, Jr. Upper Saddle River, N.J.: Prentice Hall, 153–71.

Benin, Mary, and Verna M. Keith. 1995. The social support of employed African American and Anglo mothers. *Journal of Family Issues* 16: 275–97.

Berardo, Felix M., Jeffrey Appel, and Donna H. Berardo. 1993. Age dissimilar marriages: Review and assessment. *Journal of Aging Studies* 7: 93–106.

Billingsley, S., M. Lim, and G. Jennings. 1995. Themes of long-term, satisfied marriages consummated between 1952–1967. *Family Perspective* 29: 283–95.

Bunker, B. B., J. M. Zubek, V. J. Vanderslice, and R. W. Rice. 1992. Quality of life in dual-career families: Commuting versus single-residence couples. *Journal of Marriage and the Family* 54: 399–407.

Carter, J. H. Commuter marriages. 1992. *Black Enterprise* 22: 246–50.

Chan, Anna Y., and Elaine Wethington. 1997. Are interracial marriages less satisfying? Paper presented at the Annual Conference of the National Council on Family Relations, Crystal City, Va., November.

Coburn, Joseph, Anita B. Pfeiffer, Sharon M. Simon, Patricia A. Locke, Jack B. Ridley, and Henri Mann. 1995. The American Indians. In *Educating for diversity,* edited by Carl A. Grant. Boston: Allyn and Bacon, 225–54.

Cottrell, A. B. 1993. Cross-national marriages. In *Next of kin,* edited by L. Tepperman and S. J. Wilson. Englewood Cliffs, N.J.: Prentice Hall, 96–100.

Devall, E., Martha Khodary, and Robert Steiner. 1997. The relationship between acculturation, gender roles, and marital satisfaction among Mexican-American couples. Poster session, National Council on Family Relations, Crystal City, Va., November.

Dietz, Tracy L. 1995. Patterns of intergenerational assistance within the Mexican American family. *Journal of Family Issues* 16: 344–56.

Doyle, James A. 1995. *The male experience.* Madison, Wis.: Brown and Benchmark.

Engel, J. W. 1982. *Changes in male-female relationships and family life in the People's Republic of China.* Honolulu, Hawaii: Hawaii Institute of Tropical Agriculture and Human Resources. Research Series 014.

Engel, J. W. 1984. Marriage in the People's Republic of China: Analysis of a new law. *Journal of Marriage and the Family* 46: 955–61.

Ferguson, Susan. 1995. Marriage timing of Chinese American and Japanese American women. *Journal of Family Issues* 16: 314–43.

Field, Dorothy, and Sylvia Weishaus. 1992. Marriage over half a century: A longitudinal study. In *Changing lives,* edited by M. Bloom. Columbia, S.C.: University of South Carolina Press, 269–73.

Fowers, Blaine J., and Brooks Applegate. 1995. Do marital conventionalization scales measure a social desirability response bias? A confirmatory factor analysis. *Journal of Marriage and the Family* 57: 237–41.

Garcia, Esteban Herman. 1995. The Mexican Americans. In *Educating for diversity,* edited by Carl A. Grant. Boston: Allyn and Bacon, 159–68.

Gerstel, N., and H. Gross. 1984. *Commuter marriage.* New York: Guilford Press.

Glenn, N. D. 1997. *Closed hearts, closed minds: The textbook story of marriage.* New York: Institute for American Values.

Goetting, A. 1990. Patterns of support among in-laws in the United States. *Journal of Family Issues* 11: 67–90.

Greene, Beverly. 1995. African American families. *The Phi Kappa Phi Journal* 75: 29–32.

Heaton, Tim B., and Cardell K. Jacobson. 1994. Race differences in changing family demographics in the 1980s. *Journal of Family Studies*, 290–308.

Hochschild, A., and A. Machung. 1989. *The second shift.* New York: Viking.

Huratado, Aida. 1995. Variations, combinations, and evolutions: Latino families in the United States. In *Understanding Latino families: Scholarship, policy, and practice,* edited by Ruth E. Zambrana. Thousand Oaks, Calif.: Sage, 40–61.

Huyck, M. H., and D. L. Gutmann. 1992. Thirtysomething years of marriage: Understanding experiences of women and men in enduring family relationships. *Family Perspective* 26: 249–65.

John, Robert. 1998. Native American families. In *Ethnic families in America: Patterns and variations,* 4th ed. Edited by Charles H. Mindel, Robert W. Habenstein, and Roosevelt Wright, Jr. Upper Saddle River, N.J.: Prentice Hall, 382–421.

Johnson, James H. Jr., and Walter C. Farrell Jr. 1995. Race still matters. *The Chronicle of Higher Education,* 7 July, A 48.

Jones, F. L. 1997. Individual and group effects on ethnic intermarriage: A multilevel analysis. *Australian Journal of Social Research.* 3: 17–35.

Kalmijn, Matthus. 1993. Trends in black/white intermarriage. *Social Forces,* 119–46.

Kibria, Nazli. 1995. *Family tightrope.* Princeton, N.J.: Princeton University Press.

Kitano, K. J., and H. H. L. Kitano. 1998. The Japanese-American family. In *Ethnic families in America: Patterns and variations,* 4th ed. Edited by Charles H. Mindel, Robert W. Habenstein, and Roosevelt Wright, Jr. Upper Saddle River, N.J.: Prentice Hall, 311–30.

Knox, D., T. Britton, and B. Crisp. 1997. Age discrepant relationships reported by university faculty and their students. *College Student Journal* 31: 290–93.

Landis-Klein, C., L. A. Foley, L. Nall, P. Padgett, and L. Walters-Palmer. 1995. Attitudes toward marriage and divorce held by young adults. *Journal of Divorce and Remarriage* 23: 63–73.

Lawler, E. J., and J. Yoon. 1996. Commitment in exchange relations: Test of a theory of relational cohesion. *American Sociological Review* 61: 89–108.

Lewis Jr., R., G. Yancey, and S. S. Bletzer. 1997. Racial and nonracial factors that influence spouse choice in Black/White marriages. *Journal of Black Studies* 28: 60–78.

Lopez, L. C., and M. Hamilton. 1997. Comparison of the role of Mexican-American and Euro-American family members in the socialization of children. *Psychological Reports* 80: 283–88.

Malia, Julia A., and E. Michelle Blackwell. A study of the nature of mother- and daughter-in-law relationships using the OSR NUD-IST Program. Paper presented at the Annual Conference of the National Council on Family Relations, Crystal City, Va., November.

Marano, H. E. 1992. The reinvention of marriage. *Psychology Today.* January/February, 49 passim.

Mayo, Y. 1997. Machismo, fatherhood, and the Latino family: Understanding the concept. *Journal of Multicultural Social Work* 5: 49–61.

McAdoo, H. P. 1998. African-American families. In *Ethnic families in America: Patterns and variations,* 4th ed. Edited by Charles H. Mindel, Robert W. Habenstein, and Roosevelt Wright, Jr. Upper Saddle River, N.J.: Prentice Hall, 361–81.

Michael, Robert T., John H. Gagnon, Edward O. Laumann, and Gina Kolata. 1994. *Sex in America: A definitive survey.* Boston: Little, Brown.

Mirande, A. 1991. Ethnicity and fatherhood. In *Fatherhood and families in cultural context,* edited by F. W. Bozett and S. M. H. Hanson. New York: Springer, 33–82.

National Center for Health Statistics. 1998. Births, marriages, divorces, and deaths for June, 1997. Monthly vital statistics report 46, no. 6. Hyattsville, Md.: National Center for Health Statistics. 28 January.

Nock, Steven L. 1995. Commitment and dependency in marriage. *Journal of Marriage and the Family,* 57: 503–14.

Ortiz, Vilma. 1995. The diversity of Latino families. In *Understanding Latino families: Scholarship, policy, and practice,* edited by Ruth E. Zambrana. Thousand Oaks, Calif.: Sage, 18–39.

Porter, Lawrence C., and Richard S. Wampler. 1997. Multi-generation perspective of mothers and fathers. Paper presented at the Annual Conference of the National Council on Family Relations, Crystal City, Va., November.

Rogers, Richard G. 1995. Marriage, sex, and mortality. *Journal of Marriage and the Family* 57: 515–26.

Rosenblatt, Paul C., Teri A. Karis, and Richard D. Powell. 1995. *Multiracial couples*. Thousand Oaks, Calif.: Sage.

Ruvolo, Ann. P., and J. Veroff. 1997. For better or for worse: Real-ideal discrepancies and the marital well being of newlyweds. *Journal of Social and Personal Relationships* 14: 223–42.

Sacher, J. A., and M. A. Fine. 1996. Predicting relationship status and satisfaction after six months among dating couples. *Journal of Marriage and the Family* 58: 21–32.

Serovich, J., and S. Price. 1992. In-law relationships: A role theory perspective. Paper presented at the 54th Annual Conference of the National Council on Family Relations, Orlando, Florida. Used by permission.

Shehan, C. L., F. M. Berardo, H. Vera, S. M. Carley. 1991. Women in age-discrepant marriages. *Journal of Family Issues* 12: 291–305.

Shehan, C. L., E. W. Bock, and G. R. Lee. 1990. Religious heterogamy, religiosity, and marital happiness: The case of Catholics. *Journal of Marriage and the Family* 52: 73–79.

Sousa, Lori A'lise. 1995. Interfaith marriage and the individual and family life cycle. *Family Therapy* 22: 97–104.

Sprecher, S., and Diane Felmlee. 1993. Conflict, love and other relationship dimensions for individuals in dissolving, stable, and growing premarital relationships. *Free Inquiry in Creative Sociology* 21: 115–25.

Staples, Robert. 1988. The Black American family. In *Ethnic families in America: Patterns and variations*, edited by C. H. Mindel, R. W. Habenstein, and R. Wright Jr. New York: Elsevier, 303–24.

State of Hawaii data book, 1992: A statistical abstract. State of Hawaii: Department of Business, Economic Development & Tourism.

Statistical Abstract of the United States: 1997. 117th ed. Washington, D.C.: U.S. Bureau of the Census.

Stewart, Ron. 1994. Family life satisfaction of African-American men: The importance of socio-demographic characteristics. In *The black family*, 5th ed., edited by Robert Staples. Belmont, Calif.: Wadsworth, 121–29.

Thao, T. C. 1992. Among customs on marriage, divorce, and the rights of married women. In *Cultural diversity and families*, edited by K. G. Arms, J. K. Davidson Jr., and N. B. Moore. Dubuque, Iowa: Brown and Benchmark, 54–66.

Vaillant, C. O., and G. E. Vaillant. 1993. Is the U-curve of marital satisfaction an illusion? A 40-year study of marriage. *Journal of Marriage and the Family* 55: 230–39.

Vega, William A. 1995. The study of Latino families. In *Understanding Latino families: Scholarship, policy, and practice*, edited by Ruth E. Zambrana. Thousand Oaks, Calif.: Sage, 3–17.

Vera, Hernan, Felix M. Berardo, and Joseph S. Vandiver. 1990. Age irrelevancy in society: The test of mate selection. *Journal of Aging Studies* 4, no. 1: 81–95.

Wallerstein, Judith, and Sandra Blakeslee. 1995. *The good marriage*. Boston: Houghton-Mifflin.

Walters, Lynda Henley, Patsy Skeen, Wielislawa Warzywoda-Krusynska, and Tatyana Gurko. 1997. Marital happiness in young families: Similarities and differences across countries. Paper presented at the Annual Conference of the National Council on Family Relations, Crystal City, Va., November.

Whitehead, B. D. 1997. *The divorce culture*. New York: Knopf.

Yee, B. W. K. 1992. Gender and family issues in minority groups. In *Cultural diversity and families*, edited by K. G. Arms, J. K. Davidson Sr., and N. B. Moore. Dubuque, Iowa: Brown and Benchmark, 5–10.

CHAPTER 6 Communication and Conflict Resolution

The survival of marriage is not about love, it's about skills. It's a skill to know how not to escalate a conflict if your relationship is not working.

Diane Sollee, Director, Coalition for Marriage, Family, and Couples Education

Nonverbal communication always telegraphs more of the "real" message than verbal communication.

"**M**y wife said I don't listen to her, at least I think that's what she said," quipped Laurence Peter. His statement reflects the frustration of spouses who feel that they are not understood and the blind attempt of their partners to become better communicators. Good communication is, indeed, one of the core factors that separates couples who stay together from those who part. In this chapter we review the basic principles of effective communication, the nature of conflict in relationships, and a sequence partners might go through in resolving a conflict. We begin by emphasizing that communication is both verbal and nonverbal.

Verbal and Nonverbal Communication

Communication can be defined as the process of exchanging information and feelings between two people. Although most attention is usually given to verbal content, much interpersonal communication is nonverbal. Examples of nonverbal communication include facial expressions, gestures, and posture. Researchers L'Abate and Bagarozzi (1993) reviewed the literature on nonverbal communication and marital satisfaction. Some of their findings follow:

- Symbolic interactionists emphasize that the behaviors individuals display toward each other carry symbolic messages concerning the sender's evaluations of the receiver. When Dan smiles or touches or moves close to Sharon, he is conveying a message of approval much different from what frowning, avoiding touch, and moving away from her would convey.

- The nonverbal part of a message carries more weight when the verbal and nonverbal components conflict. For example, if Juanita tells Miguel, "I love you" but crosses her arms, stands back, and looks at the floor when saying so, Miguel is likely to feel that Juanita doesn't really mean what she says.

- Dissatisfied couples tend to attribute hurtful intent to their spouses more frequently than satisfied couples. Melissa and Chad are an unhappy couple. If Melissa forgets to leave the porch light on for Chad, he is likely to assume that she did so intentionally to convey her displeasure with him and their relationship. If the couple were happy with each other, Chad would probably not interpret negatively Melissa's failure to leave the porch light on.

- The same nonverbal behavior is also subject to multiple interpretations. For example, crying may be viewed as the manipulation of one's partner, the sharing of hurt feelings, or the expression of one's openness and vulnerability. Interpersonal conflict might occur when one partner interprets crying as manipulation while the other interprets it as an expression of vulnerability.

Principles and Techniques of Effective Communication

Persons who value effective communication in their relationship follow certain principles and techniques, including the following:

1. *Make Communication a Priority.* Communicating effectively implies making communication an important priority in a couple's relationship. When communication is a priority, partners make time for communication to occur in a setting without interruptions—they are alone; they do not answer the phone; they turn the television off. Making communication a priority results in the exchange of more information between the partners, which increases the knowledge each partner has about the other. In relationships where communication is a priority, partners may be more willing to communicate about difficult but important topics.

2. *Establish and Maintain Eye Contact.* Shakespeare noted that a person's eyes are the "mirrors to the soul." Partners who look at each other when they are talking not only communicate an interest in each other but are able to gain information about the partner's feelings and responses to what is being said. Not looking at your partner may be interpreted as lack of interest and prevents you from observing nonverbal cues.

3. *Ask Open-Ended Questions.* When your goal is to find out your partner's thoughts and feelings about an issue, it is best to use open-ended questions. An open-ended question encourages your partner to give an answer that contains a lot of information. Closed-ended questions, which elicit a one-word answer such as yes or no, do not provide the opportunity for the partner to express in detail feelings and preferences. Table 6.1 provides examples of open-ended and closed-ended questions.

4. *Use Reflective Listening.* Effective communication requires being a good listener. One of the skills of a good listener is the ability to reflect back what the partner says to you as well as to demonstrate an awareness of what the partner is feeling. The technique of reflective listening involves paraphrasing or restating what the person has said to you while being sensitive to what the partner is feeling. For example, suppose you ask your partner, "How was your day?" and your partner responds, "I felt exploited today at work because I went in early and stayed late and a memo from my new boss said that future bonuses would be eliminated because of a company takeover." Listening to what your partner is both saying and

Table 6.1

Open-Ended and Closed-Ended Questions	
Open-Ended Questions	**Closed-Ended Questions**
How do you feel about living together?	Do you think living together is wrong?
How do you feel about my going out with my friends without you?	Do you think we should have a "boys/girls night out"?
How do you feel about abortion?	Do you believe in abortion?
What are your religious beliefs?	Do you believe in God?
What are your feelings about marriage?	Do you ever want to get married?

feeling, you might respond, "You feel frustrated because you really worked hard and felt unappreciated."

Reflective listening serves the following functions: (1) creates the feeling for the speaker that she or he is being listened to and is being understood and (2) increases the accuracy of the listener's understanding of what the speaker is saying. If a reflective statement does not accurately reflect what the speaker thinks and feels, the speaker can correct the inaccuracy by restating her or his thoughts and feelings.

An important quality of reflective statements is that they are nonjudgmental. For example, suppose two lovers are arguing about spending time with their respective friends and one says, "I'd like to spend one night each week with my friends and not feel guilty about it." The partner may respond by making a statement that is judgmental (critical or evaluative). Judgmental responses serve to punish or criticize someone for what he or she thinks, feels, or wants and often result in frustration and resentment. Table 6.2 provides several examples of judgmental statements and nonjudgmental reflective statements.

5. *Use "I" Statements.* "I" statements focus on the feelings and thoughts of the communicator without making a judgment on others. Because "I" statements are a clear and nonthreatening way of expressing what you want and how you feel, they are likely to result in a positive change in the listener's behavior.

In contrast, "You" statements blame or criticize the listener and often result in increasing negative feelings and behavior in the relationship. For example, suppose you are angry at your partner for being late. Rather than say, "You are always late and irresponsible" (which is a "you" statement), you might respond with, "I get upset when you are late and would feel better if you called me when you will be delayed." The latter focuses

A kiss is a lovely trick designed by nature to stop speech when words become superfluous.
Ingrid Bergman
Swedish actress

So much they talked, so little they said.
Charles Churchill
The Rosciad

Table 6.2

Judgmental and Nonjudgmental Reflective Responses to Your Partner's Statement "I'd Like to Spend One Evening Each Week with My Friends"	
Nonjudgmental Reflective Statements	**Judgmental Statements**
It sounds like you really miss your friends.	You only think about what *you* want.
You think it is healthy for us to be with our friends some of the time.	Your friends are more important to you than I am.
You really enjoy your friends and want to spend some time with them.	You just want a night out so that you can meet someone new.
You think it is important that we not abandon our friends just because we are involved.	You just want to get away so you can drink.
From your point of view, our being apart one night each week will make us even closer.	You are selfish.

on your feelings and what you want your partner to do in the future rather than blaming your partner for being late.

6. *Avoid Brutal Criticism.* Research on marital interaction has consistently shown that one brutal "zinger" can erase twenty acts of kindness (Notarius & Markman, 1994). Because intimate partners are capable of hurting each other so intensely, be careful in how you communicate disapproval to your partner.

7. *Say Positive Things about Your Partner.* We all like to hear others say positive things about us. Communication in any relationship feels better when it contains many positive references to the individuals involved. These positive references may be in the form of compliments. For example, a spouse who tells a partner, "You deserved a raise, you're the best!" or "You smell wonderful" is giving positive feedback to the partner. Positive feedback may also be in the form of appreciation. "Thanks for putting gas in the car" and "Thank you for that delicious dinner" are examples of expressions of appreciation.

8. *Tell Your Partner What You Want.* In addition to complimenting your partner or telling your partner what he or she does that pleases you, focus on what you want rather than on what you don't want. Rather than say, "Don't leave the car without gas," it may sound better to say, "Please keep at least a fourth of a tank of gas in the car." Or, rather than say, "You always leave the bathroom a wreck," an alternative would be to say, "Please hang up your towel after you take a shower."

9. *Stay Focused on the Issue.* Branching refers to going out on different limbs of an issue rather than staying focused on the issue. If you are discussing the overdrawn checkbook, stay focused on the checkbook. To remind your partner that he or she is equally irresponsible when it comes to getting things repaired or doing housework is to get off the issue of the checkbook. Stay focused.

10. *Make Specific Resolutions to Disagreements.* To prevent the same issues or problems from recurring, it is important to agree on what each partner will do in similar circumstances in the future. For example, if going to a party together results in one partner's drinking too much and drifting off with someone else, what needs to be done in the future to ensure an enjoyable evening together?

    Resolving conflicts is not easy, especially when each partner blames the other, is defensive, or avoids talking about the issue. Resolving disagreements takes time, energy, and skill. The result is a sense of pride the partners may have about their relationship and their ability to resolve conflict.

11. *Give Congruent Messages.* A message is congruent when the verbal and nonverbal behavior match. A person who says, "O.K. You're right" and smiles as he or she embraces the partner with a hug is communicating a congruent message. In contrast, the same words accompanied by leaving the room and slamming the door communicate a very different message.

12. *Share Power.* One of the greatest sources of dissatisfaction in a relationship is conflict over power (Kurdek, 1994). Power is the ability to impose one's

*Power, the most intoxicating
of all immortal drugs.*
P. C. Wren
Uniform of Glory

will on the partner and to avoid being influenced by the partner. Few spouses assert to the other that "I am more powerful than you." However, differences in power are reflected in the way couples make family decisions. Decisions about whether to have children and how many, whether to move and where, and whether to buy what and when are often made in reference to differential influence. In general, the spouse with the more prestigious occupation, higher income, and education exerts the greater influence on family decisions. But power may also take the form of love and sex. The person in the relationship who loves less and who needs sex less has enormous power over the partner who is very much in love and who is dependent on the partner for sex.

Lathrop conceptualized spouses as having different power "tools" that complement each other.

The man's power tool is material—usually money and physical strength, and logic. The woman's power tool is sex, or more correctly, relationship, and the non-rational. In a healthy, generative, creating partnership, these energies are valued by the other and result in an "income producing" partnership (1995, 96).

Power is a subtle element in communication and is expressed in numerous ways, including:

Withdrawal—(not speaking to the partner)

Guilt induction—("How could you ask me to do this?")

Being pleasant—("Kiss me and help me move the sofa.")

Negotiation—("I'll go with you to your parents if you will let me golf for a week with my buddies.")

Deception—(running up bills on charge card)

Blackmail—("I'll tell your parents you do drugs if you. . . .")

Physical abuse—(or verbal threats)

Egalitarian relationships have the greatest capacity for satisfaction because neither partner feels exploited by the other. "In a good relationship, ideally there is a balance of power" (Duncan & Rock, 1993, 51).

13. *Keep the Process of Communication Going.* Communication includes both content (verbal and nonverbal information) and process (interaction). It is important not to allow difficult content to shut down the communication process (Turner, 1997). To ensure that the process continues, the partners should focus on the fact that the sharing of information is essential and reinforce each other for keeping the process alive. For example, if your partner tells you something that you do that bothers him or her, it is important to thank him or her for telling you that rather than becoming defensive. In this way, your partner's feelings about you stay out in the open rather than hidden behind a wall of resentment. Otherwise, if you punish such disclosure because you don't like the content, subsequent disclosure will stop.

RECENT RESEARCH

Burleson and Denton (1997) studied the communication patterns of thirty distressed couples and thirty nondistressed couples and found that "communication problems may be better viewed as a symptom than a diagnosis of marital difficulties" (p. 899). Hence, if the couple are not getting along, it may be that their negative communication is a reflection of their negative emotional state. The antidote may be to improve their relationship first before positive communication can be improved.

*The same emotions are
present in both man and
woman, but in different tempo,
on which account man and
woman never cease to
misunderstand each other.*
Friedrich Nietzche
German philosopher

14. *Fight Fairly.* When an argument ensues, it is important to establish rules for fighting that will leave the partners and their relationship undamaged after the disagreement. Such fair-fighting guidelines include not calling each other names, not bringing up past misdeeds, not attacking each other, waiting twenty-four hours before saying what's on one's mind, and not beginning a heated discussion late at night. In some cases, a good night's sleep has a way of altering how a situation is viewed and may even result in the problem's no longer being an issue.

One of the most frustrating experiences in relationships occurs when one partner wants and tries to communicate but the other will not communicate. If your partner will not communicate with you, you might try the following (Duncan & Rock, 1993):

1. Do something that's a noticeable change from your previous strategies. Become less available for conversation and do not try to initiate or maintain discussion. Keep it short if a discussion does start. This not only removes but also reverses all pressure on the partner. The entire pattern changes, and the power shifts.

2. Interpret silence in a positive way: "We are so close we don't always have to be talking." "I feel good when you're quiet because I know that it means everything is all right between us." This negates any power your partner might be expressing through silence.

3. Focus less on the relationship and more on satisfying yourself. When you do things for yourself, you need less from others in the way of attention and assurance.

Intimacy, Disclosure, Honesty, and Dishonesty

Communication plays a major role in the level of emotional intimacy that individuals experience in a relationship. Such intimacy is affected by the amount of self-disclosure and honesty.

Self-Disclosure in Intimate Relationships

One aspect of intimacy in relationships is self-disclosure, which involves revealing personal information and feelings to another person about one's self. According to Waring, "self-disclosure is the single factor which most influences a couple's level of intimacy" (1988, 38). Gallmeier et al. (1997) found in a study of 360 undergraduates that women were significantly more likely than men to disclose and to expect their partners to disclose.

OTHER CULTURES Individuals in Japan are taught that quick self-disclosure in social relationships is inappropriate. They are much less likely to disclose information about themselves than individuals socialized in the United States (Nakanishi, 1986). ●

How can couples facilitate self-disclosure in their relationships? More than twenty-five years ago, R. D. Laing (1970) identified a communication method whereby couples can facilitate self-disclosure. The procedure involves the partners looking at an event or issue from four perspectives: his perspective, her perspective, his perspective of her perspective, and her perspective of his perspective. For example, suppose you and your partner are thinking about getting married. Each of you has different perspectives and feelings about marriage. If each presents his or her perspective and each identifies what he or she thinks is the perspective of the partner, the couple disclose a great deal about this issue. As a result the couple can, with full knowledge, decide what they want to do about getting married.

Honesty in Intimate Relationships

A student in the authors' class wrote:

> At this moment in my life I do not have any love relationship. I find college dating to be very hard. The guys here can lie to you about anything and you wouldn't know the truth. I find it's mostly about sex here and having a good time before you really have to get serious. That is fine, but that is just not what I am all about.

Most individuals value honesty in their relationships. "Truthtelling is . . . the foundation of authenticity, self-regard, intimacy, integrity, and joy. We know that closeness requires honesty, that lying erodes trust, that the cruelest lies are often 'told' in silence" (Lerner, 1993). Despite the importance of honesty or, to use Lerner's term, "truth-telling," "deception is part of everyday existence" (p. 9). Mothersill (1996) observed, "and as for lying, we all lie all the time" (p. 913). One anonymous saying captures the pervasiveness of dishonesty—"The secret of success is sincerity and once you can fake that you have made it."

Forms of Dishonesty and Deception

Dishonesty and deception take various forms. In addition to telling an outright lie, people may exaggerate the truth, pretend, or conceal the truth. They may put up a good front, be two-faced, or tell a partial truth. People also engage in self-deception when they deny or fail to acknowledge their own thoughts, feelings, values, beliefs, priorities, goals, and desires. When individuals are not honest with themselves, they are also not honest with others. As Lerner (1993) notes, "we can be no more honest with others than we are with the self" (p. 13).

Another form of dishonesty occurs when people withhold information or are silent about an issue. Lerner (1993) comments on the difference between lying and withholding information:

> In contrast to how we react to stated lies, we are slower to pass negative judgment on what is *withheld*. After all, no one can tell "the whole truth" all the time . . . Deception through silence or withholding may be excused, and even praised: "My daughter is lucky I never told her about her father," "The doctor was kind enough to spare her the truth about her illness."

When my love swears that she is made of truth, I do believe her, though I know she lies.
Shakespeare

Is it ever right to lie? The simple answer is that it is not. . .
Robert Hauptman
Professor of Learning Resources
St. Cloud State University

RECENT RESEARCH

Miller and Abraham (1998) analyzed data from 80 undergraduates and found that they used deception in their social relationships when their behavior violated the perceived expectancy of their partner.

I always tell the truth, even when I lie.
Al Pacino as *Scarface*, 1983

I'm going through eye-exercise therapy, strengthening my eyes. I'm supposed to . . . like rest them.
Martha Stewart
explaining why her eyes were shut during an Al Gore speech

Privacy versus Secrecy and Deception

In virtually every relationship, there are things that partners have not shared with each other about themselves or their past. There are times when partners do not share their feelings and concerns with each other. But when is withholding information about yourself an act of privacy and when is it an act of secrecy or deception? Privacy involves information that is nobody else's business. Former Mayor Ed Koch of New York was once asked why he never married. He replied, "I prefer not to talk about my personal life." When we withhold "private" information, we are creating or responding to boundaries between ourselves and other people. There may be no harm done in maintaining aspects of ourselves as private and not to be disclosed to others. Indeed, it is healthy to have and maintain boundaries between self and others. However, the more intimate the relationship, the greater our desire to share our most personal and private selves with our partner. And the greater the emotional consequences of not sharing.

We often withhold information or keep secrets in our intimate relationships for what we believe are good reasons—we believe that we are protecting our partners from anxiety or hurt feelings, protecting ourselves from criticism and rejection, and protecting our relationships from conflict and disintegration (Zagorin, 1996). But in intimate relationships, keeping secrets can block opportunities for healing, resolution, self-acceptance, and a deeper intimacy with your partner.

Extent of Lying among College Students

Lying is rampant among college student relationships. More than 85 percent of fifty undergraduates reported that they had lied to their partners (Saxe, 1991). Most of these lies (41%) were about involvement with other partners. In another study, seventy-seven college students kept diaries of their daily social interactions and reported telling two lies a day (DePaulo et al., 1996). Participants said that they did not regard their lies as serious and did not plan them or worry about being caught.

Table 6.3 presents the lies a sample of the authors' students reported having told to a current partner (Knox et al., 1993).

One of the ways in which college students deceive their partners is by failure to disclose that they have a sexually transmitted disease. It is estimated that 20 percent of college students will contract an STD while they are in college. Since the potential to harm an unsuspecting partner is considerable, should we have a national social policy regarding such disclosure?

Conflicts in Relationships

Conflict may be defined as the process of interaction that occurs when the behavior or desires of one person interfere with the behavior or desires of another. This section examines the inevitability, desirability, sources, and styles of relationship conflict.

Table 6.3

Lies University Students Reported Having Told to Current or Past Partner (N = 137)				
Lie	Male	Female	Total Number of Students Reporting Lie	Percent of Sample Reporting Lie
Number of previous partners	11%	20%	42	30.6
Had an orgasm	1	23	33	24.1
You're the best	8	8	22	16.1
It was good	2	15	23	16.8
I love you	7	8	21	15.3
I'm a virgin	3	5	11	8.0
No lies	1	6	9	6.6
You're the biggest	0	7	10	7.3
I like oral sex	4	1	8	5.8
On my period	0	4	6	4.4
I've never cheated	1	3	5	3.6
Yes, I want to	0	1	2	1.5
Age	1	2	3	2.2
I'll call	2	0.01	3	2.2
I've got a headache	1	2	3	2.2
No, I don't have AIDS or an STD	1	0.01	2	1.5
I'll pull out	2	0	2	1.5
Too tired	1	0.01	2	1.5
I'm on the pill	0	0.01	1	0.75
I don't have protection	1	0	1	0.75
You're beautiful	1	0	1	0.75

Knox, D., C. Schacht, J. Holt, and J. Turner. 1993. Adapted from Sexual lies among university students. *College Student Journal* 27: 269–72. Used by permission of *College Student Journal*.

Inevitability of Conflict

RECENT RESEARCH

Spouses who become parents may incur a new set of conflicts. Blanchard-Fields et al. (1997) examined the amount and type of interrole conflict in a sample of 244 parents. Both women and men experienced the highest degree of conflict during the peak child-rearing years.

If you are alone this Saturday evening from six o'clock until midnight, you are assured of six conflict-free hours. But if you plan to be with your partner, roommate, or spouse during that time, the potential for conflict exists. Whether you eat out, where you eat, where you go after dinner, and how long you stay must be negotiated. Although it may be relatively easy for you and your companion to agree on one evening's agenda, marriage involves the meshing of desires on an array of issues for potentially sixty years or more.

Although most men and women reach agreement on many issues before marriage, new needs and preferences will arise after marriage. Changed circumstances sometimes can create new conflicts. A wife of three years recalls:

One's Sexual Health Status: To Tell or Not to Tell?

Individuals often struggle over whether, or how, to tell a partner about their sexual health condition or history. If a person in a committed relationship becomes infected with an STD, that individual, or his or her partner, may have been unfaithful and had sex with someone outside the relationship. Thus, disclosure about an STD may also mean confessing one's own infidelity or confronting one's partner about his or her possible infidelity. (However, the infection may have occurred prior to the current relationship but gone undetected). For women in abusive relationships, telling their partners that they have an STD involves fear that their partners will react violently (Rothenberg & Paskey, 1995). Individuals who are infected with an STD and who are beginning a new relationship face a different set of concerns. Will their new partner view them negatively? Will they want to continue the relationship?

Although telling a partner about having an STD may be difficult and embarrassing, avoiding disclosure or lying about having an STD represents a serious ethical violation. The responsibility to inform a partner that one has an STD—*before* having sex with that partner—is a moral one. But there are also legal reasons for disclosing one's sexual health condition to a partner. If you have an STD and you do not tell your partner, you may be liable for damages if you transmit the disease to your partner. In over half the states, transmission of a communicable disease, including many STDs, is considered a crime. Penalties depend on whether the crime is regarded as a felony (which may involve a five-year prison term) or a misdemeanor (which may involve a fine of $100) (Davis & Scott, 1988). According to one attorney, the number of personal injury lawsuits arising from STDs is on the rise (Litigation of the 90s, 1996).

Some states and cities have partner notification laws that require health care providers to advise all persons with serious sexually transmitted diseases about the importance of informing their sex or needle-sharing partner(s). Partner notification laws may also require health care providers to either notify any partners the infected person names or forward the information about partners to the Department of Health, where public health officers notify the partner(s) that they have been exposed to an STD and schedule an STD testing appointment. The privacy of the infected individual is protected by not revealing his or her name to the partner being notified of potential infection. In cases where the infected person refuses to identify partners, standard partner notification laws require doctors to undertake notification without cooperation, if they know who the sexual partner or spouse is (Norwood, 1995).

REFERENCES

Davis, M., and Scott, R. S. 1988. *Lovers, doctors and the law.* New York: Harper & Row.

Litigation of the 90s: Personal injury suits from STDs. 1996. *American Medical News* 39, no. 7: 23.

Norwood, Chris. 1995. Mandated life versus mandatory death: New York's disgraceful partner notification record. *Journal of Community Health* 20, no. 2: 161–70.

Rothenberg, Karen H., and Stephen J. Paskey. 1995. The risk of domestic violence and women with HIV infection: Implications for partner notification, public policy, and the law. *American Journal of Public Health* 85, no. 11: 1569–76.

I am not arguing with you. I am telling you.
J. McNeill Whistler
The Gentle Art of Making Enemies

I can honestly say that before we got married, we never disagreed about anything, but things were different then. Both my husband and I got money from our parents and never worried about how much we spent on anything. Now I'm pregnant and unemployed, and Neal still acts like we've got someone to pick up the tab. He buys expensive toys like a computer and all the games and software he can carry. He thinks that because he uses VISA we can pay the monthly minimum and still live high. We're getting over our heads in debt, and we're always fighting about it.

Desirability of Conflict

Conflict can be healthy and productive for a couple's relationship. Indeed, ignoring and resigning oneself to a problem may actually increase stress levels. Although discussing conflictive issues and negotiating differences may not reduce immediate stress, the long-term outcome for the relationship is improvement (Gottman, 1994a). Two researchers found that the presence of conflict in a relationship was predictive of the couple's also doing things to help maintain their relationship (self-disclosing and discussing problems) (Sprecher & Felmlee, 1993).

By expressing your dissatisfactions, you alert each other to the need for changes in your relationship to keep your satisfactions high. One husband said he was "sick and tired of picking up his wife's clothes and wet towels from the bathroom floor." She, on the other hand, was angered by her husband's talking on the phone during mealtime. After discussing the issues, she agreed to take care of her clothes in exchange for his agreement to let the answering machine take the calls during meals. The payoff for expressing their negative feelings about each other's behavior was the agreement to stop those behaviors.

Confronting relationship conflict may be good for your health. In one study, spouses with high blood pressure who suppressed their anger at their husbands or wives were twice as likely to die earlier than spouses who talked about what upset them (Julius, 1986). Furthermore, it is critical to discuss an issue rather than to become angry, confrontational, resentful, or persistent with one's point of view. These latter expressions among spouses are associated with an increase in systolic blood pressure, which may contribute to cardiovascular disease (Brown & Smith, 1992).

Sources of Conflict

Conflict has numerous sources, some of which are easily recognized, while others are hidden inside the web of marital interaction.

Behavior The behavior of the partners can sometimes create negative feelings and set the stage for conflict. In your own relationship, you probably become upset when your partner does things you do not like (is late or tells lies). On the other hand, when your partner frequently does things that please you (is on time, is truthful) you tend to feel good about your partner and your relationship.

Cognitions and perceptions Aside from your partner's actual behavior, your cognitions and perceptions of a behavior can be a source of satisfaction or dissatisfaction. One husband complained about the fact that his wife "was messy and always kept the house in a wreck." The wife suggested to her husband that rather than focus on the messy house, he might focus on the thought that she enjoys spending time with him rather than spending time cleaning the house. Thus, the husband replaced the cognition "What a messy house" with the cognition "Isn't it wonderful that my wife would rather go fishing with me than stay home and clean house."

INSIGHT

Value differences in a relationship are not bad in and of themselves. The effect of such differences depends less on the degree of difference in what is valued than on the degree of rigidity with which each partner holds his or her values. Dogmatic and rigid thinkers, feeling threatened by value disagreement, try to eliminate varying viewpoints and typically produce more conflict. But partners who recognize the inevitability of difference usually try to accept in each other what they cannot successfully compromise (Scoresby, 1977).

The symmetrical style of conflict may involve one person being aggressive and hostile and the other person being very passive.

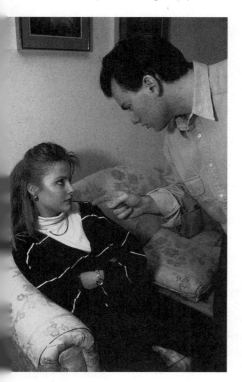

Value differences   Because you and your partner have had different socialization experiences, some of your values will be different. One wife, whose parents were both physicians, resented her mother's not being home when she was growing up. She vowed that when her own children were born, she would stay home and take care of them. But she married a man who wanted his wife to actively pursue a career and contribute money to the marriage. This is only one value conflict a couple may have. Other major value differences may be about religion (one feels religion is a central part of life; the other does not), money (one feels uncomfortable being in debt; the other has the buy-now-pay-later philosophy), and in-laws (one feels responsible for parents when they are old; the other does not).

Inconsistent rules   Partners in all relationships develop a set of rules to help them function smoothly. These unwritten but mutually understood rules include what time you are supposed to be home after work, whether you should call if you are going to be late, how often you can see friends alone, and when and how you make love. Conflict results when the partners disagree on the rules or when inconsistent rules develop in the relationship. For example, one wife expected her husband to take a second job so they could afford a new car, but she also expected him to spend more time at home with the family.

Leadership ambiguity   Unless a couple has an understanding about which partner will make decisions in which area (for example, the husband will decide over which issues to ground teenage children; the wife will decide how much money to spend on vacations), each partner may continually try to "win" a disagreement. All conflict is seen as an "I win-you lose" encounter because each partner is struggling for dominance in the relationship. "In low-conflict marriages, leadership roles vary and are flexible, but they are definite. Each partner knows most of the time who will make certain decisions" (Scoresby, 1977, 141).

Job stress   When you are scheduled to take four exams on one day, you are under a lot of pressure to prepare for them. The stress of such preparation may cause you to be irritable in interactions with your partner. A similar effect occurs when spouses are under job stress; they are less easy to get along with. When spouses are happy and satisfied with their employment, they are much more likely to report satisfaction in their marriage and in their relationships with their children (Belsky et al., 1985). One husband said:

> If I've been on the road all week and haven't made any sales, I feel terrible. And I'm on edge when the wife wants to talk to me or touch me. I seem like I get obsessed with how things are at work, and if they aren't okay, nothing else seems okay either. But if I've made a lot of sales and the commission checks are rolling in, I'm a great husband and father.

Styles of Conflict

Spouses develop various styles of conflict. If you were watching a videotape of various spouses disagreeing over the same issue, you would notice at least three styles of conflict. These styles are described in *The Marriage Dialogue* (Scoresby, 1977).

Complementary   In the *complementary style* of conflict, the wife and husband tend to behave in opposite ways: dominant-submissive, talkative-quiet, active-passive. Specifically, one person lectures the other about what should or should not occur. The other person says little or nothing and becomes increasingly unresponsive. For example, a husband was angry because his wife had left the outside lights of their house on all night. He berated her the next morning, saying she was irresponsible. She retreated in silence.

Some evidence suggests that the complementary style of conflict is more characteristic of southern husbands and southern wives than it is of northern spouses. This is a potential result of the legacy of patriarchy in the southern culture, which southern wives are more likely to accept than question (Wilson & Martin, 1988).

Symmetrical   In the *symmetrical style* of conflict, both partners react to each other in the same way. If she yells, he yells back. If one attacks, so does the other. The partners try to "win" their positions without listening to each other's point of view. In the preceding incident, the wife would blast back at the husband, perhaps pointing out that he lived there too and was equally responsible for seeing that the lights were out before going to bed.

Parallel   In the *parallel style* of conflict, both partners deny, ignore, and retreat from addressing a problem issue. "Don't talk about it, and it will go away" is the theme of this conflict style. Gaps begin to develop in the relationship, neither partner feels free to talk, and both partners believe that they are misunderstood. Both eventually become involved in separate activities rather than spending time together. In the outside-light example, neither partner said anything about the lights being left on all night, but the husband resented the fact that they were.

John Gottman (1994b) interviewed and studied over 200 couples and identified three other styles of conflict resolution:

1. *Validating*, in which couples acknowledge their differences and calmly work out a compromise to their mutual satisfaction.

2. *Volatile*, in which conflict erupts into passionate disputes. Nothing is resolved, but the conflict remains high.

3. *Conflict-avoiding*, in which couples agree to disagree but the conflict remains. Resentment may continue to fester, since nothing is resolved. INI

Geoffrey Grief (1995) provided data from 199 ex-husbands on the various patterns they used to reduce conflict in their previous marriages. The most frequently used patterns included withdrawal (84%), talking it out (74%), and fighting it out verbally.

Resolving Interpersonal Conflict

Every relationship experiences conflict. If left unresolved, conflict may create tension and distance in the relationship with the result that the partners stop talking, stop spending time together, and stop being intimate. Developing and using conflict-resolution skills is critical for the maintenance of a good relationship.

Howard Markman is head of the Center for Marital and Family Studies at the University of Denver. He and his colleagues have been studying 150 couples at yearly intervals (beginning before marriage) to determine those factors most responsible for marital success. They have found that communication skills that reflect the ability to handle conflict, which they call "constructive arguing," are the single biggest predictor of marital success over time (Marano, 1992). According to Markman:

> Many people believe that the causes of marital problems are the differences between people and problem areas such as money, sex, children. However, our findings indicate it is not the differences that are important, but how these differences and problems are handled, particularly early in marriage (Marano, 1992, 53).

There is also merit in developing and using conflict-negotiation skills before problems develop. Not only are individuals more willing to work on issues when things are going well, but they have not developed negative patterns of response that are difficult to change.

This section describes principles and techniques that have been helpful in resolving interpersonal conflict. Such principles and techniques permit a couple to manage present and future conflict by emphasizing the "process" of their interaction and negotiation and not focusing on "fixing" a specific problem (Kolevzon & Jenkins, 1992).

Address Recurring, Disturbing Issues

It is important to address issues in the relationship. A helpful ground rule is "Either one can bring up an issue at any time, but the listener has the right to say 'this is not a good time.' If the listener says, 'this is not a good time,' he or she takes the responsibility to find a good time within 24 hours" (Stanley & Trathen, 1994, 158).

Some couples are uncomfortable talking about issues that plague them. They fear that a confrontation will further weaken their relationship. Pam is jealous that Mark spends more time with other people at parties than with her. "When we go someplace together," she blurts out, "he drops me to disappear with someone else for two hours." Her jealousy is spreading to other areas of their relationship. "When we are walking down the street and he turns his head to look at another woman, I get furious." If Pam and Mark don't discuss her feelings about Mark's behavior, their relationship may deteriorate as a result of a negative response cycle: He looks at another woman, she gets angry, he gets angry at her getting angry and finds that he is even more attracted to other women, she gets angrier because he escalates his looking at other women, and so on.

To bring the matter up, Pam might say, "I feel jealous when you spend more time with other women at parties than with me. I need some help in dealing with these feelings." By expressing her concern in this way, she has identified the problem from her perspective and asked her partner's cooperation in handling it.

When discussing difficult relationship issues, it is important to avoid attacking, blaming, or being negative. Such reactions reduce the motivation of the partner to talk about an issue and thus reduce the probability of a positive outcome.

By confronting issues, couples help to keep resentment from growing in their relationships.

OTHER CULTURES Comfort in bringing up an issue to one's partner varies by culture and sex. In a comparison of 104 American undergraduates and 214 Chinese undergraduates about disagreements in an intimate relationship, Americans more than Chinese and women more than men were more comfortable bringing up problem issues with their partners (Dzindolet et al., 1996). ●

Focus on New Behaviors

I do not know what I want, but I know that I want it.
O. Henry
The Venturers

Dealing with conflict is more likely to result in resolution if the partners focus on what they want rather than what they don't want. For example, rather than tell Mark she doesn't want him to spend so much time with other women at parties, Pam might tell him that she wants him to spend more time with her at parties. Table 6.4 provides more examples of resolving conflict by focusing on "wants" rather than on "don't wants."

Summarize Partner's Perspective

We often assume that we know what our partner thinks and why our partner does things. Sometimes we are wrong. Rather than assume how our partner thinks and feels about a particular issue, we might ask our partner open-ended questions in an effort to get him or her to tell us thoughts and feelings about a particular situation. Pam's words to Mark might be, "What is it like for you when we go to parties?" "How do you feel about my jealousy?"

Once your partner has shared his or her thoughts about an issue with you, it is important for you to summarize your partner's perspective in a nonjudgmental way. After Mark has told Pam how he feels about their being at parties

Table 6.4

Example of Focusing on "Wants" Rather than "Don't Wants"	
Wants	**Don't Wants**
I would like us to go to the beach over spring break.	I don't want to go to your parents for spring break.
I would like to see the Whoopi Goldberg movie.	I don't want to see the Clint Eastwood movie.
I like New Age music.	I don't like heavy metal music.
I would like you to come to bed earlier.	I don't want you to stay up so late at night.
I really prefer to have sex in the evening when we can relax and take our time.	I don't want to have sex in the morning before going to work.

together, she might summarize his perspective by saying, "You feel that I cling to you more than I should, and you would like to feel free to be with others at parties without fear of my getting upset." (She may not agree with his view, but she knows exactly what it is—and Mark knows that she knows.)

Generate Win-Win Solutions

A win-win solution is one in which both people involved in a conflict feel satisfied with the agreement or resolution to the conflict. It is imperative to look for win-win solutions to conflicts. Solutions in which one person wins and the other person loses mean that one person does not have his or her needs met. As a result, the person who "loses" may develop feelings of resentment, anger, hurt, and hostility toward the winner and may even look for ways to get even. In this way, the winner is also a loser. In intimate relationships, one winner really means two losers.

Generating win-win solutions to interpersonal conflict often requires brainstorming. The technique of brainstorming involves suggesting as many alternatives as possible without evaluating them. Brainstorming is crucial to conflict resolution because it shifts the partners' focus from criticizing each other's perspective to working together to develop alternative solutions.

The authors and their colleagues (Knox et al., 1995) studied the degree to which 200 college students who were involved in ongoing relationships were involved in win-win, win-lose, and lose-lose relationships. Descriptions of the various relationships follow.

Win-win relationships are those in which conflict is resolved so that each partner derives benefits from the resolution. For example, suppose a couple have a limited amount of money and disagree on whether to spend it on eating out or on seeing a current movie. One possible win-win solution might be for the couple to eat a relatively inexpensive dinner and rent a movie.

An example of a *win-lose* solution would be for one of the partners to get what he or she wanted (eat out or go to a movie), with the other partner getting nothing of what he or she wanted. A *lose-lose* solution would be for the partners to neither go out to eat nor see a movie and to be mad at each other.

Over three-quarters (77.1%) of the students reported being involved in a win-win relationship, with men and women reporting similar percentages. Twenty percent of the respondents were involved in win-lose relationships. Only 2 percent reported that they were involved in lose-lose relationships. Eighty-five percent of the students in win-win relationships reported that they expected to continue their relationships, in contrast to only 15 percent of students in win-lose relationships. No student in a lose-lose relationship expected the relationship to last.

Evaluate and Select a Solution

After generating a number of solutions, the partners should evaluate each solution and select the best one. In evaluating solutions to conflicts, it may be helpful to ask the following questions:

1. Does the solution satisfy both individuals? (Is it a win-win solution?)

2. Is the solution specific? Does the solution specify exactly who is to do what, how, and when?

3. Is the solution realistic? Can both parties realistically follow through with what they have agreed to do?

4. Does the solution prevent the problem from recurring?

5. Does the solution specify what is to happen if the problem recurs?

Kurdek (1995) emphasized that conflict-resolution styles that stress agreement, compromise, and humor are associated with marital satisfaction, whereas conflict engagement, withdrawal, and defensiveness styles are associated with lower marital satisfaction. In his own study of 155 married couples, the style in which the wife engaged the husband in conflict and the husband withdrew was particularly associated with low marital satisfaction for both spouses.

Sometimes individuals drink rather than confront a partner about an issue.

Be Alert to Defense Mechanisms

Effective conflict resolution is sometimes blocked by *defense mechanisms*—unconscious techniques that function to protect individuals from anxiety and minimize emotional hurt. The following paragraphs discuss some common defense mechanisms:

Escapism is the simultaneous denial and withdrawal from a problem. The usual form of escape is avoidance. The spouse becomes "busy" and "doesn't have time" to think about or deal with the problem, or the partner may escape into recreation, sleep, alcohol, marijuana, or work. Denying and withdrawing from problems in relationships offer no possibility for confronting and resolving the problems. Even when one partner brings up an issue to discuss, the other may withdraw.

A prerequisite to resolving conflict in your relationship is feeling confident that your partner will listen to your concerns and be supportive of you. The following scale assesses the degree to which your relationship is characterized by supportive communication.

Supportive Communication Scale

This scale is designed to assess the degree to which partners experience supportive communication in their relationships. After reading each item, circle the number that best approximates your answer.

0 = strongly disagree (SD)
1 = disagree (D)
2 = undecided (UN)
3 = agree (A)
4 = strongly agree (SA)

	SD	D	UN	A	SA
1. My partner listens to me when I need someone to talk to.	0	1	2	3	**4**
2. My partner helps me clarify my thoughts.	0	1	2	3	**4**
3. I can state my feelings without him/her getting defensive.	0	1	2	3	**4**
4. When it comes to having a serious discussion, it seems we have little in common (reverse scored).	0	**1**	2	3	4
5. I feel put down in a serious conversation with my partner (reverse scored).	**0**	1	2	3	4
6. I feel it is useless to discuss some things with my partner (reverse scored).	**0**	1	2	3	4
7. My partner and I understand each other completely.	0	**1**	**2**	3	4
8. We have an endless number of things to talk about.	0	1	2	3	**4**

3.5
3.6

SCORING: Look at the numbers you circled. Reverse score the numbers for questions 4, 5, and 6. For example, if you circled a 0, give yourself a 4; if you circled a 3 give yourself a 1, etc. Add the numbers and divide by 8, the total number of items. The lowest possible score would be 0, reflecting the complete absence of supportive communication; the highest score would be 4, reflecting complete supportive communication. The average score of 94 male partners who took the scale was 3.01; the average score of 94 female partners was 3.07. Thirty-nine percent of the couples were married, 38 percent were single, 23 percent were living together. The average age was just over 24.

Source: Susan Sprecher, Sandra Metts, Brant Burleson, Elaine Hatfield, and Alicia Thompson. 1995. Domains of expressive interaction in intimate relationships: Associations with satisfaction and commitment. *Family Relations* 44: 203–10. Copyright (1995) by the National Council on Family Relations, 3989 Central Ave. NE, Suite 550, Minneapolis, MN 55421. Reprinted with permission.

NATIONAL DATA

In a national random sample of 947 adults, 43 percent of the men and 26 percent of the women reported that they withdrew when a conflict arose (Stanley & Markman, 1997).

Rationalization is the cognitive justification for one's own behavior that unconsciously conceals one's true motives. A husband worked out every evening at the health club for three hours and said that he did so to ensure the best possible health and to make business contacts. But the underlying reason for his getting out of the house every evening was to escape an unsatisfying home life. Since the idea that he was in a dead marriage was too painful and difficult for him to face, he rationalized to himself and his wife that he spent time away for his health and career.

Projection occurs when one spouse unconsciously attributes his or her own feelings, attitudes, or desires to the partner. For example, the wife who desires to have an affair may accuse her husband of being unfaithful to her. Projection may be seen in such statements as "You spend too much money" (projection for "I spend too much money") and "You want to break up" (projection for "I want to break up"). Projection interferes with conflict resolution by creating a mood of hostility and defensiveness in both partners. The issues to be resolved in the relationship remain unchanged and become more difficult to discuss.

Displacement involves shifting your feelings, thoughts, or behaviors from the person who evokes them onto someone else. The wife who is turned down for a promotion and the husband who is driven to exhaustion by his boss may direct their hostilities (displace them) onto each other rather than toward their respective employers. Similarly, spouses who are angry at each other may displace this anger onto someone else, such as the children.

By knowing about defense mechanisms and their negative impact on resolving conflict, you can be alert to them in your own relationships. When a conflict continues without resolution, one or more defense mechanisms may be operating.

GLOSSARY

communication The process of exchanging information and feelings between two people.

complementary conflict style A style in which the individuals react in opposite ways to each other. For example, he yells; she is quiet.

conflict The interaction that occurs when the behavior or desires of one person interfere with the behavior or desires of another.

defense mechanisms Unconscious techniques that function to protect individuals from anxiety and minimize emotional hurt.

displacement Shifting one's feelings, thoughts, or behaviors from the person who evokes them onto someone else who is a safer target.

lose-lose relationship A relationship in which conflict is resolved so that neither partner benefits from the resolution.

parallel conflict style A pattern of conflict in which both individuals deny, ignore, and retreat from discussing an issue in their relationship.

projection Attributing one's own thoughts, feelings, and desires to someone else while avoiding recognition that these are one's own thoughts, feelings, and desires.

rationalization The cognitive justification for one's own behavior that unconsciously conceals one's true motives.

symmetrical conflict style   A conflict style in which both individuals react to each other in the same way. For example, he yells; she yells back.

win-lose relationship   A relationship in which conflict is resolved so that one partner benefits at the expense of the other.

win-win relationship   A relationship in which conflict is resolved so that each partner derives benefits from the resolution.

SUMMARY

Talking with one's partner about problematic issues in the relationship and negotiating win-win solutions is as important to having a satisfying, durable relationship as is having similar interests and values.

Verbal and Nonverbal Communication

Individuals communicate both verbally (what they say) and nonverbally (gestures, tone, choice of words, eye contact, etc.); the nonverbal part of the message is more powerful. Dissatisfied couples attribute more hurtful intent to nonverbal messages than do satisfied couples. Nonverbal communication is always subject to multiple interpretations.

Principles and Techniques of Effective Communication

Some basic principles and techniques of effective communication include making communication a priority, maintaining eye contact, asking open-ended questions, using reflective listening, complimenting each other, and sharing the power. The latter involves letting your partner win, which increases the likelihood that your partner will be more flexible over subsequent issues.

Intimacy, Disclosure, Honesty, and Dishonesty

Intimacy in relationships is influenced by the level of self-disclosure and honesty. High levels of self-disclosure are associated with increased intimacy. Most individuals value honesty in their relationships. Honest communication is associated with trust and intimacy. Despite the importance of honesty in relationships, deception occurs frequently in interpersonal relationships. Partners sometimes lie to each other about previous sexual relationships, how they feel about each other, and how they experience each other sexually. Telling lies is not the only form of dishonesty. People exaggerate, minimize, tell partial truths, pretend, and engage in self-deception. Partners may withhold information or keep secrets in order to protect themselves and/or preserve the relationship. However, the more intimate the relationship, the greater our desire to share our most personal and private selves with our partner and the greater the emotional consequences of not sharing. In intimate relationships, keeping secrets can block opportunities for healing, resolution, self-acceptance, and a deeper intimacy with your partner.

Conflicts in Relationships

Conflict is both inevitable and desirable. Unless individuals confront and resolve issues over which they disagree, one or both may become resentful and withdraw from the relationship. Conflict may result from one partner's doing something the other does not like, having different perceptions, or having different values. Sometimes it is easier for one partner to view a situation differently or alter a value than for the other partner to change the behavior causing the distress.

Resolving Interpersonal Conflict

The sequence of resolving conflict includes deciding to address recurring issues rather than suppressing them, asking the partner for help in resolving the issue, finding out the partner's point of view, summarizing in a nonjudgmental way the partner's perspective, brainstorming for alternative win-win solutions, and selecting a plan of action. Negotiating win-win solutions reinforces the relationship. Defense mechanisms that interfere with conflict resolution include escapism, rationalization, projection, and displacement.

REFERENCES

Belsky, J., M. Perry-Jenkins, and A. C. Crouter. 1985. The work-family interface and marital change across the transition to parenthood. *Journal of Family Issues* 6: 205–20.

Blanchard-Fields, F., Y. Chen, and C. E. Hebert. 1997. Interrole conflict as a function of life stage, gender, and gender-related personality attributes. *Sex Roles: A Journal of Research* 37: 155–74.

Brown, P. C., and T. W. Smith. 1992. Social influence, marriage, and the heart: Cardiovascular consequences of interpersonal control in husbands and wives. *Health Psychology* 11: 88–96.

Burleson, B. R., and W. H. Denton. 1997. The relationship between communication skill and marital satisfaction: Some moderating effects. *Journal of Marriage and the Family.* 59: 884–902.

DePaulo, Bella M., Susan E. Kirkendol, Deborah A. Kashy, Melissa M. Wyer, and Jennifer A. Epstein. 1996. Lying in everyday life. *Journal of Personality and Social Psychology* 70, no. 5: 979–97.

DePaulo, B. M., and D. A. Kashy. 1998. Everday lies in close and casual relationships. *Journal of Personality and Social Psychology* 74: 63–79.

Duncan, B. L., and J. W. Rock. 1993. Saving relationships: The power of the unpredictable. *Psychology Today.* January/February, 26, 46–51, 86, 95.

Dzindolet, M., Xiaolin Zie, and William Meredith. 1996. Marriage and family life attitude: Comparison of Chinese and American Students. Paper presented at the Annual Meeting of the National Council on Family Relations, Kansas City, November 9. Used by permission of Dr. William Meredith, Department of Consumer and Family Science, University of Nebraska-Lincoln.

Fowers, B. J. 1998. Psychology and the good marriage. *American Behavioral Scientist* 41: 516–41.

Gallmeier, C. P., M. E. Zusman, D. Knox, and L. Gibson. 1997. Can we talk? Gender differences in disclosure patterns and expectations. *Free Inquiry in Creative Sociology* 25: 129–225.

Gottman, John. 1994a. *What predicts divorce? The relationship between marital processes and marital outcomes.* Hillsdale, N.J.: Lawrence Erlbaum.

Gottman, John. 1994b. *Why marriages succeed or fail.* New York: Simon and Schuster.

Gottman, J. M., J. Coan, S. Carrere, and C. Swanson. 1998. Predicting marital happiness and stability from newlywed interactions. *Journal of Marriage and the Family* 60: 5–22.

Grief, Geoffrey L. 1995. Single custodial fathers and their past relationships: Blueprint for the future? *Journal of Couples Therapy* 5: 55–68.

Julius, M. 1986. Marital stress and suppressed anger linked to death of spouses. *Marriage and Divorce Today* 11, no. 35: 1–2.

Keenan, J. P., G. G. Gallup, Jr., N. Goulet, and M. Kulkarni. 1997. Attributions of deception in human mating strategies. *Journal of Social Behavior and Personality* 12: 45–52.

Knox, D., C. Schacht, J. Holt, and J. Turner. 1993. Adapted from Sexual lies among university students. *College Student Journal* 27: 269–72. Used by permission of *College Student Journal.*

Knox, David, C. Schacht, J. Turner, and P. Norris. 1995. College students' preference for win-win relationships. *College Student Journal* 29: 44–46.

Knox, D., S. Hatfield, and M. E. Zusman. 1998. College student dicussion of relationship problems. *College Student Journal* 32: 19–21.

Kolevzon, M. S., and L. A. Jenkins. 1992. The relationship self-assessment inventory (RSAI): A guide to resolving conflict in couples therapy. Paper presented at the 54th Annual Conference of the National Conference on Family Relations, Orlando, Florida. Used by permission.

Kurdek, Lawrence A. 1994. Areas of conflict for gay, lesbian, and heterosexual couples: What couples argue about influences relationship satisfaction. *Journal of Marriage and the Family* 56: 923–34.

Kurdek, Lawrence A. 1995. Predicting change in marital satisfaction from husbands' and wives' conflict resolution styles. *Journal of Marriage and the Family* 57: 153–64.

L'Abate, L., and D. A. Bagarozzi. 1993. *Sourcebook of marriage and family interaction.* New York: Brunner/Mazel.

Laing, R. D. 1970. *Knots.* New York: Random House.

Lathrop, Donald D. 1995. Power tools. *Journal of Couples Therapy* 5: 95–98.

Lerner, Harriet G. 1993. *The dance of deception: Pretending and truth-telling in women's lives.* New York: HarperCollins.

Marano, H. E. 1992. The reinvention of marriage. *Psychology Today.* January/February, 49 passim.

Millar, K. U., and T. Abraham. 1998. Deceptive behavior in social relationships: Consequences of violated expectations. *Journal of Psychology* 122: 263–73.

Mothersill, M. 1996. Some questions about truthfulness and lying. *Social Research* 63: 913–29.

Nakanishi, M. 1986. Perceptions of self-disclosure in initial interaction: A Japanese sample. *Human Communication Research* 13: 167–90.

Notarius, C., and H. Markman. 1994. *We can work it out: Making sense of marital conflict.* New York: Putnam.

Payn, B., K. Tanfer, J. O. G. Billy, and W. R. Grady. 1997. Men's behavior change following infection with a sexually transmitted disease. *Family Planning Perspective* 29: 152–57.

Reiser, C. 1998. *Gendered reflections on anger.* Westport, Conn.: Greenwood Press.

Saxe, Leonard. 1991. Lying: Thoughts of an applied social psychologist. *American Psychologist* 46: 4, 409–15.

Schafer, R. B., K. A. S. Wickrama, and P. M. Keith. 1996. Self-concept disconfirmation, psychological distress, and marital happiness. *Journal of Marriage and the Family* 58: 167–77.

Scoresby, A. L. 1977. *The marriage dialogue.* Reading, Mass.: Addison-Wesley.

Sprecher, S., and Diane Felmlee. 1993. Conflict, love and other relationship dimensions for individuals in dissolving, stable, and growing premarital relationships. *Free Inquiry in Creative Sociology* 21: 115–25.

Stanley, S. M., and H. J. Markman. 1997. *Marriage in the 90s: A nationwide random phone survey.* Denver: PREP, Inc. (303-759-9931).

Stanley, S. M., and D. W. Trathen. 1994. Christian PREP: An empirically based model for marital and premarital intervention. *Journal of Psychology and Christianity* 13: 158–65.

Turner, J. 1997. Personal communication, Isla Mujures, Mexico.

Waring, Edward M. 1988. *Enhancing marital intimacy through facilitating cognitive self-disclosure.* New York: Brunner/Mazel.

Wilson, K., and P. Y. Martin. 1988. Regional differences in resolving family conflicts: Is there a legacy of patriarchy in the South? *Sociological Spectrum* 8: 197–211.

Zagorin, P. 1996. The historical significance of lying and dissimulation. *Social Research* 63: 863–912.

CHAPTER 7 Sexuality in Relationships

Remember that sex is not out there, but in here, in the deepest layer of your own being. There is not only a morning after—there are lots of days and years afterwards.

Jacob Neusner in *Words of Wisdom*

RECENT RESEARCH

Kirby et al. (1997) analyzed data on 10,600 youths in the seventh and eighth grades who were randomly assigned to Postponing Sexual Involvement (an abstinence curriculum) or not and found that youths in the PSI curriculum were just as likely to become sexually active, get pregnant, and contract an STD as those who did not take the abstinence curriculum.

Legal and religious institutions provide strict guidelines on what are regarded as appropriate and inappropriate expressions of sexual behavior.

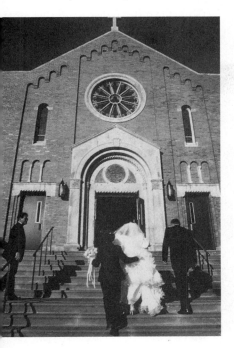

Individuals are inherently sexual. At birth, baby boys have erections and baby girls exude the precursors of vaginal lubrication. As children develop into adults, they learn the nuances of interpersonal sexuality. In this chapter we examine the sexual values that guide individual sexual behavior, the conditions of sexual fulfillment, sexuality in the middle and later years, and sexual dysfunctions of women and men. We begin by reviewing alternative sexual values.

Sexual Values

Values are guidelines for choosing something we believe to be right or moral. Raths et al. (1996) identified seven criteria of a value: (1) something chosen freely, (2) something chosen from among alternatives, (3) something chosen after thoughtful consideration of the consequences of each action, (4) something prized and cherished, (5) something affirming, (6) something that results in acting on choices, and (7) something that is repeating. *Sexual values* can be thought of as having each of these characteristics. Sexual values are operative in nonmarital, marital, heterosexual, and homosexual relationships. Three basic sexual value perspectives are absolutism, relativism, and hedonism.

Absolutism

Absolutism refers to a belief system based on the unconditional power and authority of religion, law, or tradition. Many religions teach that sexual intercourse between spouses is morally correct, and nonmarital, extramarital, and homosexual sex are sins against self, God, and community. Some religions also view masturbation, use of contraception, and abortion as sins. Some evidence suggests lawmakers in our society tend to be conservative and absolutist against sex before marriage. In 1997 Congress authorized $250 million in funds for state sex education programs that would emphasize abstinence (Ritter, 1997). This ignores the fact that such programs may not delay sexual involvement.

In the United States, persons most likely to have absolutist sexual values are women, persons over 50, married individuals, and persons affiliated with conservative Protestant denominations (Michael et al., 1994). Data from 14,396 individuals over a fourteen-year period revealed that conservative Protestants remained steadfast that premarital sex was always wrong (Petersen & Donnenwerth, 1997). Fear of becoming infected with HIV may also be the basis for an individual's adopting an absolutist sexual value perspective.

A subcategory of absolutism is *asceticism*—the belief that giving into carnal lusts is wrong and that one must rise above the pursuit of sensual pleasure to a life of self-discipline and self-denial. Asceticism is reflected in the sexual values of Catholic priests, monks, nuns, and some other celibates.

Relativism

Relativism is a value system emphasizing that sexual decisions should be made in the context of a particular situation. Whereas an absolutist might feel that it is wrong for unmarried people to have intercourse, a relativist might feel that the moral correctness of sex outside of marriage depends on the particular situation. For example, a relativist might feel that in some situations sex between casual dating partners is wrong, such as when one individual pressures the other into having sex or lies in order to persuade the other to have sex. In other cases, when there is no deception or coercion, and when the dating partners are practicing "safer sex," intercourse between casual dating partners may be viewed as acceptable.

Sexual values and choices that are based on relativism often consider the degree of love, commitment, and relationship involvement as important factors. In a sample of undergraduates who reported never having had intercourse, the most frequent reason among both females and males was that they had not been in a relationship long enough or been in love enough (Sprecher & Regan, 1996). A relativist might feel that although sex between casual dating partners is wrong, nonmarital sex between two individuals who have a committed, monogamous relationship is acceptable.

The *sexual double standard* is a view that encourages and accepts sexual expression by men more than expression by women and reflects a relativistic perspective on sexuality. In general, it is more acceptable for men than women to initiate sex, to have more sexual partners, and to have sex without love. In the United States, college men are more likely than college women to adhere to the sexual double standard.

OTHER CULTURES In a study of American and Swedish university students, U.S. university men reported more disapproval toward 18-year-old women who had many sex partners than toward 18-year-old men who had many sex partners. The sexual double standard was less evident among U.S. university women, Swedish university women, or Swedish university men (Weinberg, Lottes, & Shaver, 1995). ●

Hedonism

Hedonism suggests that the ultimate value and motivation for human behavior are the pursuit of pleasure and the avoidance of pain. The hedonist's sexual values are reflected in the creed, "If it feels good, do it." Hedonists are sensation seekers; they tend to pursue novel, exciting, and optimal levels of stimulation and arousal. Their goal is pleasure.

What percentage of U.S. adults have the sexual values of absolutism, relativism, and hedonism? Michael and his colleagues (1994) collected national data on U.S. sexual values using value categories similar to those we have discussed. The category of "traditional" (similar to absolutism) includes persons who say that their religious beliefs always guide their sexual behavior, homosexuality is always wrong, and that premarital sex, teenage sex, and extramarital sex are wrong. The "relational" category (similar to relativism) includes those who believe that sex should be part of a loving relationship but not always reserved for marriage. Finally, their "recreational" (similar to hedonism) category characterized those who feel that sexual activity and interaction do not require the context of a marital, committed, or even loving relationship—that sex need not have anything to do with love.

NATIONAL DATA Percentage of U.S. Adults Representing Three Broad Categories of Sexual Values

Category	Percentage
Traditional (Absolutism)	(33⅓%)
Relational (Relativism)	50%
Recreational (Hedonism)	25%

(From *Sex in America* by Michael et al. Copyright © 1994 by CSG Enterprise, Inc., Edward O. Laumann, Robert T. Michael, and Gina Kolata. By permission of Little, Brown and Company, pp. 232–33. Respondents could be in more than one category.)

Individuals' sexual values and choices often reflect a combination of absolutism, relativism, and hedonism. For example, an individual may be absolutist (or traditional) regarding extramarital sex, relativistic (or relational) regarding premarital sex, and hedonistic (recreational) regarding masturbation or pornography. In a survey of 2,843 U.S. adults, 77 percent reported that extramarital sex is always wrong, but only 20 percent reported that premarital sex is always wrong, suggesting that most U.S. adults have absolutist values concerning extramarital sex but relativistic values concerning premarital sex (Michael et al., 1994, 234).

Sexual Behaviors

We have examined alternative sexual value systems. We now focus on various sexual behaviors people report engaging in. Our discussion ends with a review of gender differences in sexual behavior.

Masturbation

Masturbation involves stimulating one's own body with the goal of experiencing pleasurable sexual sensations.

Alternative terms for masturbation include autoerotism, self-pleasuring, solo sex, and sex without a partner. Several older, more pejorative terms for masturbation are self-pollution, self-abuse, solitary vice, sin against nature, voluntary pollution, and onanism. The negative connotations associated with these terms are a result of various myths (e.g., masturbation causes insanity, blindness, and hair growth on the palms of the hands) about masturbation.

Oral Sex

Fellatio is oral stimulation of the man's genitals by his partner. In many states, legal statutes regard fellatio as a "crime against nature." "Nature" in this case refers to reproduction, and the "crime" is sex that does not produce babies. Nevertheless, most men have experienced fellatio.

Masturbation—it's sex with someone I love.
Woody Allen
Writer, film director

NATIONAL DATA
Among Americans aged 18–59, about 60 percent of men and 40 percent of women report that they have masturbated in the past year (Michael et al., 1994, 60).

A niggling feeling of discomfort and unease follows masturbation, even in those who do not feel guilty about it.
Charlotte Wolff
Love Between Women

NATIONAL DATA
Seventy-nine percent of men report having received oral sex at some time (Michael et al., 1994).

With regard to the "last sexual event," 28 percent of men report that the last time they had sex they experienced fellatio (Michael et al., 1994). Cohabiting males are more likely to report having experienced fellatio from their female partners during their last sexual encounter than are husbands from their wives (35% to 23%) (Michael et al., 1994).

Cunnilingus is oral stimulation of the woman's genitals by her partner. With regard to the "last sexual event," 20 percent of women report that the last time they had sex their partner performed cunnilingus on them (Michael et al., 1994). Women who are neither married nor living together report the highest frequency of cunnilingus as their last sexual event (26%, noncohabiting; 22%, cohabiting; 17%, married).

OTHER CULTURES Societies differ in regard to the degree to which they regard various sexual behaviors as appropriate. In the United States both masturbation and oral sex are regarded as appropriate behaviors for consenting adults. In countries practicing the Islamic religion, both masturbation and oral sex are not permitted, either before, during, or after marriage. ●

Vaginal Intercourse

Vaginal intercourse, or coitus, refers to the insertion of a man's penis into a woman's vagina. In a study by Reinisch et al. (1995), the age of first intercourse for their random sample of undergraduate heterosexual women (and men) was 17. Women reported six partners; men eight.

OTHER CULTURES While most university students in the U.S. report having engaged in intercourse, the percentages of 468 university students in Hong Kong and 461 university students in Shanghai who reported having had intercourse were 3 percent and 5 percent, respectively (Fan et al., 1995). Conservative sexual values still reign in these Asian cities. ●

The frequency of intercourse declines with age. Call et al. (1995) offer the following factors associated with higher and lower frequencies of intercourse:

High Frequency	Low Frequency
Age 20–45	Over age 65
Good health	Poor health
Recently married	Married long time
Recently remarried	First marriage
Currently living together	Not currently living together
Previously lived together	Never cohabited
Childfree or having older children	Having young children or being pregnant
Sterilized	Not sterilized

Figure 7.1

Frequency of Sex Last Month by Age and Marital Status

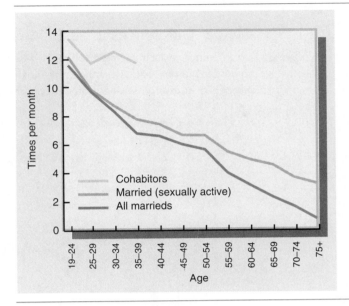

Source: Vaughn Call, Susan Sprecher, and Pepper Schwartz. The incidence and frequency of marital sex in a national sample. *Journal of Marriage and the Family*, 1995, 57, 639–652. Copyright (1995) by the National Council on Family Relations, 3989 Central Ave. NE, Suite 550, Minneapolis, MN 55421. Reprinted with permission.

> *I'm too shy to express my sexual needs except to people over the phone I don't know.*
> Gary Shandling
> Comedian

Figure 7.1 reflects the frequency of intercourse by age and marital status. Fourteen percent of men and 10 percent of women report no intercourse in the past year (Michael et al., 1994). The characteristics of those most likely to report no intercourse in the past year include being young or old and neither living with someone nor being married. "Even though married life is not seen as very erotic, it is actually the social arrangement that produces the highest rate of partnered sexual activity among heterosexuals" (Michael et al., 1994, 118). "In the United States, regular sex partners who share a common household have solved the problem of access and opportunity by marrying or cohabiting" (p. 119). In a study of ninety-one women aged 50 to 91, 41 percent reported having no sexual partner (Sydow, 1995).

NATIONAL DATA

In the past twelve months, about 10 percent of both women and men in a national sample report having engaged in anal sex (Michael et al., 1994, 140).

INSIGHT

Anal (not vaginal) intercourse is the sexual behavior associated with the highest risk of HIV infection (Brody, 1995). The potential to tear the rectum so that blood contact becomes possible presents the greatest danger. AIDS is lethal. Partners who use a condom during anal intercourse reduce their risk of infection not only from HIV but also from other STDs.

Anal Sex

While vaginal intercourse was reported as "very appealing" by 78 percent of women aged 18–44 and by 83 percent of men aged 18–44, only 1 percent of the women and 5 percent of the men reported that anal sex was "very appealing" (Michael et al., 1994, 146, 147).

Outercourse

Outercourse is sexual behavior that does not expose a partner to blood, semen, or vaginal secretions. Outercourse includes hugging, cuddling, masturbation, fantasizing, massage, and body-to-body rubbing with clothes on. In an age of HIV, outercourse is a risk-free form of sexual expression.

Gender Differences in Sexual Behavior

Do differences exist in the reported sexual behaviors of women and men? Yes. According to national data, women report thinking about sex less often, report fewer partners over the course of a year, and report experiencing orgasm less often.

NATIONAL DATA   Some Gender Differences in Sexual Behavior

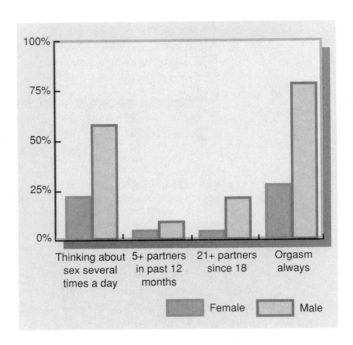

According to data from interviews with 3,432 adults, 19 percent of women compared with 54 percent of men reported thinking about sex several times a day, 2 percent of women compared with 5 percent of men reported having had five or more sexual partners in the past twelve months, 3 percent of women compared with 17 percent of men reported having had twenty-one or more sexual partners since age 18, and 29 percent of women compared with 75 percent of men reported always having had an orgasm during intercourse (from *Sex in America* by Michael et al. Copyright © 1994 by CSG Enterprise, Inc., Edward O. Laumann, Robert T. Michael, and Gina Kolata. By permission of Little, Brown and Company, pp. 102, 128, 156).

Men and women also differ in their motivations for sexual intercourse. Men report the desire for sexual pleasure, conquest, and relief of sexual tension more often than women, who emphasize emotional closeness and affection (Michael et al., 1994; Brigman & Knox, 1992). One survey of 2,365 unmarried respondents aged 18 to 30 revealed that half the men, in contrast to a quarter of the women, could not recall the first and last names of everyone they had had sex with (Rubenstein, 1993). Men are also more approving than women of sex earlier in a relationship.

Some students have a hard time understanding that the consequence of one unprotected sexual encounter may not be reversible.
American College Health Association

Both socialization and biological factors help to explain the differences in sexual behavior between women and men. Social learning theorists emphasize that men are socialized by the media and peers to think about and to seek sexual experiences. Men are also accorded social approval and called "studs" for their sexual exploits. Women, on the other hand, are more often punished and labeled "sluts" for high numbers of sexual partners. Biological factors are also involved. Men have higher testosterone levels, which are associated with more frequent and aggressive sexual behavior.

Racial and Ethnic Differences in Sexual Behavior

Are there any racial and ethnic differences in sexual behavior? Some. Michael et al. (1994) compared white, Hispanic, and African-American men and women with regard to oral sex. Their data are presented in Table 7.1, which shows that African-American men and women are much less likely to both give and receive oral sex than are white or Hispanic men and women. Michael et al. (1994) suggested:

> If others in your social network are trying oral sex, it is likely that you will have a partner who will try it with you, or that you will try it with your partner. If others in your network are willing to engage in it, you will be encouraged to try it too. But if others in your social network resist oral sex, you will be much less likely to try it, and, if you do try it, less likely to enjoy it (1994, 143).

HIV/STD Risk and Prevention

Aside from masturbation, many of the sexual behaviors individuals engage in involve the risk of contracting a sexually transmissible disease.

Persons most likely to contract an STD have multiple sex partners and are not married. This is because heterosexuals with a history of risk factors for HIV do not regularly use condoms with their primary partners.

Table 7.1

Racial and Ethnic Differences in Regard to Oral Sexual Behavior			
	Percentage White G / R	Percentage Hispanic G / R	Percentage African-American G / R
Men Giving/Receiving Oral Sex	81% / 81%	66% / 67%	51% / 66%
Women Giving/Receiving Oral Sex	75% / 78%	56% / 62%	34% / 49%

Source: Adapted from *Sex in America* by Michael et al. Copyright © 1994 by CSG Enterprise, Inc., Edward O. Laumann, Robert T. Michael, and Gina Kolata. By permission of Little, Brown and Company, p. 141.

Women are also more likely to contract an STD. For example, a woman who has one act of unprotected intercourse with a man infected with gonorrhea has up to a 90 percent chance of contracting the disease. In contrast, a man who has one act of unprotected intercourse with a woman infected with gonorrhea has about a 30 percent chance of becoming infected (SIECUS Fact Sheet, 1997). Likewise, male-to-female transmission of HIV is two to four times more efficient than female-to-male transmission (Hankins, 1996). This is due to the larger mucosal surface area exposed to the virus in women and the greater amount of virus present in semen as compared with that in vaginal secretions.

Other risk factors include being homosexual or bisexual and using alcohol/drugs. However, it is not the group individuals belong to but the behaviors they practice that put them at risk for STD infection. Table 7.2 presents various behaviors according to level of STD risk.

Individuals who learn they are infected with a sexually transmissible disease often experience psychological consequences similar to those in other life crises. These psychological reactions include shock, withdrawal from social interaction, anger (especially at the person who gave them the infection), fear, shame, and depression.

Individuals who have an STD often struggle over whether, or how, to tell a partner about their sexual health condition or history. If the infected individual is in a committed sexual relationship, acquiring an STD may imply that that individual, or his or her partner, has been unfaithful and has had sex with someone outside the relationship. Thus, disclosure about an STD may also mean confessing one's own infidelity or confronting one's partner about his or her possible infidelity. For women in abusive relationships, telling their partners that they have an STD involves fear that their partners will react violently (Rothenberg & Paskey, 1995).

Because of the serious health consequences of AIDS, STD prevention and control efforts in recent years have largely focused on preventing transmission of HIV. However, given the potentially serious health consequences of other STDs, including the increased susceptibility to HIV infection, prevention and control of all STDs are warranted.

Numerous school-based, clinic, and community programs have been designed and implemented to prevent STDs by modifying high-risk sexual behavior. One of the more controversial STD prevention strategies involves making condoms available to students in high school. This chapter's Social Policy section looks at this controversy.

Sexual Fulfillment: Some Facts

Individuals who have good sexual relationships with their partners are often aware of some basic facts about human sexuality, some of which are discussed in the following subsections.

Sexual Attitudes and Behaviors Are Learned

Whether you believe that "sex is sinful" or "if it feels good, do it," your sexual attitudes have been learned. Your parents and peers have had a major impact

The secret of a good sexual relationship is making love WITH your partner, not TO your partner.

Dianna Lowe, Kenneth Lowe
Married couple

Table 7.2

What Sexual Behaviors Are Most Risky?

Extremely High Risk

- Unprotected anal and vaginal intercourse
- Sharing needles (for drug use and body piercing)

Moderate Risk

- Unprotected oral sex on a man or woman
- Unprotected oral-anal contact (rimming)
- Unprotected fisting or intercourse using single or multiple fingers
- Sharing devices that draw blood (e.g., whips)
- Sharing unprotected sex toys (e.g., dildos)
- Allowing body fluids to come in contact with broken skin and mucous membranes

Low Risk

- Deep (French) kissing
- Anal and vaginal intercourse with a condom used correctly
- Fisting or intercourse using one or more fingers protected by a finger cot[a] or latex glove
- Oral-anal contact (rimming) with a dental dam[b]
- Oral sex on a man or woman using a condom or dental dam

No Risk

- Dry kissing
- Hugging and nongenital touching
- Massage
- Using vibrators or sex toys (not shared)
- Masturbation (alone or with partner)

[a]A *finger cot* is a mini-condom worn on the finger(s) for finger intercourse.
[b]A *dental dam* is a latex square used to cover the anus or vagina during oral sex. Household plastic wrap or a condom cut lengthwise can be used as a substitute for a dental dam.

on your sexual attitudes, but there have been other influences as well: school, church or synagogue, and the media. Your attitudes about sex would have been different if the influences you were exposed to had been different.

The same is true of sexual behavior. The words you say, the sequence of events in lovemaking, the specific behaviors you engage in, and the positions you adopt during sexual intercourse are a product of your culture and the

Should Condoms Be Available in High Schools?

By the time they graduate from high school, the majority of U.S. students have engaged in sexual intercourse. Unprotected teenage sexual activity contributes to teenage pregnancy and the high rate of STDs among adolescents, who constitute one quarter of the 12 million new STD cases each year. School programs that provide condoms to students are designed to reduce teenage pregnancy and the spread of STDs by increasing condom use through reducing teenagers' embarrassment when buying condoms, by eliminating the cost, and by improving access.

In a survey of 431 U.S. schools that have condom availability programs, nearly all offered condoms as part of a more comprehensive program involving components such as counseling, sex education, or HIV education (Kirby & Brown, 1996). In 81 percent of the schools in this survey, either active or passive parental consent was required before a student could obtain a condom. Ten percent of schools required active consent, in which students obtained written parental consent in order to receive condoms. In the 71 percent of schools that required passive consent, the school sent notices home to parents indicating that they must sign the form or contact someone at the school only if they wished to withhold consent. After parental consent, the second most common requirement for students wishing to receive condoms was counseling, which is mandatory in about half (49%) of schools with condom availability programs (Kirby & Brown, 1996). "During counseling, students are commonly informed that abstinence is the safest method of protection against STDs; they are also instructed about the proper methods of storing and using condoms" (p. 199).

In most schools with condom availability programs, condoms are free of charge. Only 3 percent of schools with such programs make condoms available to students through vending machines, at a cost of about twenty-five cents per condom. In most condom availability programs, students must ask an adult (principal, teacher, counselor, nurse, or other health worker) for condoms. Only 5 percent of the schools in the Kirby and Brown survey provided condoms in bowls or baskets.

Chances are, the high school you attended did not make condoms available to students. Only 2.2 percent of all public high schools and 0.3 percent of all

high school districts in the United States make condoms available to students (Kirby & Brown, 1996). Why are these programs not widely implemented? In some states, there are legal restrictions against such programs. Although many states require high schools to provide instruction in HIV or STD prevention, nineteen states prohibit or restrict availability of, or in some cases, information about contraceptives to students through school health and education programs (Committee on Prevention and Control of STDs, 1997). Segments of the population, as well as powerful conservative groups such as the Family Research Council and Focus on the Family strongly oppose school condom availability on the premise that giving kids condoms might seem to condone their sexual activity or encourage promiscuity. The federal government provides a million dollars for each state if an abstinence policy is promoted. Parents opposed to condom availability programs have filed suit against school districts, claiming that such programs usurped their parental rights. In 1993, a New York state appellate court ruled in favor of those opposed to New York City's school condom availability program, which was the first program in the nation to make condoms available without parental consent. The result was that the public schools implemented the program, but allowed parents an "opt-out" option whereby they could send notification to the school if they did not want their child to have access to condoms. Less than 1 percent of parents of high school students in the New York City school system have chosen that option (Mahler, 1996). More recently, the Massachusetts Supreme Judicial Court upheld a lower court ruling rejecting the parents' claim that the program violated their rights. The parents challenged the court's ruling, but the U.S. Supreme Court declined to review the case. In refusing to hear this case, "the Supreme Court has, for now, left resolution of these issues in the hands of the states" (Mahler, 1996, 77).

Parental opposition to school condom availability programs is in contrast to the recommendation by such organizations as the American School Health Association, the American College of Obstetricians and Gynecologists, the National Medical Association, and the National Institute of Medicine that condoms be made available to adolescents as part of comprehensive school health and STD prevention programs.

At present, it is uncertain whether the public will demand that public schools implement this recommendation throughout the United States (or even allow that they do so). Giving teenagers condoms contradicts moral values against nonmarital sexual relations. However, as one school official responded, "This is not a matter of morality, it is a matter of life and death" (Seligmann et al., 1991, p. 61). Researchers who collected data from 7,119 high school students in New York and 5,738 high school students in Chicago found that while condom availability does increase the use of condoms it "does not increase rates of sexual activity" (Guttmacher et al., 1997).

REFERENCES

Committee on Prevention and Control of STDs. 1997. The hidden epidemic: Confronting sexually transmitted diseases. *SIECUS Report* 25, no. 3: 4–14.

Guttmacher, Sally, Lisa Lieberman, David Ward, Nick Freudenberg, Alice Radosh, and Don Des Jarlais. 1997. Condom availability in New York City public high schools: Relationships to condom use and sexual behavior. *American Journal of Public Health* 87: 1427–33.

Kirby, Douglas B., and Nancy L. Brown. 1996. Condom availability programs in U.S. schools. *Family Planning Perspectives* 28: 196–202.

Mahler, Karen. 1996. Condom availability in the schools: Lessons from the courtroom. *Family Planning Perspectives* 28, no. 2: 75–77.

Seligmann, J., L. Beachy, J. Gordon, J. McCormick, and M. Starr. 1991. Condoms in the classroom. *Newsweek*, 9 December, 61.

NATIONAL DATA

Almost 60 percent (57%) of over 23,000 respondents in a Prodigy (a computer service) poll agreed that sex does not come naturally—that good sex is something we learn (Prodigy, 1995).

RECENT RESEARCH

Ferroni and Taffe (1997) analyzed data from 656 Western Australian women and found that psychosexual health was strongly related to a woman's communication with her partner about sexual needs.

learning history you and your partner have had. The fact that learning accounts for most sexual attitudes and behaviors is important because negative patterns can be unlearned and positive patterns can be learned.

Time and Effort Are Needed for Effective Sexual Communication

Most of us who have been reared in homes in which discussions about sex were infrequent or nonexistent may have developed relatively few skills in talking about sex. Talking about sex with our partners may seem awkward. Overcoming our awkward feelings requires retraining ourselves so that sex becomes as easy for us to talk about as the latest movie.

Spectatoring Interferes with Sexual Functioning

One of the obstacles to sexual functioning is *spectatoring*, which involves mentally observing in an evaluative way your sexual performance and that of your partner. When the researchers in one extensive study observed how individuals actually behave during sexual intercourse, they reported a tendency for sexually dysfunctional partners to act as spectators by mentally observing their own and their partners' sexual performance. For example, the man would focus on whether he was having an erection, how complete it was, and whether it would last. He might also watch to see whether his partner was having an orgasm (Masters & Johnson, 1970).

Physical and Mental Health Affect Sexual Performance

Effective sexual functioning requires good physical and mental health. Physically, this means regular exercise, good nutrition, lack of disease, and lack of fatigue. Regular exercise, whether walking, jogging, aerobics, swimming, or bicycling, is related to higher libido, sexual desire, and intimacy (Ash, 1986). Performance in all areas of life does not have to diminish with age—particularly if people take care of themselves physically (Bronte, 1989).

Low self-esteem, which may lead to depression, may also influence one's sexual desire and performance. Sometimes a vicious cycle may develop. The person feels unworthy and unloved (poor self-esteem) and may need validation through someone outside the relationship. If the partners do not talk about how they feel, they may become vulnerable to an affair, which may result in even lower self-esteem and depression, thus increasing one's lack of desire, capacity to perform, and commitment to the relationship.

Good health also implies being aware that some drugs may interfere with sexual performance. Alcohol is the most frequently used drug by American adults. Although a moderate amount of alcohol can help a person become aroused through a lowering of inhibitions, too much alcohol can slow the physiological processes and deaden the senses. Shakespeare may have said it best: "It [alcohol] provokes the desire but it takes away the performance" (*Macbeth*, act 2, scene 3). The result of an excessive intake of alcohol for women is a reduced chance of orgasm; for men, overindulgence results in a reduced chance of attaining an erection.

The reactions to marijuana are less predictable than the reactions to alcohol. While some individuals report a short-term enhancement effect, others say that marijuana just makes them sleepy. In men, chronic use may decrease sex drive because marijuana may lower testosterone levels.

Antidepressants may also depress sexual functioning. For example, Paxil and Prozac may delay ejaculation in men and interfere with orgasm in women.

Sexual Fulfillment: Some Prerequisites

In addition to having similar sexual value systems (discussed in the beginning of the chapter), other prerequisites for a sexually fulfilling relationship include the following.

Self-Knowledge and Self-Esteem

Sexual fulfillment involves knowledge about yourself and your body. Such information not only makes it easier for you to experience pleasure but also allows you to give accurate information to a partner about pleasing you. It is not possible to teach a partner what you don't know about yourself.

Sexual fulfillment also implies having a positive self-concept. To the degree that you have positive feelings about yourself and your body, you will regard yourself as a person someone else would enjoy touching, being close to, and making love with. If you do not like yourself or your body, you might wonder why anyone else would.

A Good Relationship

A guideline among therapists who work with couples who have sexual problems is to treat the relationship before focusing on the sexual issue. The sexual relationship is part of the larger relationship between the partners, and what happens outside the bedroom in day-to-day interaction has a tremendous influence on what happens inside the bedroom. The statement "I can't fight with you all day and want to have sex with you at night" illustrates the social context of the sexual experience. Oggins et al. (1993) observed in a sample of 199 African-American and 174 Caucasian couples that, particularly for women, perceptions of sexual enjoyment were associated with reports of relationship satisfaction.

A couple's sexual relationship positively influences the couple's overall relationship in several ways: (1) as a shared pleasure, a positively reinforcing event; (2) by facilitating intimacy, as many couples feel closer and share their feelings before or after a sexual experience; and (3) by reducing tension generated by the stresses of everyday living and couple interaction (McCarthy, 1982).

Open Sexual Communication

Sexually fulfilled partners are comfortable expressing what they enjoy and do not enjoy in the sexual experience. Unless both partners communicate their needs, preferences, and expectations to each other, neither is ever sure what the other wants. In essence, the Golden Rule ("Do unto others as you would have them do unto you") is not helpful, because what you like may not be the same as what your partner wants. A classic example of the uncertain lover is the man who picks up a copy of *The Erotic Lover* in a bookstore and leafs through the pages until the topic on how to please a woman catches his eye. He reads that women enjoy having their breasts stimulated by their partner's tongue and teeth. Later that night in bed, he rolls over and begins to nibble on his partner's breasts. Meanwhile, she wonders what has possessed him and is unsure what to make of this new (possibly unpleasant) behavior. Sexually fulfilled partners take the guesswork out of their relationship by communicating preferences and giving feedback. This means using what some therapists call the touch-and-ask rule. Each touch and caress may include the question "How does that feel?" It is then the partner's responsibility to give feedback. If the caress does not feel good, the partner can say what does feel good. Guiding and moving the partner's hand or body are also ways of giving feedback.

Women often expect men to be aware of women's sexual thoughts, feelings, and preferences without telling them. The following examples, from the authors' classes, are what women want men to know about women:

- It does not impress women to hear about other women in the man's past.
- If men knew what it is like to be pregnant, they would not be so apathetic about birth control.
- Most women want more caressing, gentleness, kissing, and talking *before* and *after* intercourse.
- Some women are sexually attracted to other women, not to men.

Sexual interaction communicates how the partners are feeling and acts as a barometer for the relationship. Each partner brings to a sexual encounter, sometimes unconsciously, a motive (pleasure, reconciliation, procreation, duty), a psychological state (love, hostility, boredom, excitement), and a physical state (tense, exhausted, relaxed, turned on). The combination of these factors will change from one encounter to another. Tonight the wife may feel aroused and loving and seek pleasure, but her husband may feel exhausted and hostile and have sex only out of a sense of duty. Tomorrow night, both partners may feel relaxed and have sex as a means of expressing their love for each other.

The verbal and nonverbal communication preceding, during, and after sexual interaction also may act as a barometer for the relationship. One wife said:

I can tell how we're doing by whether or not we have intercourse and how he approaches me when we do. Sometimes he just rolls over when the lights are out and starts to rub my back. Other times, he plays with my face while we talk and kisses me and waits till I reach for him. And still other times, we each stay on our side of the bed so that our legs don't even touch.

- Sometimes the woman wants sex even if the man does not. Sometimes she wants to be aggressive without being made to feel that she shouldn't be.
- Intercourse can be enjoyable without orgasm.
- Many women do not have an orgasm from penetration only; they need direct stimulation of their clitoris by their partner's tongue or finger. Men should be interested in fulfilling their partner's sexual needs.
- Most women prefer to have sex in a monogamous love relationship.
- When a woman says no, she means it. Women do not want men to expect sex every time they are alone with their partner.
- Many women enjoy sex in the morning, not just at night.
- Sex is *not* everything.
- Women need to be lubricated before penetration.
- Men should know more about menstruation.
- Many women are no more inhibited about sex than men.
- Women do not like men to roll over, go to sleep, or leave right after orgasm.
- Intercourse is more of a love relationship than a sex act for many women.
- The woman should not always be expected to supply a method of contraception. It is also the man's responsibility.
- Women tend to like a loving, gentle, patient, tender, and understanding partner. Rough sexual play can hurt and be a turnoff.
- Men should always have a condom with them and initiate putting it on.

Men in the authors' classes also made a list of things they wish women knew about men and sex:

- Men do not always want to be the dominant partner; women should be aggressive.
- Men want women to enjoy sex totally and not be inhibited.
- Men enjoy tender and passionate kissing.
- Men really enjoy fellatio and cunnilingus.
- Women need to know a man's erogenous zones.
- Oral sex is good and enjoyable, not bad and unpleasant.
- Many men enjoy a lot of romantic foreplay and slow, aggressive sex.
- Men cannot keep up intercourse forever. Most men tire more easily than women.
- Looks are not everything.
- Women should know how to enjoy sex in different ways and different positions.
- Women should not expect a man to get a second erection right away.
- Many men enjoy sex in the morning.
- Pulling the hair on a man's body can hurt.
- Many men enjoy sex in a caring, loving, exclusive relationship.
- It is frustrating to stop sex play once it has started.

- Women should know that not all men are out to have intercourse with them. Some men like to talk and become friends.

These respective comments by men and women about the other sex emphasize the importance of being direct with one's partner about sexual desires. Open sexual communication is vital to a sexually fulfilling relationship.

Addressing Safer Sex Issues

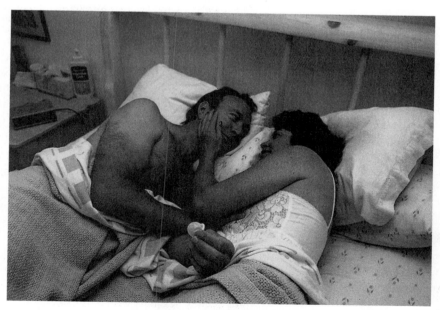

Sexuality in an age of HIV and STD infections demands talking about safer sex issues with a new potential sexual partner. Two researchers asked 252 university women and 207 university men whether it was "very likely" they would bring up a variety of issues with new potential sexual partners. The issues and percentages follow (Gray & Saracino, 1991, 261).

Issue	Percentage
Discuss use of condoms before having intercourse.	54
Ask to have a monogamous relationship.	49
Insist on using a condom before intercourse.	40
Ask how many sexual partners he/she has had.	27

These percentages reflect the ambivalence college students have over safer sex issues. While over half say that they would discuss condoms, only 40 percent report that they would insist on using a condom. And while almost half would ask for sexual exclusivity, only a fourth would ask about previous sexual partners. Only 5 percent of this sample reported it was "very likely" that both they and their partners would be tested for HIV before having sexual intercourse (Gray & Saracino, 1991, 262).

RECENT RESEARCH

Almost one-third (32.7%) of 438 never-married undergraduate women reported that they "rarely" asked a new sex partner about his total number of previous sex partners. Those who did ask were self-confident young women whose sexual self-esteem was intact (Davidson & Moore, 1997).

The difference between sex and death is that with death you can do it alone and no one is going to make fun of you.
Woody Allen
Writer/Director

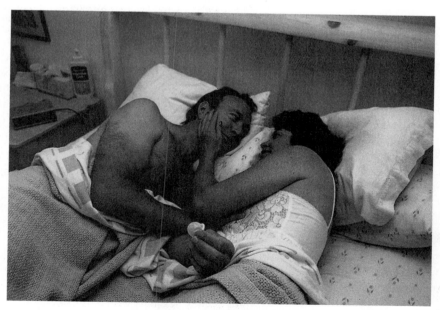

Condom use helps to protect against HIV/STD infection.

Although most worldwide HIV infection occurs through heterosexual transmission, in the United States, HIV infection remains the most threatening STD for male homosexuals and bisexuals. Men who have sex with men account for more cases of AIDS in the United States than persons in any other transmission category (Council on Scientific Affairs, 1996). Data from 280 homosexual men who participated in a longitudinal study in upstate New York revealed that in spite of their knowledge about how HIV is transmitted, many continue to take chances of sexual transmission. Love is a central reason:

> In all the years of the study, we have noted that men in committed relationships and men just falling in love take chances for HIV infection that they did not take with more casual partners. Men were more willing to have unprotected anal sex even if they did not know the loved man's sero-status (Ames et al., 1995, 68).

The researchers emphasized the need to make condom use an integral part of commitment, rather than a sign of distrust.

Homosexual women are not without risk of contracting HIV. "Female-to-female transmission of HIV can occur through exposure to cervical and vaginal secretions of an HIV-infected woman. The amount of shedding from these secretions likely increases the risk of HIV exposure" (Council on Scientific Affairs, 1996, p. 1355). In addition, lesbians and bisexual women may also be at risk for HIV if they have sex with high-risk male partners and they inject drugs (Norman, 1996).

Realistic Expectations

To achieve sexual fulfillment, expectations must be realistic. A couple's sexual needs, preferences, and expectations may not coincide. It is unrealistic to assume that your partner will want to have sex with the same frequency and in the same way that you do on all occasions. It may also be unrealistic to expect the level of sexual interest and frequency of sexual interaction in long-term relationships to remain consistently high.

Debunking Sexual Myths

INSIGHT

Sexual fulfillment means not asking things of the sexual relationship that it cannot deliver. Failure to develop realistic expectations will result in frustration and resentment.

Sexual fulfillment also means not being victim to sexual myths. Some of the more common myths include that sex equals intercourse and orgasm, that women who love sex don't have values, and that the double standard is dead. Zilbergeld (1992) identified some sexual myths unique to men. Some of these include that men are totally liberated and comfortable with sex, that real men don't have sex problems, and that men are always interested in and ready for sex. He also challenged the myth that good sex is spontaneous, with no planning. Indeed, individuals with busy and hectic schedules most often plan their lovemaking if it is to occur at all. See Table 7.3 for other sexual myths.

A Healthy Attitude toward Sex

SIECUS, the Sex Information and Education Council of the United States, identifies healthy sexuality as "consensual, non-exploitative, honest, pleasurable and protected" (Painter, 1995, 4d). In addition, healthy sexuality includes a positive

Table 7.3

Common Sexual Myths

Masturbation is sick.

Masturbation will make you go blind and grow hair on your palm.

Sex education makes children promiscuous.

Sexual behavior usually ends soon after age 60.

People who enjoy pornography end up committing sexual crimes.

Most "normal" women have orgasms from penile thrusting alone.

Extramarital sex always destroys a marriage.

Extramarital sex will strengthen a marriage.

Simultaneous orgasm with one's partner is the ultimate sexual experience.

My partner should enjoy the same things that I do sexually.

A man cannot have an orgasm unless he has an erection.

Most people know a lot of accurate information about sex.

Using a condom ensures that you won't get HIV.

Most women prefer a partner with a large penis.

Few women masturbate.

Women secretly want to be raped.

An erection is necessary for good sex.

An orgasm is necessary for good sex.

The only way to get rid of temptation is to yield to it.
Oscar Wilde
(1854–1900)

attitude toward sex. Traumatic sexual experiences such as rape or child sexual abuse may create an aversion toward sex. Any sexual advance or contact can cause the individual to become anxious and fearful. Such negative reactions to sex are best dealt with through sex therapy.

People can be viewed on a continuum from feeling uncomfortable and wanting to avoid sex (*erotophobia*) to enjoying and wanting to be involved in sex (*erotophilia*). The erotophobic may feel nauseated by erotic material; the erotophilic will feel aroused by it.

Sexuality in the Middle Years

Age is largely socially defined. Cultural definitions of "old" and "young" vary from society to society, from time to time, and from person to person. For example, the older a person, the less likely that person is to define a particular age as old. In a national study of over 2,503 men and women aged 18 to 75, 30 percent of those under 25 responded that "old" is between 40 and 64 years of age. But among those over the age of 65, only 8 percent reported that 65 was old (Clements, 1993). Similarly, with an average life expectancy of 20 years in ancient Greece or Rome, one was old at 18.

The U.S. Census Bureau regards middle age as having reached 45. Using the family life cycle, middle age is that time of life that extends from when the last child leaves home until retirement or the death of either spouse. Changes in sexuality are different for women and for men during this period.

Women and Menopause

Menopause is the primary physical event for middle-aged women. Defined as the permanent cessation of menstruation, menopause is caused by the gradual decline of estrogen produced by the ovaries. It occurs around age 50 for most women but may begin much earlier or later. Signs that the woman may be nearing menopause include decreased menstrual flow and a less predictable cycle. After twelve months with no period, the woman is said to be through menopause. During this time the woman should use some form of contraception. Women with irregular periods may remain at risk of pregnancy up to twenty-four months following their last menstrual period.

The term *climacteric* is often used synonymously with menopause. But menopause refers only to the time when the menstrual flow permanently stops, while climacteric refers to the whole process of hormonal change induced by the ovaries, pituitary gland, and hypothalamus. Reactions to such hormonal changes may include hot flashes, in which the woman feels a sudden rush of heat from the waist up. Hot flashes are often accompanied by an increased reddening of the skin surface and a drenching perspiration. Other symptoms may include heart palpitations, dizziness, irritability, headaches, backache, and weight gain. However, it is important to emphasize that not all women experience these changes.

For 85 percent of women, the symptoms associated with decreasing levels of estrogen will stop within one year of their final period. "But for those who are in too much misery to wait it out, estrogen can do wonders" (Wallis, 1995, 51).

To minimize the effects of decreasing levels of estrogen, some physicians recommend estrogen replacement therapy (ERT), particularly to control hot flashes, night sweats, and vaginal dryness. Researchers do not agree on the advisability of women taking such therapy, also referred to as HRT (hormone replacement therapy) or postmenopausal hormone therapy. Although HRT reduces the risk of heart attacks, colon cancer, and osteoporosis and keeps the skin looking plumped up and moist, it is associated with an increased risk of breast and ovarian cancer (Folsom et al., 1995). Women who are at high risk for breast or uterine cancer or who have clotting problems should not take estrogen. Twenty-four percent of the participants in Baber's (1997) study reported using hormone replacement therapy. Those on HRT were more likely to report experiencing more intense orgasms and increased interest in sexual activity than those not using hormones.

RECENT RESEARCH

Baber (1997) studied 304 midlife women (mean age = 47.6) of different menopausal status and found no significant differences in regard to the degree of conflict reported, their emotional and sexual satisfaction in their relationships, and their interest in sexuality.

OTHER CULTURES A cross-cultural look at menopause suggests that a woman's reaction to this phase of her life may be related to the society in which she lives. For example, among Chinese women, fewer menopausal symptoms have been observed. Researchers have suggested that this may occur because older

These physiological changes in the middle-aged man, along with psychological changes, have sometimes been referred to as male menopause. During this period, the man may experience nervousness, hot flashes, insomnia, and lack of interest in sex. But these changes most often occur over a long period of time, and the anxiety and depression some men experience seem to be as much related to their life situation (lack of career success) as to hormonal alterations.

A middle-aged man who is not successful in his career is often forced to recognize that he will never achieve what he had hoped but will carry his unfulfilled dreams to the grave. This knowledge may be coupled with his awareness of diminishing sexual vigor. For the man who has been taught that masculinity is measured by career success and sexual prowess, middle age may be particularly difficult.

women in China are highly respected, as are older people generally. Researchers Karen Matthews and Nancy Avis conducted a longitudinal study of 541 U.S. women as they progressed through menopause and found that the negative expectations our society has of the menopausal years "may cause at least some of the problems women experience" (quoted in Adler, 1991, 14). ●

Men and the Decrease in Testosterone

The profound hormonal changes and loss of reproductive capacity that occur in women during menopause do not occur in men (Keogh, 1990). However, production of testosterone usually begins to decline around age 40 and continues to decrease gradually until age 60, when it levels off. A 20-year-old man usually has about twice as much testosterone in his system as a 60-year-old man (Young, 1990). The decline is not inevitable but is related to general health status.

The consequences of lowered testosterone include (1) more difficulty in getting and maintaining a firm erection, (2) greater ejaculatory control, with the possibility of more prolonged erections, (3) less consistency in achieving orgasm, (4) fewer genital spasms during orgasm, (5) a qualitative change from an intense, genitally focused sensation to a more diffused and generalized feeling of pleasure, and (6) an increase in the length of the refractory period, the period after orgasm during which the man is unable to ejaculate again or have another erection.

Sexuality in the Later Years

Middle-aged women and men eventually become elderly women and men. In the United States today, one is not usually considered elderly until he or she reaches age 65.

Sexual interest, capacity, and behavior wane as a couple age.

Sexuality of Elderly Men

Table 7.4 describes the physiological changes that elderly men experience during the sexual response cycle.

Four hundred and forty-seven men aged 30 to 99 revealed that sexual interest, activity, and ability decline with age. Sexual interest declined from a mean of 4.4 (5 = extremely interested, 4 = very interested) in men aged 30 to 39 to 2.0 (2 = slightly interested, 1 = not interested) in men aged 90 to 99 (Mulligan & Moss, 1991). Note that while sexual interest declined, it did not stop. No respondent reported a total absence of interest.

Just as interest declined, so did intercourse frequency. It dropped from a mean of once a week in those 30 to 39 years of age to once a year in those 90 to 99 years of age (Mulligan & Moss, 1991). Part of the decline among the 90- to 99-year-old group may be attributed to the lack of a spouse or sexual partner. The frequency, rigidity, and duration of erections decreased dramatically with age. Men aged 90 to 99 reported a mean of 1.9 in terms of rigidity of erection with 1 = flaccid erection never lasting long enough for intercourse and 5 = extremely rigid and always lasting long enough for intercourse. (In contrast,

Table 7.4

Physiological Sexual Changes in Elderly Men	
Phases of Sexual Response	**Changes in Men**
Excitement Phase	As men age, it takes them longer to get an erection. While the young man may get an erection within 10 seconds, elderly man may take several minutes (10 to 30). During this time, he usually needs intense stimulation (manual or oral). Unaware that the greater delay in getting erect is a normal consequence of aging, men who experience this for the first time may panic and have erectile dysfunction.
Plateau Phase	The erection may be less rigid than when the man was younger, and there is usually a longer delay before ejaculation. This latter change is usually regarded as an advantage by both the man and his partner.
Orgasm Phase	Orgasm in the elderly male is usually less intense, with fewer contractions and less fluid. However, orgasm remains an enjoyable experience, as over 70 percent of older men in one study reported that having a climax was very important when having a sexual experience.
Resolution Phase	The elderly man loses his erection rather quickly after ejaculation. In some cases, the erection will be lost while the penis is still in the woman's vagina and she is thrusting to cause her own orgasm. The refractory period is also increased. Whereas the young male needs only a short time after ejaculation to get an erection, the elderly man may need considerably longer.

Source: From *Health Dynamics: Attitudes and Behaviors* by W. Boskin, G. Graf, and V. Kreisworth. Copyright © 1990 West Publishing Co. By permission of Brooks/Cole Publishing Company, Pacific Grove, CA, a division of International Thomson Publishing Inc.

You have to enjoy getting older.

Clint Eastwood

Actor

INSIGHT

The enjoyment of sex reported by these men did not change with age. The men's satisfaction with their own sexuality remained substantially the same. In another study of sixty-one elderly men (average age 71) both with and without sexual partners, sexual satisfaction was rated at an average of 6.3 on a scale where 1 = no satisfaction and 10 = extremely high satisfaction (Mulligan & Palguta, 1991).

men aged 30 to 39 reported a mean rigidity of 4.5). Even though these 90-year-olds achieved an erection, only 21 percent of them were able to achieve orgasm.

Other research supports the view that sexual behavior declines with age. Schiavi et al. (1990) observed that of sixty-five healthy, married men aged 45 to 74, those who were 65 to 74 reported less sexual desire (thought about sex less frequently and could comfortably go without sex for longer periods of time), engaged in intercourse less often, masturbated less frequently, and had fewer orgasms than the men aged 45 to 65. The older group also reported lower arousal, fewer erections, and more difficulty becoming aroused. Getting an erection was their most frequently reported sexual problem. Indeed, fear of failure is the greatest sexual problem for elderly men.

Sexuality of Elderly Women

Table 7.5 describes the physiological changes elderly women experience during the sexual response cycle.

Fewer studies have been conducted on the sexuality of elderly women. However, Bretschneider and McCoy (1988) collected data from 102 white women and 100 white men ranging in age from 80 to 102. Some of the findings follow:

Table 7.5

Physiological Sexual Changes in Elderly Women	
Phases of Sexual Response	**Changes in Women**
Excitement Phase	Vaginal lubrication takes several minutes or longer, as opposed to ten to thirty seconds. Both the length and the width of the vagina decrease. Considerably decreased lubrication and vaginal size are associated with pain during intercourse. Some women report decreased sexual desire and unusual sensitivity of the clitoris.
Plateau Phase	Little change occurs as the woman ages. During this phase, the vaginal orgasmic platform is formed and the uterus elevates.
Orgasm Phase	Elderly women continue to experience and enjoy orgasm. Of women aged 60 to 91, almost 70 percent reported that having an orgasm made for a good sexual experience (Starr & Weiner, 1981). With regard to their frequency of orgasm now as opposed to when they were younger, 65 percent said "unchanged," 20 percent "increased," and 14 percent "decreased."
Resolution Phase	Defined as a return to the preexcitement state, the resolution phase of the sexual response phase happens more quickly in elderly than in younger women. Clitoral retraction and orgasmic platform disappear quickly after orgasm. This is most likely a result of less pelvic vasocongestion to begin with during the arousal phase.

Source: From *Health Dynamics: Attitudes and Behaviors* by W. Boskin, G. Graf, and V. Kreisworth. Copyright © 1990 West Publishing Co. By permission of Brooks/Cole Publishing Company, Pacific Grove, CA, a division of International Thomson Publishing Inc.

1. Thirty-eight percent of the women and 66 percent of the men reported that sex was currently important to them.

2. Thirty percent of the women and 62 percent of the men reported they had sexual intercourse sometimes.

3. Of those with sexual partners, 64 percent of the women and 82 percent of the men said that they were at least mildly happy with their partners as lovers.

4. Forty percent of the women and 72 percent of the men reported they currently masturbated.

5. Touching and caressing without sexual intercourse were the most frequently engaged-in behaviors by women (64 percent) and by men (82 percent).

These findings suggest that declines in sexual enjoyment and frequency are greater for women in the later years than for men.

Other studies confirm that sexual frequency decreases with age. Indeed, about 95 percent of women and 55 percent of men over the age of 80 report not having had sex with anyone in the past twelve months. The following figure illustrates that for the elderly, particularly widows, few sexual partners are available. This accounts for declines in sexual enjoyment and frequency and is the greatest sexual complaint of elderly women.

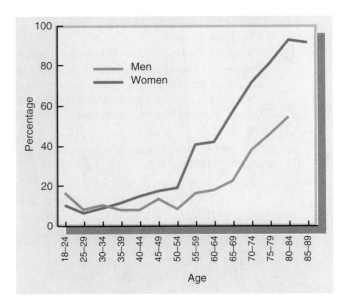

Source: From Sex in America by Michael et al. Copyright © 1994 by CSG Enterprise, Inc., Edward O. Laumann, Robert T. Michael, and Gina Kolata. By permission of Little, Brown and Company.

Sexual Dysfunctions Reported by Women

Though decreasing sexual interest, behavior, and capacity are associated with increasing age, it is not uncommon for spouses and lovers of any age to report having sexual problems. Sex therapists refer to such problems as sexual dysfunctions. Such dysfunctions may be classified by time of onset, situations in which they occur, and cause. A *primary sexual dysfunction* is one that a person has always had. A *secondary sexual dysfunction* is one that a person is currently experiencing, after a period of satisfactory sexual functioning. For example, a woman who has never had an orgasm with any previous sexual partner has a primary dysfunction, whereas a woman who has been orgasmic with previous partners but not with a current partner has a secondary sexual dysfunction.

A *situational dysfunction* occurs in one context or setting and not in another, while a *total dysfunction* occurs in all contexts or settings. For example, a man who is unable to become erect with one partner but who can become erect with another has a situational dysfunction.

Finally, a sexual dysfunction can be classified according to whether it is caused primarily by biological (or organic) factors, such as insufficient hormones or physical illness, or by psychosocial or cultural factors, such as negative learning, guilt, anxiety, or an unhappy relationship.

The national sex study reported by Michael et al. (1994) listed the following most frequent sexual problems along with the percentage of women reporting them:

- lack of interest in sex (33%)
- inability to achieve orgasm (23%)

INSIGHT

In most cases, sexual dysfunctions are caused by more than one factor. In addition, factors are often contributory rather than causal. Each by itself may not ensure the development of a sexual problem, but a problem may result from a complex interaction of factors.

McCabe (1997) compared 84 sexually dysfunctional women with 102 sexually functional women and found that the former "reported lower levels of total intimacy and social, sexual, and recreational intimacy, but not emotional or intellectual intimacy" (p. 287).

INSIGHT

Although we will discuss lack of interest in sex first because it was identified by women as their greatest sexual dysfunction, keep in mind that a male bias may be operative, since "interest in sex" has become identified as the national norm and represents what is thought to be normal for everyone. Hence, we might ask, if women were not reared in a society where male norms of sexual performance and interest prevailed, to what degree would women list "lack of interest" as a sexual concern?

If increased intimacy is the goal of sex therapy, interventions must be designed with this goal in mind and sex therapists need to ensure consistency between desired outcomes and treatment strategies.

Clark Christensen
Marriage therapist

- unpleasurable sex (21%)
- difficulty lubricating (18%)
- pain during intercourse (13%)

We will examine the potential causes and alternative solutions for the top three problems.

Lack of Interest in Sex

Also referred to as hypoactive sexual desire, lack of interest in sex may be caused by one or more factors, including restrictive upbringing, relationship dissatisfaction (e.g., anger or resentment toward one's sexual partner), nonacceptance of one's sexual orientation, learning a passive sexual role, and physical factors such as stress, illness, drug use, and fatigue. In addition, abnormal hormonal states have been shown to be associated with low sexual desire.

The treatment for lack of interest in sex depends on the underlying cause or causes of the problem. The following are some of the ways in which lack of sexual desire can be treated:

1. *Improve relationship satisfaction.* Treating the relationship before treating the sexual problem is standard therapy in treating any sexual dysfunction, including lack of interest in sex. A prerequisite for being interested in sex with a partner, particularly from the viewpoint of a woman, is to be in love and feel comfortable and secure with the partner. Couple therapy focusing on a loving egalitarian relationship becomes the focus of therapy.

2. *Identify and implement conditions for satisfying sex.* Bass (1985) suggested that many women who believe they have a low sexual drive have mislabeled the problem. In many cases, the "real" problem (according to Bass) is not that the woman has low sexual desire but rather that she has not identified or implemented the conditions under which she experiences satisfactory sex. Bass tells his clients who believe they have a low sex drive that "just as their desire to eat is only temporarily diminished when confronted with certain unappetizing foods, so too their sexual desire is only temporarily inhibited through their failure to identify and implement their requirements for enjoyable sex" (1985, 62).

3. *Practice sensate focus.* Sensate focus is a series of exercises developed by Masters and Johnson used to treat various sexual dysfunctions. Sensate focus may also be used by couples who are not experiencing sexual dysfunction but who want to enhance their sexual relationship.

   In doing the *sensate focus* exercise, the couple (in the privacy of their bedroom) remove their clothing and take turns touching, feeling, caressing, and exploring each other in ways intended to provide sensual pleasure. In the first phase of sensate focus, genital touching is not allowed. The person being touched should indicate whether he or she finds a particular touching behavior unpleasant, at which point the partner will stop or change what is being done.

   During the second phase of sensate focus, the person being touched is instructed to give positive as well as negative feedback (i.e., to indicate what is enjoyable as well as what is unpleasant). During the third phase, genital touching can be included, without the intention of producing orgasm. The goal of progressing through the three phases of sensate focus is to help the partners learn to give and receive pleasure by promoting trust and communication and by reducing anxiety related to sexual performance.

4. *Be open to reeducation.* Reeducation involves being open to examining and reevaluating the thoughts, feelings, and attitudes learned in childhood. The goal is to redefine sexual activity so that it is viewed as a positive, desirable, healthy, and pleasurable experience.

5. *Consider other treatments.* Other treatments for lack of sexual desire include rest and relaxation. This is indicated where the culprit is chronic fatigue syndrome (CFS), the symptoms of which are overwhelming fatigue, low-grade fever, and sore throat. Other treatments for lack of sexual desire include hormone treatment and changing medications (if possible) in cases where medication interferes with sexual desire. In addition, sex therapists often recommend that people who are troubled by a low level of sexual desire engage in masturbation as a means of developing positive sexual feelings. Therapists also recommend the use of sexual fantasies. Women who do not have sexual fantasies or who report feeling guilty about having them report higher levels of sexual dissatisfaction (Cado & Leitenberg, 1990).

Inability to Achieve Orgasm

Orgasmic difficulty, also referred to as inhibited female orgasm, or orgasmic dysfunction, occurs when a woman is unable to achieve orgasm after a period of continuous stimulation. Difficulty achieving orgasm can be primary, secondary, situational, or total. Situational orgasmic difficulties, in which the woman is able to experience orgasm under some circumstances but not others, are the most common. Many women are able to experience orgasm during manual or oral clitoral stimulation but are unable to experience orgasm during intercourse (i.e., in the absence of manual or oral stimulation).

Biological factors associated with orgasmic dysfunction can be related to fatigue, stress, alcohol, and some medications, such as antidepressants and antihypertensives. Sixty-one percent of sixty-five married women identified fatigue as an important reason why they had difficulty achieving an orgasm (Davidson & Darling, 1988). Diseases or tumors that affect the neurological system, diabetes, and radical pelvic surgery (e.g., for cancer) may also impair a woman's ability to experience orgasm.

Psychosocial and cultural factors associated with orgasmic dysfunction are similar to those related to lack of sexual desire. Causes of orgasm difficulties in women include restrictive childrearing and learning a passive female sexual role. Guilt, fear of intimacy, fear of losing control, ambivalence about commitment, and spectatoring may also interfere with the ability to experience orgasm. Other women may not achieve orgasm because of their belief in the myth that women are not supposed to enjoy sex.

NATIONAL DATA

Twenty-three percent of women report trouble experiencing orgasm for at least one of the past 12 months (Michael et al., 1994, 126).

INSIGHT

Women who have difficulty achieving orgasm tend to report having experienced their first menstruation later than women who report having no difficulty. Raboch and Raboch suggested "that biological factors are important for insufficient orgasmic capacity" (1992, 118). In other words, women may have different biological capacities for orgasm.

In describing a woman's lack of arousal, it is important to consider that the problem may not be the woman's inability to become aroused but her partner's failure to provide the kind of stimulation required for arousal to occur or her failure to let her partner know what it takes.

A woman who does not achieve orgasm because of lack of sufficient stimulation is not considered to have a sexual dysfunction. In one study, 64 percent of the women who did not experience orgasm during sexual intercourse said that the primary reason was lack of noncoital clitoral stimulation. The type of stimulation most effective in inducing orgasm was manual and oral stimulation and manipulation of the clitoral and vaginal area (Darling, Davidson, & Cox, 1991).

Not all therapists agree that masturbation is beneficial for the pair-bonded nonorgasmic woman. Schnarch (1991, 1993) suggests that masturbation focuses the individual on personal, individualistic happenings, when intimacy with the partner is the more appropriate focus.

The essence of sexual intimacy lies not in mastering specific sexual skills or reducing performance anxiety or having regular orgasms but in the ability to allow one's self to deeply know and to be deeply known by the partner (Schnarch, 1993, 43).

Clinicians are admonished to encourage partners to explore and experiment with each other and to stimulate each other (Christensen, 1995, 97).

Relationship factors, such as anger and lack of trust, can also produce orgasmic dysfunction. For some women, the lack of information can result in orgasmic difficulties (e.g., some women do not know that clitoral stimulation is important for orgasm to occur). Some women might not achieve orgasm with their partners because they do not tell their partners what they want in terms of sexual stimulation out of shame and insecurity (Kelly, Strassberg, & Kircher, 1990).

Because the causes for primary and secondary orgasm difficulties vary, the treatment must be tailored to the particular woman. Treatment can include rest and relaxation, change of medication, or limiting alcohol consumption prior to sexual activity. Sensate focus exercises might help a woman explore her sexual feelings and increase her comfort with her partner. Treatment can also involve improving relationship satisfaction and teaching the woman how to communicate her sexual needs.

Masturbation is a widely used treatment for women with orgasm difficulties. LoPiccolo and Lobitz (1972) developed a nine-step program of masturbation for women with orgasm difficulties. The rationale behind masturbation as a therapeutic technique for a nonorgasmic woman is that masturbation is the technique that is most likely to produce orgasm and can enable her to teach/show her partner what she needs. Masturbation gives the individual complete control of the stimulation, provides direct feedback to the woman of the type of stimulation she enjoys, and eliminates the distraction of a partner. Kinsey et al. (1953) reported that the average woman reached orgasm in 95 percent or more of her masturbatory attempts. In addition, the intense orgasm produced by masturbation leads to increased vascularity in the vagina, labia, and clitoris, which enhances the potential for future orgasms.

Unpleasurable Sex

Sex that is not pleasurable may be both painful and aversive. Pain during intercourse, or *dyspareunia*, occurs in about 10 percent of gynecological patients and may be caused by vaginal infection, lack of lubrication, a rigid hymen, or an improperly positioned uterus or ovary. Because the causes of dyspareunia are often medical, a physician should be consulted. Sometimes surgery is recommended to remove the hymen.

Dyspareunia may also be psychologically caused. Guilt, anxiety, or unresolved feelings about a previous trauma, such as rape or childhood molestation, may be operative. Therapy may be indicated.

Some women report that they find sex aversive. Sexual aversion, also known as sexual phobia and sexual panic disorder, is characterized by the individual's wanting nothing to do with genital contact with another person. The immediate cause of sexual aversion is an irrational fear of sex. Such fear may result from negative sexual attitudes acquired in childhood or sexual trauma such as rape or incest. Some cases of sexual aversion may be caused by fear of intimacy or hostility toward the other sex.

Treatment for sexual aversion involves providing insight into the possible ways in which the negative attitudes toward sexual activity developed, increasing the communication skills of the partners, and practicing sensate focus. Understanding the origins of the sexual aversion may enable the individual to view change as possible. Through communication with the partner and through sensate focus exercises, the individual may learn to associate more positive feelings with sexual behavior.

Sexual Dysfunctions Reported by Men

Men also report sexual problems. The national sex study reported by Michael et al. (1994) listed the following most frequent sexual problems along with the percentage of men reporting them:

- Rapid ejaculation (28%)
- Lack of interest in sex (15%)
- Loss of erection (11%)

Rapid Ejaculation

Also referred to as premature ejaculation, rapid ejaculation is defined as persistent or recurrent ejaculation with minimal sexual stimulation before, upon, or shortly after penetration and before the person wishes it (American Psychiatric Association, 1994). Rapid ejaculation or failing to control the timing of his ejaculation is a man's most common sexual dysfunction, which can lead to other sexual dysfunctions, such as female inorgasmia, low sexual desire, and sexual aversion.

Whether a man ejaculates too soon is a matter of definition, depending on his and his partner's desires. Some partners define a rapid ejaculation in positive terms. One woman said she felt pleased that her partner was so excited by her that he "couldn't control himself." Another said, "The sooner he ejaculates, the sooner it's over with, and the sooner the better." Other women prefer that their partner delay ejaculation. Thirty-one percent of 709 female nurses reported that their partners ejaculated before they had an orgasm, and 23 percent wanted their partners to delay their ejaculation (Darling et al., 1991). Some women regard a pattern of rapid ejaculation as indicative of selfishness in their partner. This feeling can lead to resentment and anger.

OTHER CULTURES On the island of Inis Beag, off the coast of Ireland, men are expected to ejaculate as fast as they can. By doing so it is believed that they spare the woman as much unpleasantness as possible by getting the sex over with as quickly as possible. Only the men are thought to have sexual needs and to enjoy sex (Messenger, 1971). ●

While there is no agreement about the cause of rapid ejaculation (Grenier & Byers, 1995), biological causes of rapid ejaculation are increasingly being regarded as more significant than previously thought (Assalian, 1994). Some men are thought to have a constitutionally hypersensitive sympathetic nervous system that predisposes them to rapid ejaculation. "Further research is needed in

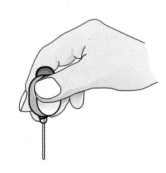

the area of physiological processes of ejaculation and the role of brain neurotransmitters in human sexual response" (Assalian, 1994, 3).

A procedure used for treating rapid ejaculation is the squeeze technique, developed by Masters and Johnson. The partner stimulates the man's penis manually until the man signals that he feels the urge to ejaculate. At the man's signal, the partner places her thumb on the underside of his penis and squeezes hard for three to four seconds. The man will lose his urge to ejaculate. After thirty seconds, the partner resumes stimulation, applying the squeeze technique again when the man signals. The important rule to remember is that the partner should apply the squeeze technique whenever the man gives the slightest hint of readiness to ejaculate. (The man can also use the squeeze technique during masturbation to teach himself to delay his ejaculation.)

Another technique used to delay ejaculation is known as the pause technique, also referred to as the stop-start technique. This technique involves the man's stopping penile stimulation (or signaling his partner to stop stimulation) at the point that he begins to feel the urge to ejaculate. After the period of pre-ejaculatory sensations subsides, stimulation resumes. This process may be repeated as often as desired by the partners.

Success of the squeeze technique and stop-start technique is disputed. While success rates of Masters and Johnson (1970) are often interpreted to be 98 percent, subsequent research has shown rates closer to 60 percent. And most of the gains are lost at follow-up. "Since it is not entirely clear why the intervention works in the first place, it is difficult to identify why the treatment gains were lost over time," observe Grenier and Byers (1995, 465). Even pharmacological interventions that have been employed have not been grounded in studies using control groups and long-term follow-up. Grenier and Byers (1995) suggest that future research should be launched to provide more definitive data on the treatment of this pervasive dysfunction.

Still another method of increasing the delay of ejaculation is for the man to ejaculate often. In general, the greater the number of ejaculations a man has in one twenty-four-hour period, the longer he will be able to delay each subsequent ejaculation. The man's relationship with his partner is also important. The more comfortable, relaxed, and anxiety-free he is, the longer he will delay ejaculation.

Lack of Interest in Sex

Since lack of interest in sex experienced by men is similar to lack of interest in sex experienced by women, the same causes and treatments apply as previously discussed.

Loss of Erection

Also referred to as erectile dysfunction, loss of erection involves the man's inability to get and maintain an erection. Like other sexual dysfunctions, erectile dysfunction can be primary, secondary, situational, or total. Occasional, isolated episodes of the inability to attain or maintain an erection are not considered dysfunctional; these are regarded as normal occurrences. To be classified

Forty percent of 197 homosexual men and 26 percent of 62 heterosexual men reported that difficulty getting an erection had been a problem (Rosser et al., 1997).

Viagra is the biggest breakthrough in the treatment for male sexual dysfunction in 30 years.
Robert Kolodny
Behavioral Medicine Institute
New Canaan, Conn.

The mind messes up more shots than the body.
Tommy Bolt
Professional golfer

INSIGHT

Some men are not accustomed to satisfying their partner in any other way (cuddling, cunnilingus, digital stimulation) than through the use of an erect penis. Most of the women in one study (86 percent) who had male partners with erectile dysfunction reported that their partner never engaged in any sexual activities other than intercourse (Carroll & Bagley, 1990). However, when these women were asked, "What is your favorite part of sexual behavior?" 60 percent said foreplay and 3 percent said afterplay; only 37 percent said sexual intercourse was their favorite part of sexual interaction.

as an erectile dysfunction, the erection difficulty should last continuously for a period of at least three months. "The age when a man may begin to experience erection difficulty is unpredictable but the incidence rises rapidly for men in their mid-50s and by age 70, at least half of the male population experience erectile difficulties" (Mulcahy, 1995).

Up to 70 percent of all cases of erectile dysfunction may be caused by physiological conditions (Mulcahy, 1995). Such biological causes include blockage in the arteries, diabetes, neurological disorders, heavy smoking, alcohol/other drug abuse, chronic disease (kidney or liver failure), pelvic surgery, and neurological disorders.

Psychosocial factors associated with erectile dysfunction include fear (e.g., of unwanted pregnancy, intimacy, HIV or other STDs), guilt, and relationship dissatisfaction. For example, the man who is having an extradyadic sexual relationship may feel guilty. This guilt may lead to difficulty in achieving or maintaining an erection in sexual interaction with the primary partner and/or the extradyadic partner.

Anxiety may also inhibit the man's ability to create and maintain an erection. One source of anxiety is performance pressure, which may be self-imposed or imposed by a partner. In self-imposed performance anxiety, the man constantly "checks" (mentally or visually) to see that he is erect. Such self-monitoring (also referred to as spectatoring, discussed earlier in the chapter,) creates anxiety, since the man fears that he may not be erect.

Partner-imposed performance pressure involves the partner's communicating to the man that he must get and stay erect to be regarded as a good lover. Such pressure usually increases the man's anxiety, thus ensuring no erection. Whether self- or partner-imposed, the anxiety associated with performance pressure results in a vicious cycle—anxiety, erectile difficulty, embarrassment, followed by anxiety, erectile difficulty, and so on.

Performance anxiety may also be related to alcohol use. After consuming more than a few drinks, the man may initiate sex but may become anxious after failing to achieve an erection (too much alcohol will interfere with erection). Although alcohol may be responsible for his initial "failure," his erection difficulties continue because of his anxiety.

Treatment of erectile dysfunction (like treatment of other sexual dysfunctions) depends on the cause(s) of the problem. When erection difficulties are caused by psychosocial factors, treatment may include improving the relationship with the partner and/or removing the man's fear, guilt, or anxiety (i.e., performance pressure) about sexual activity. These goals may be accomplished through couple counseling, reeducation, and sensate focus exercises. A sex therapist would instruct the man and his partner to temporarily refrain from engaging in intercourse so as to remove the pressure to attain or maintain an erection. During this period, the man is encouraged to pleasure his partner in ways that do not require him to have an erection (e.g., cunnilingus, manual stimulation of partner). Once the man is relieved of the pressure to perform and learns alternative ways to satisfy his partner, his erection difficulties (if caused by psychosocial factors) often disappear.

Treatment for erectile dysfunction related to biological factors can include modification of medication, alcohol, or other drugs, or hormone treatment.

INSIGHT

Some sex therapists feel that too much emphasis in male sexual therapy is placed on the performing penis. One therapist suggested:

For every dollar devoted to perfecting the phallus, let's insist that a dollar be devoted to assisting women with their complaints about partner impairments in kissing, tenderness, talk, hygiene, and general eroticism. Too many men still can't dance, write love poems, erotically massage the clitoris, or diaper the baby and let Mom get her rest. The fundamental problem is with the human sexual response cycle model of sexual relations. If we continue to work within this barren conceptualization, we will have nowhere to go but towards maximizing mechanical, compartmentalized sexual components (Tiefer, 1994, 8).

Surgery to improve the blood flow in the penis is also an effective treatment for some men with erectile difficulty.

Another option for treating biologically caused erectile dysfunction is a penile prosthesis (or penile implant). These may be an inflatable prosthesis or a permanent semirigid rod. In a study of twenty-seven men (who had had an implant) and their partners, the majority of the men (72 percent) and their partners (65 percent) would recommend penile implants for men with erectile dysfunction (McCarthy & McMillan, 1990). However, restoration of erectile competence may not improve the overall relationship between the man and his partner.

Another treatment alternative for erectile dysfunction is injection therapy, whereby the patient injects premixed solutions of papaverine or prostaglandin E combined with phentolamine into the corpora cavernosa of the penis via a syringe. The injection results in a firm erection five to ten minutes after the injection and lasts about fifteen minutes (Levitt & Mulcahy, 1995). In a study of forty-two men who used papaverine hydrochloride and phentolamine mesylate, the quality of erections, sexual satisfaction, and frequency of intercourse were all improved (Althof et al., 1991). Disadvantages of injection therapy include discomfort and bruising associated with self-injections, sustained or recurring erections, fibrotic nodules, and abnormal liver function. A new procedure has been developed to put drops of alprostadil at the tip of the penis to induce an erection. A pellet inserted into the urethra will also induce an erection. New medications such as sildenafil (marketed as Viagra) are taken an hour before sex to help with an erection (LeLand, 1997).

An alternative to penile prostheses, injections, and various medications for the treatment of erectile dysfunction is the use of a vacuum device, which produces an erection that lasts for thirty minutes. Vacuum devices contain a chamber large enough to fit over the erect penis, a pump, connector tubing, and tension rings.

When the pump is activated, negative pressure is created within the system, which pulls blood into the penis to produce either erectile augmentation or an erection-like state. After adequate tumescence is achieved, the tension band is guided from the chamber to the base of the penis to produce entrapment of blood (Witherington, 1991, 73).

Of twenty men evaluated for erectile dysfunction for whom vacuum erection, erection devices, or constriction bands were recommended, only four experienced improvement in their erections by using the specific suggested method (Shuetz-Mueller et al., 1995). Because of anxiety, social shame, and cognitive rigidity, most of the men did not follow what was recommended, stopped sexual activity, or tried the intracavernosal injection method.

Other medical treatments include the use of trazodone in combination with yohimbine. Although how this combination works is not clear, some men report developing "abnormally prolonged, rigid erections" (Leslie, 1994).

OTHER CULTURES Cultures differ in their treatment of sexual dysfunctions. In China, men with erectile dysfunction are regarded as "suffering from deficiency of Yang elements in the kidney" and are treated with a solution prepared with wa-

Student Sexual Risks Scale

The following self-assessment allows you to evaluate the degree to which you may be at risk for engaging in behavior that exposes you to HIV. Safer sex means sexual activity that reduces the risk of transmitting the AIDS virus. Using condoms is an example of safer sex. Unsafe, risky, or unprotected sex refers to sex without a condom, or to other sexual activity that might increase the risk of AIDS virus transmission. For each of the following items, check the response that best characterizes your opinion.

A = Agree
U = Undecided
D = Disagree

	A	U	D
1. If my partner wanted me to have unprotected sex, I would probably give in.	✓		
2. The proper use of a condom could enhance sexual pleasure.		✓	
3. I may have had sex with someone who was at risk for HIV/AIDS.			✓
4. If I were going to have sex, I would take precautions to reduce my risk of HIV/AIDS.	✓		
5. Condoms ruin the natural sex act.			✓
6. When I think that one of my friends might have sex on a date, I ask him/her if he/she has a condom.	✓	✓	
7. I am at risk for HIV/AIDS.			✓
8. I would try to use a condom when I had sex.	✓		
9. Condoms interfere with romance.			✓
10. My friends talk a lot about safer sex.	✓		
11. If my partner wanted me to participate in risky sex and I said that we needed to be safer, we would still probably end up having unsafe sex.	✓		✓
12. Generally, I am in favor of using condoms.	✓		
13. I would avoid using condoms if at all possible.			✓
14. If a friend knew that I might have sex on a date, he/she would ask me whether I was carrying a condom.	✓		
15. There is a possibility that I have HIV/AIDS.			✓
16. If I had a date, I would probably not drink alcohol or use drugs.			✓
17. Safer sex reduces the mental pleasure of sex.			✓
18. If I thought that one of my friends had sex on a date, I would ask him/her if he/she used a condom.	✓		
19. The idea of using a condom doesn't appeal to me.			✓
20. Safer sex is a habit for me.		✓	
21. If a friend knew that I had sex on a date, he/she wouldn't care whether I had used a condom or not.			✓

	A	U	D
22. If my partner wanted me to participate in risky sex and I suggested a lower-risk alternative, we would have the safer sex instead.	✓✓	—	—
23. The sensory aspects (smell, touch, etc.) of condoms make them unpleasant.	—	—	✓✓
24. I intend to follow "safer sex" guidelines within the next year.	✓✓	—	—
25. With condoms, you can't really give yourself over to your partner.	—	—	✓✓
26. I am determined to practice safer sex.	✓✓	—	—
27. If my partner wanted me to have unprotected sex and I made some excuse to use a condom, we would still end up having unprotected sex.	—	—	✓✓
28. If I had sex and I told my friends that I did not use condoms, they would be angry or disappointed.	✓	✓	—
29. I think safer sex would get boring fast.	—	—	✓✓
30. My sexual experiences do not put me at risk for HIV/AIDS.	✓✓	—	—
31. Condoms are irritating.	—	—	✓✓
32. My friends and I encourage each other before dates to practice safer sex.	—	✓✓	—
33. When I socialize, I usually drink alcohol or use drugs.	✓✓	—	—
34. If I were going to have sex in the next year, I would use condoms.	✓	✓	—
35. If a sexual partner didn't want to use condoms, we would have sex without using condoms.	—	✓	✓
36. People can get the same pleasure from safer sex as from unprotected sex.	✓✓	—	—
37. Using condoms interrupts sex play.	—	—	✓✓
38. It is a hassle to use condoms.	—	—	✓✓

68
70

(To be read after completing the scale)

SCORING: Begin by giving yourself eighty points. Subtract one point for every undecided response. Subtract two points every time that you agreed to odd-numbered items or to item number 38. Subtract two points every time you disagreed with even-numbered items 2 through 36.

Interpreting Your Score

Research shows that students who make higher scores on the SSRS are more likely to engage in risky sexual activities, such as having multiple sex partners and failing to consistently use condoms during sex. In contrast, students who practice safer sex tend to endorse more positive attitudes toward safer sex, and tend to have peer networks that encourage safer sexual practices. These students usually plan on making sexual activity safer, and they feel confident in their ability to negotiate safer sex even when a dating partner may press for riskier sex. Students who practice safer sex often refrain from using alcohol or drugs, which may impede negotiation of safer sex, and often report having engaged in lower-risk activities in the past. How do you measure up?

(Continued on following page)

(Below 15) Lower risk

(Of 200 students surveyed by DeHart and Birkimer, 16 percent were in this category.) Congratulations! Your score on the SSRS indicates that relative to other students your thoughts and behaviors are more supportive of safer sex. Is there any room for improvement in your score? If so, you may want to examine items for which you lost points and try to build safer sexual strengths in those areas. You can help protect others from HIV by educating your peers about making sexual activity safer.

(15 to 37) Average risk

(Of 200 students surveyed by DeHart and Birkimer, 68 percent were in this category.) Your score on the SSRS is about average in comparison with those of other college students. Though it is good that you don't fall into the higher-risk category, be aware that "average" people can get HIV, too. In fact, a recent study indicated that the rate of HIV among college students is ten times that in the general heterosexual population. Thus, you may want to enhance your sexual safety by figuring out where you lost points and work toward safer sexual strengths in those areas.

(38 and above) Higher risk

(Of 200 students surveyed by DeHart and Birkimer, 16 percent were in this category.) Relative to other students, your score on the SSRS indicates that your thoughts and behaviors are less supportive of safer sex. Such high scores tend to be associated with greater HIV-risk behavior. Rather than simply giving in to riskier attitudes and behaviors, you may want to empower yourself and reduce your risk by critically examining areas for improvement. On which items did you lose points? Think about how you can strengthen your sexual safety in these areas. Reading more about safer sex can help, and sometimes colleges and health clinics offer courses or workshops on safer sex. You can get more information about resources in your area by contacting the CDC's HIV/AIDS Information Line at 1-800-342-2437.

Source: DeHart, D. D., and Birkimer, J. C. 1997. The Student Sexual Risks Scale (modification of SRS for popular use; facilitates student self-administration, scoring, and normative interpretation). Developed specifically for this text by Dana D. DeHart, College of Social Work at the University of South Carolina; John C. Birkimer, University of Louisville. Used by permission of Dana DeHart.

ter and several chemicals designed to benefit kidney function. They may also be given acupuncture therapy (Shikai, 1990, 198). ●

As we end this section on the treatment of sexual dysfunctions, it is important to point out that sex therapy has its downside. Indeed, rather than improving a person's sexuality and a couple's happiness, it can have negative outcomes. McCarthy noted some of the unwanted effects of sex therapy as

increased sexual self-consciousness, increased anticipatory and performance anxiety, an acute problem becoming a chronic sexual dysfunction, therapy resulting in the dissolution of what had been a marginally functional relationship, increased feelings of personal deficit, and vulnerability resulting from shared sexual secrets being used against the person, especially in divorce proceedings (1995, 36).

The buyer should beware, and couples who decide to seek sex therapy should see only a credentialed therapist. The American Association of Sex Educators, Counselors, and Therapists (AASECT), maintains a list of certified sex therapists throughout the country. The association's address is 435 N. Michigan Ave., Suite 1717, Chicago, IL 60611-4067. Phone: 312-644-0828.

GLOSSARY

absolutism   Sexual value system based on the unconditional allegiance to the authority of science, law, tradition, or religion.

asceticism   The belief that giving into carnal lusts is wrong and one must rise above the pursuit of sensual pleasure to a life of self-discipline and self-denial.

climacteric   Refers to the whole process of hormonal change induced by the ovaries (e.g., hot flashes).

dyspareunia   Pain during intercourse.

erectile dysfunction   A man's inability to get and maintain an erection.

erotophilia   Enjoying and wanting to be involved in sex.

erotophobia   Feeling uncomfortable and wanting to avoid sex.

hedonism   Sexual value system emphasizing the pursuit of pleasure and the avoidance of pain.

inhibited female orgasm   Inability to achieve an orgasm after a period of continuous stimulation.

menopause   Permanent cessation of menstruation caused by the gradual decline of estrogen produced in the ovaries.

outercourse   Sexual behavior that does not expose a partner to blood, semen, or vaginal secretions.

primary sexual dysfunction   A sexual dysfunction that the person has always had.

rapid ejaculation   Persistent or recurrent ejaculation with minimal sexual stimulation before, upon, or shortly after penetration and before the partner wishes it.

relativism   Sexual value system emphasizing that decisions should be made in the context of the situation (hence, values are relative).

secondary sexual dysfunction   A sexual dysfunction the person is currently experiencing, which has been preceded by a period of satisfactory sexual functioning.

sensate focus   An exercise whereby the partners focus on pleasuring each other in nongenital ways.

sexual double standard   The view that encourages and accepts sexual expression by men more than by women, it reflects a relativistic perspective on sexuality.

situational dysfunction   Sexual dysfunction that occurs in one situation (e.g., with one partner) but not in another.

spectatoring   Mentally observing one's own sexual performance and that of one's partner, which usually interferes with sexual performance.

total dysfunction   Sexual dysfunction that occurs in all contexts and settings.

SUMMARY

Sexuality may enhance an already good relationship.

Sexual Values

Sexual values are moral guidelines for sexual behavior. Three sexual value perspectives are absolutism ("rightness" is defined by an official code of morality), relativism ("rightness" depends on the situation—who does what, with whom, in what context), and hedonism ("if it feels good, do it").

Sexual Behaviors

Masturbation is defined as sexual self-stimulation with the goal of giving one's self pleasure. Sixty percent of men and 40 percent of women report having masturbated in the past year. About three-fourths of both women and men report receiving oral sex, and over 80 percent of both sexes have engaged in vaginal intercourse before marriage. Anal sex is a relatively infrequent heterosexual sexual behavior; only about 10 percent of both women and men in a national sample reported having engaged in it. In regard to gender differences in sexuality, women report thinking about sex less often, report having fewer partners over the course of a year, and report experiencing orgasm less often. There are also racial differences in sexual behavior. For example, African-American men report engaging in cunnilingus less often than white men; African-American women report engaging in fellatio less often than white women.

HIV/STD Risk and Prevention

Persons most likely to contract an STD have multiple sex partners, are not married, are women, are homosexual, and/or use alcohol/drugs. However, it is one's sexual behavior that exposes him or her to the risk, not the group of which that person is a member. The highest exposure behavior involves unprotected anal and vaginal intercourse.

Individuals who learn they are infected with a sexually transmissible disease often react with shock, withdrawal from social interaction, anger (especially at the person who gave them the infection), fear, shame, and depression. Telling one's partner that one has an STD is difficult.

Numerous school-based, clinic, and community programs have been designed and implemented to prevent STDs by modifying high-risk sexual behavior. One of the more controversial STD prevention strategies involves making condoms available to students in high school. But because of legal restrictions and parental opposition, only 2.2 percent of all public high schools and 0.3 percent of all high school districts in the United States make condoms available to students.

Sexual Fulfillment: Some Facts

The sexual attitudes people have and the behaviors they engage in are learned. Since most have learned anxiety and discomfort in talking about sex, time and effort are needed to improve one's sexual communication. Spectatoring should also be minimized, since it may have a negative effect on sexual functioning.

Sexual Fulfillment: Some Prerequisites

Fulfilling sexual relationships involve self-knowledge, self-esteem, a good nonsexual relationship, open sexual communication, safer sex practices, and making love "with" not "to" one's partner. Other variables include realistic expectations ("my partner will not always want what I want") and not buying into sexual myths ("masturbation is sick"). Basic knowledge about sexuality is also helpful—sexual functioning is influenced by physical and mental health.

Sexuality in the Middle Years

Sexuality is also affected by the aging process. Women experience menopause around age 50, and some experience hot flashes, irritability, and headaches. Hormone replacement therapy to relieve the symptoms is an alternative that a woman must consider carefully with her physician. Men in the middle years are experiencing decreasing levels of testosterone, which translates into more difficulty in getting and maintaining an erection and a greater amount of time between erections.

Sexuality in the Later Years

Sexuality after age 90 is accompanied by decreases in a desire for sexual activity. Both women and men report less frequent intercourse and masturbation as they age. Men over 90 also report less ability to achieve orgasm. Nevertheless, about 30 percent of women and twice that percentage of men in their later years report being interested in sex.

Sexual Dysfunctions Reported by Women and Men

The most frequently reported sexual dysfunctions in women are lack of interest in sex, inability to achieve orgasm, and unpleasurable sex. Men report the sexual dysfunctions of rapid ejaculation, lack of interest in sex, and loss of erection. Treatments include resolving relationship problems, replacing negative views of sexuality with positive and healthy views, sensate focus exercises, and masturbation.

REFERENCES

Adler, T. 1991. Women's expectations are menopause villains. *Monitor,* July, 14.

Althof, Stanley E., Louisa A. Turner, Stephen D. Levine, Candace Risen, Elroy Kursh, Donald Bodner, and Martin Resnick. 1989. Why do so many people drop out of autoinjection therapy for impotence? *Journal of Sex and Marital Therapy,* 15: 121–29.

American Council on Education and University of California. 1997. *The American freshman: National norms for fall, 1997.* Los Angeles: Los Angeles Higher Education Research Institute.

American Psychiatric Association. 1987. *Diagnostic and statistical manual of mental disorders.* 3rd ed. rev. Washington, D.C.: American Psychiatric Association.

American Psychiatric Association. 1994. *Diagnostic and statistical manual of mental disorders.* 4th ed. Washington, D.C.: American Psychiatric Association.

Ames, L. J., A. B. Atchinson, and D. T. Rose. 1995. Love, lust, and fear: Safer sex decision making among gay men. *Journal of Homosexuality* 30: 53–73.

Ash, P. 1986. Healthy sexuality and good health. *Sexuality Today* 9, no. 24: 1.

Assalian, Pierre. 1994. Premature ejaculation: Is it really psychogenic? *Journal of Sex Education and Therapy* 20: 1–4.

Baber, Kristine. 1997. Women's midlife sexuality. Paper presented at the Annual Conference of the National Council on Family Relations, Crystal City, Va., November.

Bass, B. A. 1985. The myth of low sexual desire: A cognitive behavioral approach to treatment. *Journal of Sex Education and Therapy* 11: 61–64.

Bretschneider, Judy G., and Norma L. McCoy. 1988. Sexual interest and behavior in healthy 80 to 102 year olds. *Archives of Sexual Behavior* 17: 109–129.

Brigman, Bonnie, and David Knox. 1992. University student's motivations to have intercourse. *College Student Journal* 26: 406–8.

Brody, Stuart. 1995. Lack of evidence for transmission of human immunodeficiency virus through vaginal intercourse. *Archives of Sexual Behavior* 24: 383–94.

Bronte, Lydia. 1989. *Head first: The biology of hope.* New York: Dutton.

Brown, L. K., S. M. Kessel, K. J. Lourie, H. H. Ford, and L. P. Lipsitt. 1997. Influence of sexual abuse on HIV-related attitudes and behaviors in adolescent psychiatric inpatients. *Journal of the American Academy of Child and Adolescent Psychiatry* 36: 316–22.

Cado, S., and H. Leitenberg. 1990. Guilt reactions to sexual fantasies during intercourse. *Archives of Sexual Behavior* 19: 49–63.

Call, Vaughn, Susan Sprecher, and Pepper Schwartz. 1995. The incidence and frequency of marital sex in a national sample. *Journal of Marriage and the Family* 57: 639–52.

Carroll, J. L., and D. H. Bagley. 1990. Evaluation of sexual satisfaction in partners of men experiencing erectile failure. *Journal of Sex and Marital Therapy* 16: 70–78.

Christensen, Clark. 1995. Prescribed masturbation in sex therapy: A critique. *Journal of Sex and Marital Therapy* 21: 87–99.

Clements, Mark. 1993. What we say about aging. *Parade Magazine*. 12 December, 4–5.

Council on Scientific Affairs. 1996. Health care needs of gay men and lesbians in the United States. *Journal of the American Medical Association* 275: 1354–59.

Darling, C. A., J. K. Davidson, and R. P. Cox. 1991. Female sexual response and the timing of partner orgasm. *Journal of Sex and Marital Therapy* 17: 3–21.

Davidson, J. K., and C. A. Darling. 1988. The stereotype of single women revisited: Sexual practices and sexual satisfaction among professional women. *Health Care for Women International* 9: 317–36.

Davidson, J. K., and N. B. Moore. 1997. Communicating with new sex partners: Questions that make a difference. Paper presented at the Annual Conference of the National Council on Family Relations, Crystal City, Va., November.

Fan, M. S., J. H. Hong, M. L. Ng, L. K. C. Lee, P. K. Lui, and Y. H. Choy. 1995. Western influences on Chinese sexuality: Insights from a comparison of the sexual behavior and attitudes of Shanghai and Hong Kong freshmen at universities. *Journal of Sex Education and Therapy* 21: 158–66.

Ferroni, P., and J. Taffe. 1997. Women's emotional well-being: The importance of communicating sexual needs. *Sexual and Marital Therapy* 12: 127–38.

Folsom, Aaron R., Pamela J. Mink, Thomas A. Sellers, Ching-Ping Hong, and John D. Potter. 1995. Hormonal replacement therapy and morbidity and mortality in a prospective study of postmenopausal women. *American Journal of Public Health* 85: 1128–32.

Gray, L. A., and M. Saracino. 1991. College students' attitudes, beliefs, and behaviors about AIDS: Implications for family life educators. *Family Relations* 40: 258–63.

Grenier, Guy, and E. Sandra Byers. 1995. Rapid ejaculation: A review of conceptual, etiological, and treatment issues. *Archives of Sexual Behavior* 24: 447–70.

Gross, Kevin H., and M. L. Morris. 1997. College students' sexual behavior expectancies. Paper presented at the Annual Conference of the National Council on Family Relations, Crystal City, Va., November.

Grunseit, A., S. Kippax, P. Aggleton, M. Baldo, and Gary Slutkin. 1997. Sexuality education and young people's sexual behavior: A review of studies. *Journal of Adolescent Research* 12: 421–53.

Hankins, C. 1996. Sexual transmission of HIV to women in industrialized countries. *World Health Statistics Quarterly* 49: 106–12.

Hynie, M., J. E. Lydon, and A. Taradash. 1997. Commitment, intimacy, and women's perceptions of premarital sex and contraceptive readiness. *Psychology of Women Quarterly* 21: 447–64.

Kelly, M. P., D. S. Strassberg, and J. R. Kircher. 1990. Attitudinal and experiential correlates of anorgasmia. *Archives of Sexual Behavior* 19: 165–77.

Keogh, E. J. 1990. The male menopause. *Australian Family Physician* 19: 833–40.

Kinsey, A. C., W. B. Pomeroy, C. E. Martin, and P. H. Gebhard. 1953. *Sexual behavior in the human female.* Philadelphia: W. B. Saunders.

Kirby, D., M. Korpi, R. P. Barth, and H. H. Cagampang. 1997. The impact of the postponing sexual involvement curriculum among youths in California. *Family Planning Perspectives* 29: 100–108.

Knox, D., and M. Zusman. 1998. Original data collected for this text.

Kreiger, John N., R. W. Coombs, A. C. Collier, D. D. Ho, S. O. Ross, J. E. Zeh, and L. Corey. 1995. Intermittent shedding of human immunodeficiency virus in semen: Implications for sexual transmission. *Journal of Urology* 154: 1035–40.

LeLand, J. 1997. A pill for impotence? *Newsweek.* 17 November, 62–68.

Leslie, S. W. 1994. 1994. *Impotence: Current diagnosis and treatment.* Birmingham, Ala.: Vet-Co, Inc.

Levitt, Eugene E., and John J. Mulcahy. 1995. The effect of intracavernosal injection of papaverine hydrochloride on orgasm latency. *Journal of Sex and Marital Therapy*, 21: 39–56.

LoPiccolo, J., and C. Lobitz. 1972. The role of masturbation in the treatment of orgasmic dysfunction. *Archives of Sexual Behavior* 2: 163–71.

Masters, W .H., and V. E. Johnson. 1970. *Human sexual inadequacy.* Boston: Little, Brown.

McCabe, M. P. 1997. Intimacy and quality of life among sexually dysfunctional men and women. *Journal of Sex & Marital Therapy* 23: 276–90.

McCarthy, Barry W. 1982. Sexual dysfunctions and dissatisfactions among middle-years couples. *Journal of Sex Education and Therapy* 8: 9–12.

McCarthy, Barry W. 1995. Learning from unsuccessful sex therapy patients. *Journal of Sex Therapy* 21: 31–39.

McCarthy, J., and S. McMillan. 1990. Patient/partner satisfaction with penile implant surgery. *Journal of Sex and Marital Therapy* 16: 25–37.

McLaughlin, C. S., C. Chen, E. Greenberger, and C. Biermeier. 1997. Family, peer, and individual correlates of sexual experience among Caucasian and Asian American late adolescents. *Journal of Research on Adolescence* 7: 33–53.

Messenger, J. C. 1971. Sex and repression in an Irish folk community. In *Human sexual behavior: Variations in the ethnographic spectrum*, edited by D. S. Marshall and R. C. Suggs. New York: Basic Books, 3–37.

Michael, Robert T., John H. Gagnon, Edward O. Laumann, and Gina Kolata. 1994. *Sex in America: A definitive survey.* Boston: Little, Brown.

Miller, B. C., M. C. Norton, T. Curtis, E. J. Hill, P. Schvaneveldt, and M. H. Young. 1997. The timing of first intercourse among adolescents: Family, peer and other antecedents. *Youth and Society* 29: 54-83.

Mulcahy, John J. 1995. Sexual function and aging. *Foundation Focus*, Winter, 1–3.

Mulligan, Thomas, and C. Renne Moss. 1991. Sexuality and aging in male veterans: A cross-sectional study of interest, ability, and activity. *Archives of Sexual Behavior* 20: 17–25.

Mulligan, T., and R. F. Palguta, Jr. 1991. Sexual interest, activity, satisfaction among male nursing home residents. *Archives of Sexual Behavior* 20: 199–204.

Norman, A. D. et al. 1996. Lesbian and bisexual women in small cities—At risk for HIV. *Public Health Reports* 111: 347–52.

Oggins, J., D. Leer, and J. Veroff. 1993. Race and gender differences in black and white newlyweds' perceptions of sexual and marital relations. *Journal of Sex Research*, 30: 152–60.

Painter, Kim. 1995. Some advocate mature teen sex. *USA Today.* 22 June, 4D.

Petersen, L. R., and G. V. Donnenwerth. 1997. Secularization and the influence of religion on beliefs about premarital sex. *Social Forces* 75, no. 3: 1071–89.

Prodigy. 1995. On-line sexual survey. 16 October.

Raboch, J., and J. Raboch. 1992. Infrequent orgasms in women. *Journal of Sex and Marital Therapy* 18: 114–20.

Raths, L., H. Merrill, and S. Simon. 1996. *Values and teaching.* Columbus, Ohio: Charles E. Merrill.

Reinisch, June M., Craig A. Hill, Stephanie A. Sanders, and Mary Ziemba-Davis. 1995. High-risk sexual behavior at a midwestern university: A confirmatory survey. *Family Planning Perspectives* 27: 79–82.

Ritter, John. 1997. Federal funds refuel a sex-ed fire: Money keyed to message of abstinence only. *USA Today.* 14 July 3A.

Rosser, B. R. S., M. E. Metz, W. O. Bockting, and T. Buroker. 1997. Sexual difficulties, concerns, and satisfaction in homosexual men: An empirical study with implications for HIV prevention. *Journal of Sex Therapy* 23: 61–73.

Rothenberg, Karen H., and Stephen J. Paskey. 1995. The risk of domestic violence and women with HIV infection: Implications for partner notification, public policy, and the law. *American Journal of Public Health* 85, no. 11: 1569–76.

Rubenstein, C. 1993. Generation sex. *Mademoiselle.* June, 130–37.

Sawyer, Robin G., E. D. Schulken, and P. J. Pinciaro. 1997. A sexual victimization in sorority women. *College Student Journal* 31: 387–95.

Schiavi, R. C., P. Schreiner-Engle, and J. Mandeli. 1990. Healthy aging and male sexual function. *American Journal of Psychiatry* 147: 766.

Schnarch, D. 1993. Inside the sexual crucible. *Family Therapy Networker* 17: 40–49.

Schover, L. R., and S. B. Jensen. 1988. *Sexuality and chronic illness: A comprehensive approach.* New York: Guilford Press.

Shuetz-Mueller, D., L. Tiefer, and A. Melman. 1995. Follow-up of vacuum and nonvacuum constriction devices as treatments for erectile dysfunction. *Journal of Sex and Marital Therapy* 21: 229–38.

Shikai, X. 1990. Treatment of impotence in traditional Chinese medicine. *Journal of Sex Education and Therapy* 16: 198–200.

SIECUS Fact Sheet. 1997. Sexually transmitted diseases in the United States. *SIECUS Report* 25, no. 3: 22–24.

Smith, C. A. 1997. Factors associated with early sexual activity among urban adolescents. *Social Work* 42: 334–46,

Sprecher, S., and Regan, P. C. 1996. College virgins: How men and women perceive their sexual status. *Journal of Sex Research* 33: 3–15.

Starr, B., and M. Weiner. 1981. *The Starr-Weiner report on sex and sexuality in the mature years.* New York: Stein and Day.

Sydow, Kirsten Von. 1995. Unconventional sexual relationships: Data about German women ages 50 to 91 years. *Archives of Sexual Behavior* 24: 271–90.

Tiefer, Lenore. 1994. Might premature ejaculation be organic? The perfect penis takes a giant step forward. *Journal of Sex Education and Therapy* 20: 7–8.

Tschann, J. M., and N. E. Adler. 1997. Sexual self-acceptance, communication with partner, and contraceptive use among adolescent females: A longitudinal study. *Journal of Research on Adolescence* 7: 413–30.

Wallis, Claudia. 1995. The estrogen dilemma. *Time.* 26 June, 46–53.

Weinberg, Martin S., Ilsa L. Lottes, and Frances M. Shaver. 1995. Swedish or American heterosexual college youth: Who is more permissive? *Archives of Sexual Behavior* 24, no. 4: 409–37.

Witherington, R. 1991. Vacuum devices for the impotent. *Journal of Sex and Marital Therapy* 17: 69–80.

Young, W. R. 1990. Changes in sexual functioning during the aging process. In *Sexology: An independent field,* edited by F. J. Bianoco and R. Hernandez Serrano. New York: Elsevier Science Publishers, 121–28.

Zilbergeld, B. 1992. *The new male sexuality.* New York: Bantam.

Parenting is a two-way street. As you take them by the hand, they take you by the heart.

Judy Ford, *Wonderful Ways to Love a Child*

"Life changing," "most important," and "best thing that ever happened" are phrases parents use to describe the impact of having children on their lives. Many who hear of such delight look forward to having their own children. Indeed, of all women in the United States between the ages of 18 and 34, 91 percent expect to have children (*Statistical Abstract of the United States: 1997*, Table 107). Over 70 percent of a random sample of all first-year students in U.S. colleges and universities noted that "raising a family" was an "essential" or a "very important" life goal (American Council on Education and University of California, 1997). Among Chinese students (in contrast to American students), children are considered necessary in a marriage (Dzindolet et al., 1996).

In this chapter we review the various methods of contraception that may be used to control the timing and number of children in a family. We also look at the effect of children on individuals and couples and various issues associated with parenting (myths and principles). We begin by examining social influences and individual motivations for having children.

Do You Want to Have Children?

As noted above, most individuals report that they want to have children. This desire has its roots in social influences and individual motivations.

Social Influences to Have Children

Our society tends to encourage childbearing, an attitude known as *pronatalism*. Our family, friends, religion, and government help to develop positive attitudes toward parenthood. Cultural observances also function to reinforce these attitudes.

Family   Our experience of being reared in families encourages us to have families of our own. Our parents are our models. They married; we marry. They had children; we have children. Some parents exert a much more active influence. "I'm 73 and don't have much time. Will I ever see a grandchild?" asked the mother of an only child. Other remarks parents have made include "If you don't hurry up, your younger sister is going to have a baby before you do," "We're setting up a trust fund for your brother's child, and we'll do the same for yours," "Did you know that Nash and Marilyn (the adult child's contemporaries) just had a daughter?" "I think you'll regret not having children when you're old," and "Don't you want a son to carry on your name?"

Friends   Our friends who have children influence us to do likewise. After sharing an enjoyable weekend with friends who had a little girl, one husband wrote to the host and hostess, "Lucy and I are always affected by Karen—she is such a good child to have around. We haven't made up our minds yet, but our desire to have a child of our own always increases after we leave your home." This couple became parents sixteen months later.

Religion Couples who choose to be childfree are less likely than couples with children to adhere to any set of religious beliefs. Religion may be a powerful influence on the decision to have children. Catholics are taught that having children is the basic purpose of marriage and gives meaning to the union. Although many Catholics use contraception and reject their church's emphasis on procreation, some internalize the church's message. One Catholic woman said, "My body was made by God, and I should use it to produce children for Him. Other people may not understand it, but that's how I feel." Mormanism and Judaism also have a strong family orientation.

Some countries are now deliberately reducing incentives and support for having children as a way of curbing population growth.
Brent Miller
Family life specialist

Government The tax structures imposed by our federal and state governments support parenthood. Married couples without children pay higher taxes than couples with children, although the reduction in taxes is not sufficient to offset the cost of rearing a child and is not large enough to be a primary inducement to have children.

Governments in other countries have encouraged or discouraged childbearing in different ways. In the 1930s, as a mark of status for women contributing to the so-called Aryan race, Adolf Hitler bestowed the German Mother's Cross—a gold cross for eight or more children, a silver cross for six or seven, and a bronze cross for four or five—on Nazi Germany's most fertile mothers.

China, on the other hand, has a set of incentives to encourage families to have only one child. Couples who have only one child are given a "one-child glory certificate," which entitles them to special priority housing, better salaries, a 5 percent supplementary pension, free medical care for the child, and an assured place for the child in school. If the couple have more than one child, they may lose their jobs, be assigned to less desirable housing, and be required to pay the government back for any benefits they have received. India had a policy similar to China's, but it was repealed after its passage led the people to force Indira Gandhi out of office.

Cultural observances Our society reaffirms its approval of parents every year by identifying special days for Mom and Dad. Each year on Mother's Day and Father's Day (and now Grandparents' Day) parenthood is celebrated across the nation with cards, gifts, and embraces. There is no cultural counterpart (e.g., Childfree Day) for persons choosing not to have children.

OTHER CULTURES In premodern societies (before industrialization and urbanization) and in some parts of China (despite China's one-child policy) and Korea today, spouses try to have as many children as possible, because children are an economic asset (Jones, Tepperman, & Wilson, 1995). At an early age, children work as farmhands or for wages, which they give to their parents. Children are also expected to take care of their parents when the parents are elderly. In China and Korea, the eldest son is expected to take care of his aging parents by earning money for them and by marrying and bringing his wife into their home to physically care for his parents. The wife's parents need their own son and daughter-in-law to provide old-age insurance. Beyond their economic value

Freedom and family commitment are mutually exclusive. You do surrender some of your freedom when you undertake the care of children and promise fidelity to a spouse.

Sarah McLanahan

Family life specialist

A baby costs $474 a month. How much do you have in your pocket?

Planned Parenthood Poster

and old-age insurance, having numerous children is regarded as a symbol of virility for the man, a source of prestige for the woman, and a sign of good fortune for the couple. In modern societies, having large numbers of children is less valued. ●

Individual Motivations for Having Children

Individual motivations for having children include the desire to love and nurture a child, to experience love and companionship with one's child, to fulfill one's desired role identity as a parent, to experience the final rite of passage into adulthood, to increase one's status among siblings/peers, and to have a sibling for an existing child. These are generally regarded as positive motivations, but there are also negative motivations, such as wanting to escape one's parents by getting pregnant, to spite one's parents, to hold a partner, to possess something, and to prove something.

Although Americans recognize the individual right to have a child, there is debate over how old is too old to have children. This chapter's Social Policy addresses this issue.

Lifestyle Changes Associated with Having Children

Couples who have children experience major changes in their lifestyles. Daily routines become focused around the needs of the children. Living arrangements change to provide space for another person in the household. Some parents change their work schedules to allow them to be home more. Food shopping and menus change to accommodate the appetites of children.

When Groat et al. (1997) studied the evaluations of 412 young adult white and African-American respondents (married as well as single) who had at least one child, about 30 percent mentioned regrets of parenthood, including "the responsibilities of having a child take up too much time," "my children cause too much tension and worry," "raising children is a nerve-wracking job," and "I wish I had waited longer to have my first child" (p. 572).

But there are compensating rewards for having a baby. Thirty percent of Groat et al.'s (1997) respondents saw many rewards of parenthood, including "children have made me feel important," "having children has given me a sense of personal accomplishment," "now we are a real family," and "my life would be empty without children."

About 20 percent of the respondents were indifferent to the rewards, and about 20 percent were ambivalent about the regrets. Mothers were more positive than fathers, and spouses were more positive than singles about parenthood.

Financial Costs of Children

Increased financial costs are also associated with parenthood. New parents are often shocked at the relentless expenses. Medical costs for prenatal care/birth, checkups, disposable diapers, crib, clothes, carseat, baby food/bottles, etc. can easily reach $3,000. Also included are the costs associated with missing work, day care, etc. Demographers note that families from higher social classes spend more money on rearing their children than do lower- or middle-class families.

Children after Fifty: How Old Is Too Old?

Arceli Keh of Highland, California, became a mother at the age of 63 (Hellmich, 1997); Tony Randall became a father at the age of 77; and Strom Thrumond was in his 80s when he became a father. Situations like these are becoming more common, and questions are now being asked about the appropriateness of elderly individuals becoming parents. Indeed, children may now be born to an older woman whose eggs were frozen in her youth (Marcus, 1997). Should policies be established regulating the upper limit at which a person becomes a parent? Our society is mostly in agreement that having a child as a teenager has negative consequences for the child, parents, and society. But what about the other end of the age spectrum?

When Tony Randall's daughter, Julia, begins college, her father will be 95; Arceli Keh will be 81 when her daughter, Cynthia, begins college. "So what?" says Arceli, "My mother is in her 80s and she's helping me with Cynthia" (Peterson & Hellmich, 1997). While this grandmother may enjoy rearing another generation, other grandparents may feel forced into the role of becoming a parent in the later years.

There are advantages and disadvantages of having a child as an elderly parent. The primary developmental advantage for the child of retirement-aged parents is the attention the parents can devote to their offspring. Not distracted by their careers, these parents have more time to nurture, play with, and teach their children. There is abundant evidence that children benefit cognitively, emotionally, and socially from concentrated attention during their early developmental years (Ramsburg, 1997; Carnegie Corporation of New York, 1994).

Elderly parents are also less likely to divorce. The median age of women who divorce is 33; of men 36 (*Statistical Abstract of the United States: 1997*, Table 149). Two researchers (Amato & Keith, 1991) analyzed data from 81,000 people in thirty-seven studies and concluded that "divorce (or permanent separation) has broad negative consequences for quality of life in adulthood" (p. 54). Children who experience divorce are more likely to experience depression and lower life satisfaction, and to have a divorce in their own marriages.

The primary disadvantage of having a child in the later years is that the parents are likely to die before or shortly after, the child reaches adulthood. The daughter of James Dickey, the famous southern writer, lamented the fact that her late father (to whom she was born when he was in his fifties) would not be present at her graduation or wedding.

There are also medical concerns for both the mother and the baby during pregnancy in later life. Although maternal and fetal outcomes for women delivering babies after the age of 45 are "generally good" (Dildy et al., 1996), there is an increased risk of morbidity and mortality for the mother. These risks are typically a function of chronic disorders that go along with aging, such as diabetes, hypertension, and cardiac disease (Cunningham et al., 1997). Stillbirths, congenital malformations, and infant mortality are also higher in women in their late thirties and forties (Fretts et al., 1995; Cunningham et al., 1997). However, a normal ultrasonography can significantly reduce the risk of Down syndrome and any chromosome abnormality (Bianco et al., 1996). After reviewing the studies on pregnancy and childbirth in women over 35, Cunningham et al. (1997) note:

> Women should realistically appraise the risks of pregnancy later in life, but should not necessarily fear delaying childbirth. Pregnancy after 35 is increasingly common in our society, and improved obstetrical care has made advanced maternal age compatible with successful pregnancy for the great majority of such women (p. 577).

Garrison et al. (1997) studied parenthood in sixty-nine couples, the majority of whom became parents for the first time after the woman was 35. The average age of the women was 50 and the average age of the men was 48. Of delayed parenthood, the authors concluded:

> With few exceptions, parents who delayed childbearing were more satisfied, less stressed, and reported better functioning than their nondelaying counterparts. This conclusion leads credence to updated life cycle theories in which it is presumed that adults who delay childbearing may be better prepared and adapt more easily to parenthood. It may be that parents who delay parenthood have different expectations and that these expectations moderate their experiences and the reporting of these experiences (p. 288).

Given that medical outcomes are usually manageable, government regulations in regard to how young a woman must be at pregnancy are not likely. In addition, control would be difficult. Fearing that she would not be accepted into the University of Southern California's Program for Assisted Reproduction, Arceli Keh did not tell them her real age. Instead, she told them she was 50.

REFERENCES

Amato, P. R. and Keith, B. 1991. Parental divorce and adult well-being. *Journal of Marriage and the Family* 53: 43–58.

Carnegie Corporation of New York. 1994. *Starting points for young children.* New York: Carnegie Corporation of New York.

Cunningham, F. G. et al., eds. 1997. *William's obstetrics.* 20th ed. Stamford, Conn.: Appleton and Lange.

Bianco, A., J. Stone, L. Lapinski Lynch, B. Berkowitz, and R. L. Berkowitz. 1996. Pregnancy outcome at age 40 and older. *Obstetrics and Gynecology* 87: 917–22.

Dildy, G. A., G. M. Jackson, G. K. Flowers, B. T. Oshiro, N. W. Varner, and S. L. Clark. 1996. Very advanced maternal

age: Pregnancy after age 45. *American Journal of Obstetrics and Gynecology* 175: 668–74.

Fretts, R. C. et al. 1995. Increased maternal age and the risk of fetal death. *New England Journal of Medicine* 333: 953–57.

Garrison, M. E., L. B. Blalock, J. J. Zarski, and P. B. Merritt. 1997. Delayed parenthood: An exploratory study of family planning. *Family Relations* 46: 281–90.

Hellmich, N. 1997. Oldest new mom is 63. *USA Today.* 24 April, 1A.

Marcus, M. B. 1997. New life from frozen eggs. *U.S. News and World Report.* 27 October.

Peterson, K. S. and Hellmich, N. No shortage of opinion on 63-year-old mom. 1997 *USA Today.* April 25., 1A.

Ramsburg, D. 1997. Brain development in young children: The early years ARE learning years. *Parent News.* No. 3 (April), 1–3.

Statistical Abstract of the United States: 1997. 116th ed. Washington, D.C.: U.S. Bureau of the Census.

RECENT RESEARCH

Schoen et al. (1997) analyzed data from 4,358 respondents in a national survey in regard to motivating factors for parenthood and found that persons for whom relationships created by children are important are more likely to intend to have a child than persons for whom such relationships are not important.

NATIONAL DATA   According to the United States Department of Agriculture, the estimated cost of rearing a child by family income group from birth through age 17 (does not include college) is as follows: low-income group—$170,920; middle-income group—$231,140; high-income group—$334,590 (Schwiesow & Staimer, 1994, A1).

College expenses significantly increase the cost of rearing a child.

NATIONAL DATA   The costs of a child's attending a state-supported and private college for four years are $29,804 and $89,876 respectively. These costs include tuition and required fees, board, and dorm (*Statistical Abstract of the United States:* 1997, Table 293).

According to a *U.S. News and World Report* feature article (Longman, 1998), the total cost of raising one child in a middle class family is nearly $1.5 million ($1,455,581.00). This figure includes the cost of housing, food, transportation, clothing, health care, day care/education, college, and wages lost due to a parent dropping out of the workforce to provide childcare.

OTHER CULTURES   The economic value of children varies greatly across societies and historical time periods. In rural, developing countries (as in early U.S. history), parents viewed children as economic assets because children provided valuable work to help sustain their family. However, with industrialization, the role of children changed in the United States from worker to consumer. Child labor and compulsory education laws took children out of factories and put them in schools. ●

Youth is a wonderful thing. What a crime to waste it on children.

George Bernard Shaw

Irish playwright

INSIGHT

Couples who put off having children are similar to couples who make it clear they do not want children, ever. Both groups tend to be white, highly educated, career-oriented city dwellers (Jones et al., 1995, 114).

I would have made a terrible parent.

Katherine Hepburn

Actress

Regardless of the lifestyle changes and economic costs, most people are motivated to have children. The Personal Application on p. 254 allows you to assess your own motivations for having children.

The Childfree Alternative

Jay Leno and his wife are an example of a childfree couple. They are not alone.

NATIONAL DATA Of all women in the United States between the ages of 18 and 34, 9 percent report that they do not expect to have a child (*Statistical Abstract of the United States: 1997*, Table 107).

With two exceptions, the percentage of couples electing to remain childfree has remained relatively stable in the past fifty years. These two periods were during the 1930s depression (when there were more childfree women) and during the 1950–1965 baby boom (when there were fewer childfree women) (Jones et al., 1995).

Women today who want to remain childfree tend to evidence greater interest in a career, to have an egalitarian relationship with their spouse, and to value freedom from the constraints of having children. Generally, wives in childfree marriages are more committed to remaining childfree than are their husbands (Jones et al., 1995, 113).

Wanting to remain childfree is also related to racial and ethnic background. Whites are most likely to consider marriage without children. African-Americans, Native Americans, Mexican-Americans, and Asian-Americans are more likely to be family-oriented and to consider children an important part of family life (O'Hare et al., 1991; O'Hare & Felt, 1991; Ahlburg & De Vita, 1992). However, being upwardly mobile may sometimes override ethnic influence. African-Americans who are striving to achieve social mobility are more likely to be voluntarily childfree (Boyd, 1989).

OTHER CULTURES Beets et al. (1997) surveyed 1,775 Dutch adults in 1987 and 1,257 from the same sample again in 1991 and found that only 10 percent reported that they preferred not to have children. ●

How happy are the marriages of couples who elect to remain childfree compared with the marriages of couples who opt for having children? While marital satisfaction declines across time for all couples whether or not they have children (MacDermid, Huston, & McHale, 1990), children tend to lessen marital satisfaction by decreasing spousal time together, spousal interaction, and agreement over finances. In addition, couples who have children report greater satisfaction before their children are born and after the children leave home. The greatest drop in marital satisfaction is during the time the children are teenagers. Childfree couples do not experience this roller coaster ride (Glenn, 1991).

Infertility

The birth of seven babies to Bobbi and Kenny McCaughey in 1997 brought the issue of infertility to national attention. The McCaugheys were *infertile*, a condition

INSIGHT

Is the childfree lifestyle for you? If you get your primary satisfactions from interacting with adults and from your career and if you require an atmosphere of freedom and privacy, perhaps the answer is yes. But if your desire for a child is at least equal to your desire for a satisfying adult relationship, career, and freedom, the answer may be no. The childfree alternative is particularly valuable to persons who would find the demands of parenthood an unnecessary burden and strain.

The reason we have so many children is that I'm deaf. When we would go to bed, my wife would ask, "Do you want to go to sleep or what?" And I would always ask "What?"

Joe Hancock

Married 24 years

defined as the inability to achieve a pregnancy after at least one year of regular sexual relations without birth control, or the inability to carry a pregnancy to a live birth. The use of fertility drugs resulted in the multiple pregnancy/births. A team of researchers (Van Balen et al., 1997) studied 131 infertile couples and found that "medical help" (as opposed to adoption, foster parenting, focusing on other goals) was the option chosen by 80 percent of the couples in response to their infertility.

Infertility problems may be attributed to the man (e.g., low sperm production, poor sperm motility) or woman (e.g., blocked fallopian tubes, endocrine imbalances). Sexually transmissible diseases may also have negative effects on both the man and the woman. More general causes of infertility include women's waiting until they are older to conceive, environmental hazards (enzymes injected into meat products, toxic waste), and medications.

Not being able to get pregnant is a psychological crisis, a grief experience, and an event that involves various new choices. Couples feel blocked in their ability to achieve a cultural ideal (parenthood) and may blame each other for the infertility. They may grieve over the biological baby they will never have and become depressed. A lower quality of life is often reported (Weaver et al., 1997). Some couples seek fertility specialists, who, depending on the cause of the infertility, may use in vitro fertilization, fertility drugs, or ovum transfer. The latter involves taking the fertilized egg from a woman impregnated by the husband's sperm in a surrogate woman and transferring it to the wife. Treatment of infertility is expensive ($30,000 plus), time-consuming (several months to a year), and frustrating. Only about 20 percent of infertile couples who seek the services of a fertility clinic end up with a live birth. Those who are successful because of the use of fertility drugs may have multiple pregnancies with low-birth-weight babies who often have future health problems.

OTHER CULTURES   Laban and Gwako (1997) studied the effect of a wife's conjugal power over family size among married women in rural Kenya and found that the higher the status of the wife, the more likely she was to use contraception and to have fewer children. ●

Contraception

Once individuals have decided on whether and when they want children, contraception becomes important. All contraceptive practices have one of two common purposes: to prevent the male sperm from fertilizing the female egg or to keep the fertilized egg from implanting itself in the uterus. In this section, we look at the various methods of contraception.

Hormonal Contraceptives

Hormonal contraceptives currently available to women include the pill, Norplant, and Depo-Provera.

Birth control pill   The birth control pill is the most commonly used method of all the nonsurgical forms of contraception. Although a small percentage of women who take the pill still get pregnant, it remains a very desirable birth control option.

NATIONAL DATA   Twenty percent of never-married women, 16 percent of currently married women, and 15 percent of formerly married women aged 15–44 use the pill as their method of contraception (*Statistical Abstract of the United States: 1997*, Table 110).

There are basically two types of birth control pills—the combination pill and the minipill. The combination pill is taken for twenty-one days, beginning on the fifth day after the start of the menstrual flow. Three or four days after the last pill is taken, menstruation occurs, and the 28-day cycle begins again. To eliminate the problem of remembering when to begin taking the pill every month, some physicians prescribe a low-dose combination pill for the first twenty-one days and a placebo (sugar pill) or an iron pill for the next seven days. In this way, the woman takes a pill every day.

The second type of oral contraceptive, the minipill, contains the same progesterone found in the combination pill but in much lower doses. The minipill contains no estrogen. Like the progesterone in the combination pill, the progesterone in the minipill provides a hostile environment for sperm and inhibits implantation of a fertilized egg in the uterus. In general, the minipill is somewhat less effective than other types of birth control pills and has been associated with a higher incidence of irregular bleeding.

Neither the combination pill nor the minipill should be taken unless prescribed by a physician who has detailed information about the woman's previous medical history. Contraindications—reasons for not prescribing birth control pills—include hypertension, impaired liver function, known or suspected tumors that are estrogen-dependent, undiagnosed abnormal genital bleeding, pregnancy at the time of the examination, and a history of poor blood circulation or blood clotting. The major complications associated with taking oral contraceptives are blood clots and high blood pressure. Also, the risk of heart attack is increased in women over age 30, particularly those who smoke or have other risk factors. Women over 40 should generally use other forms of contraception because the side effects of contraceptive pills increase with the age of the user. Infertility problems have also been noted in women who have used the combination pill for several years without the breaks in pill use recommended by most physicians.

Although the long-term negative consequences of taking birth control pills are still the subject of research, short-term negative effects are experienced by 25 percent of all women who use them. These side effects include increased susceptibility to vaginal infections, nausea, slight weight gain, vaginal bleeding between periods, breast tenderness, headaches, and mood changes (some women become depressed and experience a loss of sexual desire).

Finally, women should be aware that pill use is associated with an increased incidence of chlamydia and gonorrhea. One reason for the association of pill use and a higher incidence of STDs is that sexually active women who use the pill sometimes erroneously feel that because they are protected from

INSIGHT

In spite of the negative consequences associated with birth control pill use, numerous studies involving hundreds of thousands of women show that the overall risk of pill use is lower than the risk of full-term pregnancy and giving birth.

Immediate health benefits are also derived from taking birth control pills. Oral contraceptives tend to protect the woman against breast tumors, ovarian cysts, rheumatoid arthritis, and inflammatory diseases of the pelvis. They also regularize the woman's menstrual cycle, reduce premenstrual tension, and may reduce menstrual cramps and blood loss during menstruation. Finally, oral contraceptives are convenient, do not interfere with intercourse, and, most important, provide highly effective protection against pregnancy.

Whether to use birth control pills remains a controversial issue. Some women feel it harms their bodies to take birth control pills; others feel it harms their bodies not to take them. Whatever a woman's choice, it should be made in conjunction with the physician who knows her medical history. The physician should also alert the woman to the fact that any use of antibiotics while taking the pill may cancel the contraceptive effects of the pill.

INSIGHT

Oral contraceptives and Depo-Provera were designed for use by women. In recent years, researchers have tried to develop hormonal contraceptives for use by men. Dr. Ronald Swerdloff (1995) of the Harbor–UCLA Medical Center noted, "We have been able to show that reversible, safe, male hormonal contraception is a possibility." He and his colleagues injected 370 males once a day initially and, subsequently, once a week with synthetic testosterone. The effect was to lower sperm production so that males could have regular intercourse with pregnancy resulting only 1.4 percent of the time. Men who share the responsibility for contraception are most likely to use a male injectable contraceptive (Ringheim, 1995). However, further testing and refinement will take another ten years before a male hormonal contraceptive is commercially available (Alexander, 1995). In this regard, it is noteworthy that hormonal pills for women were commercially available long before many of the problems related to them had been solved.

RECENT RESEARCH

Lindberg et al. (1997) found that among a nationally representative sample of men aged 17–22, almost a fourth (23%) reported experiencing at least one condom break during the previous 12 months. Of all condoms used, 2.5 percent had broken.

getting pregnant, they are also protected from contracting STDs. The pill provides no protection against STDs; a condom must be worn.

Norplant In the early 1990s, the FDA approved the use of Norplant, a long-acting reversible hormonal contraceptive consisting of six thin flexible silicone capsules (36mm in length) implanted under the skin of the upper arm. As a result of its association with silicone and reported side effects (migraine headaches, shortness of breath, and weight gain), recent use has declined dramatically. Daily sales of Norplant have plummeted from 800 to 60. And out of one million women who have tried Norplant, 50,000 have retained lawyers to sue the company that manufactures it because they were not adequately warned of its potential problems (Cohen, 1995).

Depo-Provera An FDA-approved alternative to Norplant is Depo-Provera, a synthetic compound similar to progesterone injected into the woman's arm or buttock. Depo-Provera protects a woman against pregnancy for three months by preventing ovulation. It has been used by 30 million women worldwide since it was introduced in the late 1960s. For women who get their shots every three months, the failure rate is less than 1 percent.

Side effects of Depo-Provera include menstrual spotting, irregular bleeding, and some heavy bleeding the first few months of use. Mood changes, headaches, dizziness, and fatigue have also been observed. Some women report a weight gain of three to five pounds. Also, after stopping the injections, it takes an average of eighteen months before the woman will become pregnant at the same rate as women who have not used Depo-Provera. The cost of injectable Depo-Provera for five years is about $140 per year.

Male Condom

The condom is currently the only form of male contraception. The condom is a thin sheath, made of latex or polyurethane. Unlike condoms made of latex, the polyurethane condom can be used with sterile lubricants; it avoids the latex allergy some people experience, blocks the HIV virus and other sexually transmitted diseases, and allows for greater sensitivity during intercourse. Condoms made of natural membranes (sheep intestinal lining) are not recommended because they are not effective in preventing HIV transmission.

NATIONAL DATA Fourteen percent of never-married women, 13 percent of currently married women, and 10 percent of formerly married women aged 15–44 use the male condom as their method of contraception (*Statistical Abstract of the United States: 1997*, Table 110).

The condom works by being rolled over and down the shaft of the erect penis before intercourse. When the man ejaculates, the sperm are caught inside the condom. When used in combination with a spermicidal lubricant that is placed on the inside of the reservoir tip of the condom as well as a spermicidal or sperm-killing agent that the woman inserts inside her vagina, the condom is a highly effective contraceptive.

Like any contraceptive, the condom is effective only when used properly. It should be placed on the penis early enough to avoid any seminal leakage into

*The female condom is huge—
now you can carry your wallet
in your condom.*

Jay Leno
The Tonight Show

*The female condom allows the
woman more control over
pregnancy and avoiding STDs
including HIV.*

the vagina. In addition, polyurethane or latex condoms with a reservoir tip are preferable, as they are less likely to break. Finally, the penis should be withdrawn from the vagina immediately after ejaculation, before the man's penis returns to its flaccid state. If the penis is not withdrawn and the erection subsides, semen may leak from the base of the condom into the vaginal lips. Alternatively, when the erection subsides, the condom will come off when the man withdraws his penis if he does not hold onto the condom. Either way, the sperm will begin to travel up the vagina to the uterus and fertilize the egg.

In addition to furnishing extra protection, spermicides also provide lubrication, which permits easy entrance of the condom-covered penis into the vagina. If no spermicide is used and the condom is not of the prelubricated variety, a sterile lubricant (such as K-Y Jelly) may be needed. Vaseline or other kinds of petroleum jelly *should not* be used with condoms because vaginal infections and/or condom breakage may result.

Female Condom

The *female condom* resembles a man's condom except that it fits in the woman's vagina to protect her from pregnancy, HIV infection, and other STDs. The vaginal condom is a large, lubricated, polyurethane adaptation of the male version. It is about seven inches long and has flexible rings at both ends. It is inserted like a diaphragm, with the inner ring fitting behind the pubic bone against the cervix; the outer ring remains outside the body and encircles the labial area (see Figure 8.1). Like the male version, the female condom is not reusable. Female condoms have been approved by the FDA and are being marketed under the brand names Femidom and Reality. The one-size-fits-all device is available without a prescription and sells for about $2.25 (per condom).

The vaginal condom is durable and does not tear like latex male condoms, but it is trickier to use. The actual effectiveness rate (against STDs) of the vaginal condom has not been sufficiently studied. A major advantage of the female condom is that like the male counterpart, it protects against transmission of the

Figure 8.1
The Female Condom

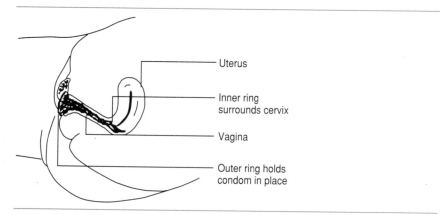

The 6 1/2-inch-long sheath lines the vagina, held in place by two plastic rings.

HIV virus and other STDs. Placement may occur up to eight hours before use, allowing greater spontaneity (Stifel & Anderson, 1997). Women and men who have used the vaginal condom are generally satisfied with the device (Gregersen & Gregersen, 1990).

In one study of fifty-two women aged 18–57, 79 percent reported that they had used the female condom at least once. Of those who reported use, 73 percent of the respondents and 44 percent of their partners preferred the female condom to the male condom (Gollub, Stein, & El-Sadr, 1995).

Intrauterine Device (IUD)

The intrauterine device, or IUD, is a small object that is inserted by a physician into the woman's uterus through the vagina and cervix (see Figure 8.2). The device is thought to prevent implantation of the fertilized egg in the uterine wall. As a result of infertility and miscarriage associated with IUDs and subsequent lawsuits against manufacturers by persons who were damaged by the device, use in the United States is now minimal. However, current studies suggest that the newer IUDs do not have rates of pelvic inflammatory disease (PID), ectopic pregnancy, or resultant infertility as high as previously suspected (Stifel & Anderson, 1997).

NATIONAL DATA   Less than one-half of 1 percent of never-married women, one percent of currently married women, and .05 percent of formerly married women aged 15–44 use the IUD as their method of contraception (*Statistical Abstract of the United States: 1997*, Table 110).

Diaphragm

Use of the diaphragm is not much greater than use of the IUD.

Happiness is having a large, loving, caring, close-knit family in another city.
George Burns
Comedian

Figure 8.2
The IUD

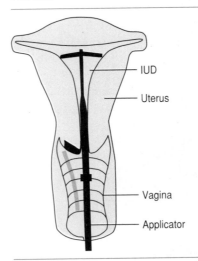

IUD

Uterus

Vagina

Applicator

NATIONAL DATA Less than one-half of 1 percent of never-married women, 2 percent of currently married women, and 1 percent of formerly married women aged 15–44 use the diaphragm as their method of contraception (*Statistical Abstract of the United States: 1997,* Table 110).

The diaphragm is a shallow rubber dome attached to a flexible, circular steel spring. Varying in diameter from two to four inches, the diaphragm covers the cervix and prevents sperm from moving beyond the vagina into the uterus. This device should always be used with a spermicidal jelly or cream.

To obtain a diaphragm, a woman must have an internal pelvic examination by a physician or nurse practitioner, who will select the appropriate size diaphragm and instruct the woman on how to insert it. The woman will be told to apply one teaspoonful of spermicidal cream or jelly on the inside of the diaphragm and around the rim before inserting it into the vagina (no more than two hours before intercourse). The diaphragm must also be left in place for six to eight hours after intercourse to permit any lingering sperm to be killed by the spermicidal agent.

After the birth of a child, a miscarriage, abdominal surgery, or the gain or loss of ten pounds, a woman who uses a diaphragm should consult her physician or health practitioner to ensure a continued good fit. In any case, the diaphragm should be checked every two years for fit.

A major advantage of the diaphragm is that it does not interfere with the woman's hormonal system and has few, if any, side effects. Also, for those couples who feel that menstruation diminishes their capacity to enjoy intercourse, the diaphragm may be used to catch the menstrual flow for a brief time.

On the negative side, some women feel that use of the diaphragm with the spermicidal gel is messy and a nuisance. For some, the use of the gel may produce an allergic reaction. Furthermore, some partners feel that the gel makes oral genital contact less enjoyable. Finally, if the diaphragm does not fit properly or is left in place too long (more than twenty-four hours), pregnancy or toxic shock syndrome can result.

Cervical Cap

The cervical cap is a thimble-shaped contraceptive device made of rubber or polyethylene that fits tightly over the cervix and is held in place by suction. Like the diaphragm, the cervical cap, which is used in conjunction with spermicidal cream or jelly, prevents sperm from entering the uterus. Cervical caps have been widely available in Europe for some time and were approved for marketing in the United States in 1988. The cervical cap cannot be used during menstruation, since the suction cannot be maintained. The effectiveness, problems, risks, and advantages are similar to those of the diaphragm.

Vaginal Spermicides

A spermicide is a chemical that kills sperm. Vaginal spermicides come in several forms, including foam, cream, jelly, and suppository. In the United States, the active agent in most spermicides is nonoxynol–9, which has been shown to kill many organisms that cause sexually transmitted diseases (including HIV).

Creams and gels are intended for use with a diaphragm. Suppositories are intended for use alone or with a condom. Foam is marketed for use alone but can also be used with a diaphragm or condom.

Spermicides must be applied before the penis penetrates the vagina (appropriate applicators are included when the product is purchased) no more than twenty minutes before intercourse. While foam is effective immediately, suppositories, creams, or jellies require a few minutes to allow the product to melt and spread inside the vagina (package instructions describe the exact time required). Each time intercourse is repeated, more spermicide must be applied. Spermicide must be left in place for at least six to eight hours after intercourse; douching or rinsing the vagina should not be done during this period.

One advantage of using spermicides is that they are available without a prescription or medical examination. They also do not manipulate the woman's hormonal system and have few side effects. A major noncontraceptive benefit of some spermicides is that they offer some protection against the transmission of sexually transmitted diseases, including HIV. However, spermicides should never be depended upon alone to be effective in preventing STD transmission.

Periodic Abstinence

Also referred to as natural family planning, rhythm method, and fertility awareness, *periodic abstinence* involves refraining from sexual intercourse during the one to two weeks each month when the woman is thought to be fertile (see Figure 8.3).

NATIONAL DATA Less than 1 percent of never-married women, 2 percent of currently married women, and 1 percent of formerly married women aged 15–44 use periodic abstinence as their method of contraception (*Statistical Abstract of the United States: 1997,* Table 110).

Women who use periodic abstinence must know their time of ovulation and avoid intercourse just before, during, and immediately after that time. Calculating the fertile period involves three assumptions: (1) ovulation occurs on day 14 (plus or minus two days) *before the onset of the next menstrual period;* (2) sperm remain viable for two to three days; and (3) the ovum survives for twenty-four hours.

Nonmethods: Withdrawal and Douching

Because withdrawal and douching are not effective in preventing pregnancy, we call them "nonmethods" of birth control.

Also known as coitus interruptus, withdrawal is the practice whereby the man withdraws his penis from the vagina before he ejaculates. The advantages of coitus interruptus are that it requires no devices or chemicals, and it is always available. The disadvantages of withdrawal are that it does not provide protection from STDs, it may interrupt the sexual response cycle and diminish the pleasure for the couple, and it is very ineffective in preventing pregnancy.

Withdrawal is not a reliable form of contraception for two reasons. First, a man can unknowingly emit a small amount of preejaculatory fluid (which is stored in the prostate or penile urethra or in the Cowper's glands), which may

INSIGHT

The calendar method of predicting the "safe" period may be unreliable for two reasons. First, the next month the woman may ovulate at a different time than any of the previous eight months. Second, sperm life varies; they may live long enough to meet the next egg in the fallopian tubes.

Figure 8.3
The Natural Family Planning Method

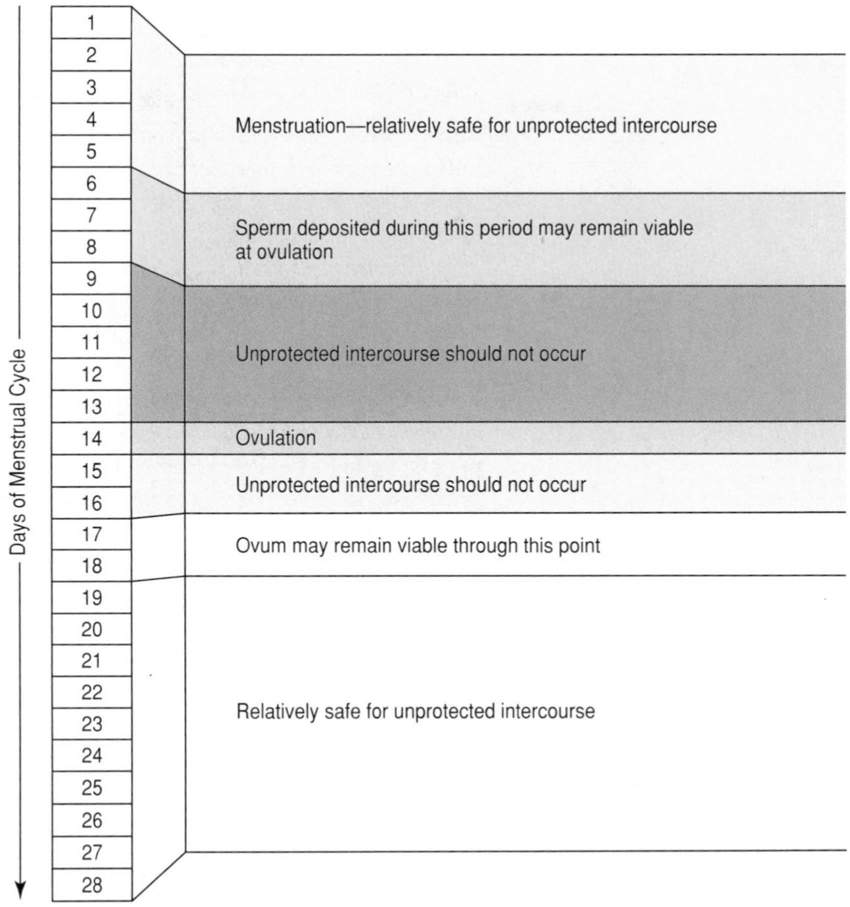

Days of Menstrual Cycle	
1	
2	
3	
4	Menstruation—relatively safe for unprotected intercourse
5	
6	
7	Sperm deposited during this period may remain viable
8	at ovulation
9	
10	
11	Unprotected intercourse should not occur
12	
13	
14	Ovulation
15	Unprotected intercourse should not occur
16	
17	Ovum may remain viable through this point
18	
19	
20	
21	
22	
23	Relatively safe for unprotected intercourse
24	
25	
26	
27	
28	

Most women's cycles are not a consistent twenty-eight days, but may vary anywhere from twenty-one to thirty-five days.

contain sperm. One drop can contain millions of sperm. In addition, the man may lack the self-control to withdraw his penis before ejaculation, or he may delay his withdrawal too long and inadvertently ejaculate some semen near the vaginal opening of his partner. Sperm deposited there can live in the moist vaginal lips and make their way up the vagina.

Though some women believe that douching is an effective form of contraception, it is not. Douching refers to rinsing or cleansing the vaginal canal. After intercourse, the woman fills a syringe with water or a spermicidal agent and flushes (so she assumes) the sperm from her vagina. But in some cases, the fluid will actually force sperm up through the cervix. In other cases, a large number of sperm may already have passed through the cervix to the uterus, so the douche may do little good. Sperm may be found in the cervical mucus within ninety seconds after ejaculation.

RECENT RESEARCH

Gold et al. (1997) assessed the beliefs of 167 physicians who had an expertise in adolescent health. Twenty-nine percent believed that repeated use of emergency contraception could pose health risks.

In effect, douching does little to deter conception and may even encourage it. In addition, douching is associated with an increased risk for pelvic inflammatory disease and ectopic pregnancy.

Emergency Contraception

Also referred to as *postcoital contraception, emergency contraception* refers to various types of morning-after pills that are used primarily in three circumstances: when a woman has unprotected intercourse; when a contraceptive method fails (such as condom breakage or slippage); and when a woman is raped. Emergency contraception methods should be used only in emergencies—those times when unprotected intercourse has occurred and medication can be taken within seventy-two hours of exposure.

Combined estrogen-progesterone The most common morning-after pills are the combined estrogen-progesterone oral contraceptives routinely taken to prevent pregnancy. In higher doses, they serve to prevent ovulation, fertilization of the egg, or transportation of the egg to the uterus. They may also make the uterine lining inhospitable to implantation. Known as the "Yuzpe method" after the physician who proposed it, this method involves ingesting four tablets of combined estrogen-progesterone. The first two tablets are taken within seventy-two hours of unprotected intercourse; two more tablets are taken twelve hours later. Side effects of combined estrogen-progesterone emergency contraception pills (sold under the trade name Ovral) include nausea, vomiting, and breast tenderness. The pregnancy rate is 1.2 percent if combined estrogen-progesterone is taken within twelve hours of unprotected intercourse, 2.3 percent if taken within forty-eight hours, and 4.9 percent if taken within forty-eight to seventy-two hours (Rosenfeld, 1997).

Mifepristone (RU-486) *Mifepristone,* also known as RU-486, is a synthetic steroid that effectively inhibits implantation of a fertilized egg. Given in a single 600-mg dose within seventy-two hours after unprotected intercourse, it makes the endometrium unsuitable for implantation. Side effects of RU-486 may include nausea, vomiting, and breast tenderness. The pregnancy rate associated with RU-486 is 1.6 percent, which suggests that RU-486 is an effective means of emergency contraception (Rosenfeld, 1997).

Postcoital methods of contraception are controversial; some people regard them as a form of abortion, while others regard them as a means of reducing the need for abortion. Recent reviewers of research on new methods of emergency contraception (von Hertzen & Van Look, 1996) urged that as soon as researchers are able to establish the lowest effective dose in mifepristone, this method of emergency contraception should be more available in clinical practice, and will be helpful in avoiding unwanted pregnancies and unnecessary abortions.

According to one expert, the wide availability and knowledge of emergency contraception in the Netherlands have contributed to lower teenage pregnancy and abortion rates than are found in the United States and the United Kingdom (Haspels, 1994).

Table 8.1 presents data on the effectiveness of various contraceptive methods for preventing pregnancy and protecting against sexually transmitted dis-

Table 8.1

Methods of Contraception and Sexually Transmitted Disease Protection from the Woman's Perspective						
	Estimated Effectivness Against Sexually Transmitted Disease	Contraceptive Effectiveness				
Method		High[a]	Average[a]	Benefits	Disadvantages	Cost[b]
Condom	30–60%	98%	88%	Entails male responsibility; offers high level of protection; is inexpensive	Difficult to negotiate; entails lack of control for woman; may be seen as interrupting sex; may imply unfaithfulness	$0.50
Female condom	Insufficiently studied	Insufficiently studied	85%[c]	Offers high level of protection against sexually transmitted disease/HIV	Is visible; is expensive; requires negotiation	$2.25
Film[d]	50%	99%	79%	Is easy to use; requires no negotiation	Requires 15 minutes' waiting time; must be applied within one hour of intercourse	$1.00
Suppository	50%	99%	79%	Is easy to use; is inexpensive; requires no negotiation	Requires 15 minutes' waiting time; must be inserted within one hour of intercourse	$0.30
Foam	50%	99%	79%	Is available OTC;[e] requires no waiting time after insertion; requires no negotiation	Requires applicator	$0.50
Jelly/cream	50%	98%	79%	Is available OTC; inexpensive; requires no negotiation	Requires applicator; must be applied within one hour of intercourse	$5.00 per tube

[a]Highest observed effectiveness; "typical" user effectiveness.
[b]Per act of intercourse based on average cost to consumer.
[c]Use-effectiveness rate presented by Food and Drug Administration hearings on Reality, January 31, 1992.
[d]Marketed as VCF (vaginal contraceptive film) in the United States and as C-film in United Kingdom.
[e]Over the counter.

Table 8.1

Method	Estimated Effectivness Against Sexually Transmitted Disease	Contraceptive Effectiveness		Benefits	Disadvantages	Cost[b]
		High[a]	Average[a]			
Cervical cap	50–70% for cervical pathogens, 0% for others	98%	82%	Is comfortable; can be used repeatedly over 2 + days; is cheaper over reproductive life; requires no waiting time after insertion; no UTIs;[g] rarely requires refitting; may require no negotiation (if not felt by partner); offers excellent cervical protection with low nonoxynol-9 use	Must be fitted; 20%–40% women not able to be fitted;[f] requires initial outlay of $100–$150; requires vaginal spermicide for best protection against sexually transmitted disease	$0.10
Diaphragm	50–75% for cervical pathogens	99%	82%	Requires no waiting time after insertion; fits nearly all women; may require no negotiation (if not felt by partner); is cheaper over reproductive life	Must be removed after 10–12 hours; may need to be refitted; carried increased risk of UTIs for some women; requires initial outlay of $50–$75	$0.10
Withdrawal	Insufficiently studied			Requires no purchases	Is not controlled by woman; is highly user dependent	0
Pill	None	96%	99%	Is convenient; is removed from sex act; affords high contraceptive efficacy	Offers no protection against sexually transmitted disease/ HIV; may raise risk; expensive	$12–$24 /month
Intrauterine device	None	99%	96%	Is not user dependent	Carries high initial cost; entails risk of PID associated with insertion	$150–$300 per insertion

[f]Rate depends on criteria for a good fit, and practitioner criteria have varied widely.

[g]Urinary tract infection.

Source: M. J. Rosenberg and E. L. Gollub. Commentary: Methods women can use that may prevent sexually transmitted diseases, including HIV. *American Journal of Public Health,* 1992, 82, no. 11, 1473–1478. Copyright 1992 American Public Health Association. Reprinted with permission.

eases. Table 8.1 also describes the benefits, disadvantages, and cost of various methods of contraception. Not included in the chart is the obvious and most effective form of birth control, abstinence, which also has the lowest cost and eliminates the risk of HIV and STD infection from intercourse.

Sterilization

Unlike the temporary and reversible methods of contraception just discussed, sterilization is a permanent surgical procedure that prevents reproduction. Sterilization may be a contraceptive method of choice when the woman should not have more children for health reasons or when individuals are certain about their desire to have no more children or to remain childfree. Most couples complete their intended childbearing in their late 20s or early 30s, leaving more than fifteen years of continued risk of unwanted pregnancy. Because of the risk of pill use at older ages and the lower reliability of alternative birth control methods, sterilization has become the most popular method of contraception among married women who have completed their families.

Slightly more than half of all sterilizations are performed on women. Although male sterilization is easier and safer than female sterilization, women feel more certain they will not get pregnant if they are sterilized. "I'm the one that ends up being pregnant and having the baby," said one woman. "So I want to make sure that I never get pregnant again."

Female sterilization   Although a woman may be sterilized by removal of her ovaries (*oophorectomy*) or uterus (*hysterectomy*), these operations are not normally undertaken for the sole purpose of sterilization, because the ovaries produce important hormones (as well as eggs) and because both procedures carry the risks of major surgery. But sometimes there is another medical problem requiring hysterectomy.

NATIONAL DATA   Women aged 15–44 who use surgical sterilization as their method of contraception (*Statistical Abstract of the United States: 1997,* Table 110).

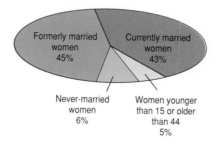

The usual procedures of female sterilization are the salpingectomy and a variant of it, the laparoscopy. *Salpingectomy,* also known as tubal ligation, or tying the tubes (see Figure 8.4), is often performed under a general anesthetic while the woman is in the hospital just after she has delivered a baby. An incision is made in the lower abdomen, just above the pubic line, and the fallopian tubes are brought into view one at a time. A part of each tube is cut out, and

Figure 8.4
Female Sterilization: Tubal Sterilization

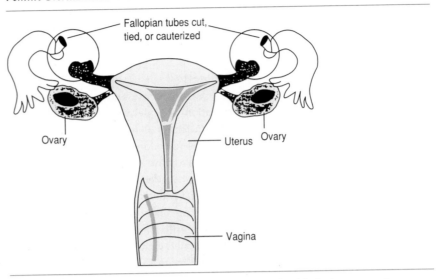

Fallopian tubes cut, tied, or cauterized

Ovary

Uterus

Ovary

Vagina

The fallopian tubes are interrupted surgically—cut and tied or blocked—to prevent passage of the eggs from the ovaries to the uterus.

INSIGHT

Though the cost of female sterilization appears high, the cost of using the pill for five years is $1,784. Over time, female sterilization is actually cheaper (Trussell et al., 1995).

the ends are tied, clamped, or cauterized (burned). The operation takes about thirty minutes. About 700,000 such procedures are performed annually. The cost is around $2,554.

A less expensive and quicker (about fifteen minutes) form of salpingectomy, which is performed on an outpatient basis, is the *laparoscopy*. Often using local anesthesia, the surgeon inserts a small, lighted viewing instrument (laparoscope) through the woman's abdominal wall just below the navel through which the uterus and the fallopian tubes can be seen. The surgeon then makes another small incision in the lower abdomen and inserts a special pair of forceps that carry electricity to cauterize the tubes. The laparoscope and the forceps are then withdrawn, the small wounds are closed with a single stitch, and small bandages are placed over the closed incisions. (Laparoscopy is also known as "the band-aid operation.")

As an alternative to reaching the fallopian tubes through an opening below the navel, the surgeon may make a small incision in the back of the vaginal barrel (vaginal tubal ligation).

These procedures for female sterilization are over 95 percent effective, but sometimes they have complications. In rare cases, a blood vessel in the abdomen is torn open during the sterilization and bleeds into the abdominal cavity. When this happens, another operation is necessary to find the bleeding vessel and tie it closed. Occasionally, injury occurs to the small or large intestine, which may cause nausea, vomiting, and loss of appetite. The fact that death may result, if only rarely, is a reminder that female sterilization is surgery and, like all surgery, involves some risks.

In addition, although some female sterilizations may be reversed, a woman should become sterilized only if she does not want to have a biological child.

NATIONAL DATA

Twelve percent of married men aged 20–39 have had a vasectomy (Forste, Tanfer, & Tedrow, 1995).

Male sterilization   Vasectomies are the most frequent form of male sterilization. They are usually performed in the physician's office under a local anesthetic. *Vasectomy* involves the physician making two small incisions, one on either side of the scrotum, so that a small portion of each vas deferens (the sperm-carrying ducts) can be cut out and tied closed. Sperm are still produced in the testicles, but since there is no tube to the penis, they remain in the epididymis and eventually dissolve. The procedure takes about fifteen minutes and costs about $800. The man can leave the physician's office within a short time. Men most likely to seek a vasectomy are in their 30s or older, have been married over eight years, and already have three children (Forste et al., 1995).

Since sperm do not disappear from the ejaculate immediately after a vasectomy (some remain in the vas deferens above the severed portion), a couple should use another method of contraception until the man has had about twenty ejaculations. The man is then asked to bring a sample of his ejaculate to the physician's office for examination under a microscope for a sperm count. In about 1 percent of the cases, the vas deferens grows back and the man becomes fertile again. In other cases, the man may have more than two tubes, which the physician was not aware of.

A vasectomy does not affect the man's desire for sex, ability to have an erection or an orgasm, amount of ejaculate (sperm comprise only a minute portion of the seminal fluid), health or chance of prostate cancer. Although in some instances a vasectomy may be reversed, a man should get a vasectomy only if he does not want to have a biological child.

Transition to Parenthood

We just want parents to give more weight to the family bond.
Sylvia Hewlett
Cornel West
The War Against Parents

Individuals who use contraception effectively are more able to control when they become parents. *Transition to parenthood* refers to that period of time from the beginning of pregnancy through the first few months after the birth of a baby. Developmentalists conflict, and symbolic interactionist theorists combine to view the transition to parenthood as a new stage in the family life cycle that involves a new allocation of resources for the couple and new definitions of each other.

We are the least child- and family-oriented society in the world.
T. Berry Brazelton
Pediatrician

As a childless married couple, the family was a dyad, a two-person system, in which each party had the other's undivided attention, affection, time, and energy. When the baby appears, the dyad becomes a triad, a three person system, and now scarce emotional resources must be shared with the new arrival. One spouse may feel great losses in this transition. "When I used to come home, my spouse had time and energy for me, we'd talk to each other; now there is nothing left for me at the end of the day." The other partner thinks, "When he came home, he used to recognize my existence, spend time with me, talk to me, and now he makes a beeline for the baby and spends hours playing and cuddling with the baby, ignoring me as if I wasn't even there, as if I didn't exist!" Each may feel abandoned and neglected as time, energy, attention, and affection are turned into the new arrival (Winton, 1995, 23).

*The "new" father is more
involved in the physical care of
children.*

Transition to Motherhood

Although pregnancy and childbirth are sometimes thought of as a painful ordeal, some women describe the experience as fantastic, joyful, and unsurpassed. Their expectations of motherhood often influence their reality.

A strong emotional bond between the mother and her baby usually develops early, so that mother and infant resist separation. Sociobiologists suggest that there is a biological basis for the attachment between a mother and her offspring. The mother alone carries the fetus in her body for nine months, lactates to provide milk, and produces oxytocin—a hormone from the pituitary gland during the expulsive stage of labor that has been associated with the onset of maternal behavior in lower animals.

Postpartum feelings Not all mothers feel joyous after childbirth. Emotional bonding may be temporarily impeded by a mild depression, characterized by irritability, crying, loss of appetite, and difficulty in sleeping. From 50 percent to 70 percent of all new mothers experience "baby blues"—transitory symptoms of depression twenty-four to forty-eight hours after the baby is born. About 10 percent experience *postpartum depression*—a more severe reaction than baby blues (Kraus & Redman, 1986).

Postpartum depression is believed to be a result of the numerous physiological and psychological changes occurring during pregnancy, labor, and delivery. Although the woman may become depressed in the hospital, she more often experiences these feelings within the first month after returning home with her baby. Most women recover within a short time; some (about 5 percent) seek therapy to speed their recovery. To minimize "baby blues" and postpartum depression, one must recognize that having misgivings about the new infant is normal and appropriate. In addition, the woman who has negative feelings about her new role as mother should elicit help with the baby from her family so that she can continue to keep up her social contacts with friends.

Choosing priorities For some women, motherhood is the ultimate fulfillment; for others, the ultimate frustration. Priorities must be established. When forced to choose between her job and family responsibilities (the babysitter does not show up, the child is sick or hurt, etc.), the employed woman and mother (unlike the man and father) generally prioritizes the role of mother over the role of employee (Grant et al., 1990; LaRossa, 1988; Mischel & Fuhr, 1988). Women report higher levels of anger than men because of the inequities in child care and household responsibilities (Ross & Van Willigen, 1996). Nevertheless most women, particularly mothers in their first marriage, report that motherhood is a profoundly happy time (Demo & Acock, 1996).

Transition to Fatherhood

Husbands vary widely in their transition to fatherhood. While some are enthralled at the birth of their children and actively participate in child care, others feel ill prepared and spend more time at work. The latter is in reference to their primary role identity of father-providers. Blankenhorn (1995) emphasizes

RECENT RESEARCH

Neville (1997) studied 60 families in which both parents were either younger than 26 or older than 29 when they began childbearing and whose child was between the ages of 3 and 5 at the time of the study. Younger fathers were more likely to engage their children in physical play while older fathers were more likely to engage their children verbally.

RECENT RESEARCH

In a national sample of 1,601 adults, parents report higher levels of psychological distress than people without children younger than 18 living at home. Children themselves do not increase distress, but they are associated with increased economic hardship and difficulty arranging child care (Bird, 1997).

RECENT RESEARCH

Interview data from a probability sample of 3,407 white and African-American adults in 21 cities revealed that those who were parents were less happy in their relationships than those without children (Tucker, 1997).

RECENT RESEARCH

Reimann (1997) examined the division of labor of lesbian couples in their transition to parenthood and observed that conflict erupted whenever one partner perceived the other as not doing equal domestic work.

that this role has far-reaching effects on children: "as fatherlessness spreads, the economic difference between America's have and have-nots will increasingly revolve around a basic question: Which of us had fathers?" (p. 45).

The importance of the father in the lives of his children goes beyond his economic contribution (Knox, 1998; Popenoe 1996; Lambert, 1995). Children from intact homes or those in which fathers maintained an active involvement in their lives after divorce tend to:

- make good grades
- have good health
- report satisfied durable marriages as adults
- report higher life satisfaction
- have higher education levels
- have stable jobs
- have lower reported child sex abuse
- be less involved in crime
- have a strong work ethic
- have a strong moral conscience
- have higher incomes as adults
- form close friendships
- have fewer premarital births

Transition from a Couple to a Family

Researchers disagree on the effect of children on marital happiness.

Children decrease marital happiness Some research suggests that parenthood decreases marital happiness. A team of researchers (Lavee et al., 1996) studied 287 intact couples who had children living at home and concluded that the stress associated with rearing children affected both their psychological well-being and perceived marital quality negatively. The more economically stressed, the greater the effect. Cowan and Cowan (1992) followed seventy-two expectant couples and twenty-four childfree couples for ten years. They noted a decrease in relationship satisfaction among parents as a result of unfulfilled expectations, different patterns of engagement into the role of parent, and different perceptions of their role as lovers/partners. Both partners were surprised that the baby did not bring them closer together. The husbands viewed themselves less in the role of the parent than did the wives, and the wives viewed themselves less in the role of lover than did the husbands. The greater the discrepancies, the greater the unhappiness.

Most couples experience a pattern of decreased happiness that bottoms out during the teen years and gradually improves. When the children leave home, parents typically report increased relationship satisfaction, though it is not as high as pre-baby levels.

Regardless of how children affect the feelings spouses have about their marriage, spouses report more commitment to their relationship once they have children (Stanley & Markman, 1992). Figure 8.5 illustrates that the more children a couple have, the more likely the couple will stay married. A primary reason for this increased commitment is the desire on the part of both parents to provide a stable family context for their children. In addition, parents of dependent children may keep their marriage together to maintain continued access to and a higher standard of living for their children. Finally, people (especially mothers) with small children feel more pressure to stay married

Cowan and Cowan (1992) have developed a program designed to assist couples in preparing for the changes that occur in the transition from spouse to parent. The following are some of their suggestions:

1. Share expectations. Discuss private notions about one's ideal family. Also share any anxieties about the new role and its impact on one's self, the relationship, and work.

2. Make time for talk, sex, and togetherness. Since the demands of the new baby will take time away from the couple, it is important to make time for communication, sex, and doing things together "even if the laundry or dinner dishes have to wait" (p. 78).

3. Don't be afraid of conflict. "Regard a fight as information that something is wrong in the relationship. The trick is not to worry that you are having a struggle, or to avoid a fight" (p. 78).

4. Talk with a friend or coworker. Talking with others who have experienced the transition from spouse to parent helps alleviate one's feelings of aloneness and isolation. Be careful, however, not to divulge information that may result in the spouse's feeling betrayed. Some mental health centers offer parenting groups to discuss parenting concerns.

Figure 8.5
Percentage of couples getting divorced by number of children

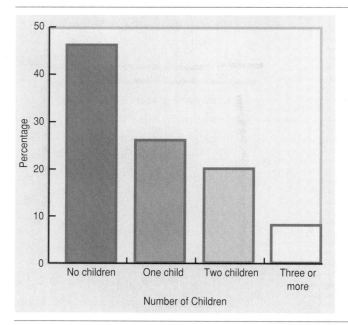

regardless of how unhappy they may be. Hence, while children may decrease happiness, they increase stability, since pressure exists to avoid divorce.

Children do not affect marital happiness   Some research suggests that children neither increase nor decrease marital happiness. In a study comparing married couples who had children with those who did not, the researchers observed that, over time, the spouses in both groups reported declines in love feelings, marital satisfaction, doing things together, and positive interactions. The parents were no less happy in their marriages than the childfree couples. The researchers concluded that "the transition to parenthood is not an inescapable detriment to marital quality" (MacDermid, Huston, & McHale, 1990, 485).

While the degree of change in marital satisfaction due to the birth of a baby remains uncertain, changes in the division of labor are more certain. Kluwer (1997) studied the division of labor of 293 Dutch couples at three time intervals (during pregnancy, six months postpartum, and two years postpartum) and found that wives took on a larger share of housework and child care and a smaller amount of paid labor than husbands across time. A pattern of wives' demanding more help, husbands' withdrawing, and mutual avoidance of conflict was also observed across time.

Parenting

The role of parenthood is like none other and is one for which there is little preparation. A review of the facts about parenthood follow.

Facts about Parenthood

There are at least five basic facts about parenthood that often go unnoticed.

Parenthood is only one stage in an individual's life Parents of newly adult children often lament, "Before you know it, your children are grown and gone." Although parents of infants sometimes feel that the sleepless nights will never end, they do end. Unlike the marriage relationship, the parent-child relationship inevitably moves toward separation. Just as the marital partners were alone before their children came, they will be alone again after their children leave. Except for occasional visits with their children and possibly with grandchildren, the couple will return to the childfree lifestyle.

Typical parents are in their early 50s when their last child leaves home. This means that spouses in first marriages have over twenty years together after their children leave home. Parenthood might best be described as only one stage in marriage and in life. Assuming individuals marry at age 24, have two children at three-year intervals, and die at 77, they will have children living with them about 30 percent of their lifetime and 40 percent of their marriage. One parent said:

> We had three kids, and I loved taking care of all of them. I think the happiest time in my life was when my husband and I would wake up in the morning and they would all be there. But that's changed now. They are married and have moved several states away. I know they still love me and they call to stay in touch, but I rarely see them anymore (authors' files).

Parents are only one influence Although parents often take the credit—and the blame—for the way their children turn out, they are only one among many influences on child development. From a macro (societal) perspective, "even the most well-intentioned and committed parents cannot make the schools less competitive, the streets less violent, and the media less vulgar" (Elkind, 1995, 28).

A micro (individual) perspective emphasizes the influence of peers, siblings, teachers, and television viewing as continuous influences. Although parents are the first significant influence, peer influence becomes increasingly important and remains so into the college years. During this time, children are likely to mirror the values and behaviors of their friends and agemates.

Siblings also have an important and sometimes lasting effect on each other's development. Siblings are social mirrors and models (depending on the age) for each other. They may also be sources of competition for each other and be jealous of each other.

Television, replete with MTV and "parental discretion advised" movies, is a major means of exhibiting language, values, and lifestyles to children that may be different from those of the parents. One father had Home Box Office and Showtime disconnected because he did not want his children seeing the movies and specials on those channels. Another parent went through the television guide each week and marked the programs he would not allow his children to watch. The guide was left on top of the television, and the children were to look at what programs had been approved before they turned the TV on.

To minimize the danger to children who surf the Internet, parents should establish some rules. These include instructing them to:

1. never give out personal information such as address, phone number, or anything that could be used to locate them;

2. never give their passwords to anyone, even their best friends or somebody who says they're with the on-line service staff;

3. avoid using crude, offensive, or threatening language that could get them and the family in trouble.

Another concern parents have of external influence is the Internet. Jupiter Communications Research predicts that by the year 2002, 75 percent of all teenagers will be on-line (Thomas, 1997). Though parents invite their children to conduct research and write term papers using the Internet, they are wary of the sex web sites and chat rooms. Parental supervision and teen privacy are potential conflict issues. The following Internet addresses provide help about Internet Issues: www.netparents.org and www.missingkids.org

Children are also influenced by different environmental situations. An only daughter adopted into an urban, Catholic, upper-class family will be exposed to a different environment than a girl born into a rural, Southern Baptist, working-class family with three male children. Some of the important environmental variables that influence children include geographic location, family size, the family's social class, religion, and racial or ethnic background.

Genetic makeup and physiological wiring in the individual child are also important influences on behavior. Reiss (1995) observed the genetic influence on cognitive functioning, personality differences, and psychopathology. Such genetic processes not only affect children but also affect the way parents respond to children.

OTHER CULTURES   Parents in other societies are also concerned about the various influences on their children. Isla Mujures is a small island off the coast of Mexico near Cancun. A study of parental concerns on the island revealed a desire that their children get a good education and avoid becoming violent (Knox & Schacht, 1997). ●

Think of the tragedy of teaching children not to doubt.
Clarence Darrow

Parenthood demands change as children grow up   The demands of parenthood change as the children move through various developmental stages. Infants, toddlers, preschoolers, preadolescents (8–11), new adolescents (12–15), middle to late adolescents (15–18), and young adults (18–22) all exhibit different behaviors and require different emotional, social, and psychological resources from parents. Over time, the child is growing from a state of total dependence to one of total independence. While developmental theorists emphasize the importance of both parents and offspring in negotiating this transition, conflict theorists note how parents and young adults compete for power and resources. Each often wants to control the other in terms of power, and they often have different ideas about how resources (e.g., money, space) should be allocated. Aquilino (1997) found that conflict between parents and children decreases when the children move away and live independently.

As children grow into adulthood, it is an important developmental task for parents to let their children become independent and to redirect their energy and attention away from their children to their own lives.

Each child is different   Children differ in their tolerance for stress, in their capacity to learn, in their comfort in social situations, in their interests, and in innumerable other ways. Parents soon become aware of the specialness of each child—of her or his difference from every other child they know and from children they have read about. Parents of two or more children are often amazed

Now kids have access to money, drugs, alcohol, and weapons. Things have changed and they have opportunities to do these terrible things.

Carol Freeman

Psychologist

at how children who have the same parents can be so different. Some differences between children may be due to differences in biology and differential gender role socialization.

Children also differ in their mental and physical health. Over 250,000 babies are born each year with birth disabilities, including 20,000 severely retarded, 16,000 with a profound hearing loss, and 8,000 with cerebral palsy. Mental and physical disabilities of children present emotional and financial challenges to their parents.

Parenting styles differ   Parenting styles vary along a continuum from authoritarian to permissive. Authoritarian parents demand obedience from their children and severely punish disobedience. They expect their children to be responsive to their demands, but they are also responsive to the demands of their children. These parents also explain their rules and provide rationales for rules and regulations. Permissive parents allow their children to do as they please. There is little conflict, because permissive parents defer to their children's wishes. Democratic parents are between the two extremes. They try to negotiate differences with their children and involve their children in the process. Authoritative parenting (regardless of the outcome assessed, yields the most favorable outcome) (Lamborn et al., 1991).

Folklore about Childrearing

Since a society cannot survive without new socialized individuals to replace the dying members, it is imperative that having and rearing children be sufficiently romanticized so as to attract a sufficient number of spouses into the full-time role of parents. LeMasters and DeFrain (1989) suggested that certain myths permeate the culture to increase the acceptability of the parenting role. Some of these are the following.

Myth 1: Rearing children is always fun   Would-be parents see television commercials of young parents and children and are led to believe that playing in the park with their 4-year-old is what childrearing is all about. Parenthood is portrayed as being a lot of fun. The truth is somewhat different from the folklore:

> Rearing children is hard work; it is often nerve-racking work; it involves tremendous responsibility; it takes all the ability one has (and more); and once you have begun, you cannot quit when you feel like it (LeMasters & DeFrain, 1989, 22, 23).

Myth 2: Good parents inevitably produce good kids   It is often assumed that children who turn out "wrong"—who abuse drugs, steal, and the like—have parents who really did not do their job. We tend to blame parents when children fail. But good parents have given both their emotional and material resources to their children and the children have not turned out well. One mother said:

> We live in one of the finer suburbs of our city, our children went to the best schools, and we spent a lot of time with them as a family (camping,

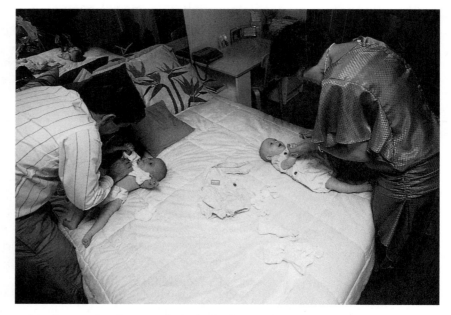

While this couple is happy with the birth of twins, they report that taking care of two infants is not "always fun."

going to the beach, skiing). But our son is now in prison. He held up a local grocery store one night and got shot in the leg. We've stopped asking ourselves what we did wrong. He was 23 and drifted into friendships with a group of guys who just decided they would pull a job one night (authors' files).

Myth 3: Love is the essential key to effective childrearing   Parents are taught that if they love their children enough, their children will turn out okay. Love is seen as the primary ingredient that, if present in sufficient quantities, will ensure a successful child. But most parents love their children dearly and want only the best for them. Love is not enough and does not guarantee desirable behavior. One parent said:

We planned our children in courtship, loved them before they got here, and have never stopped loving them. But they are rude, irresponsible, and hardly speak to us. We are frustrated beyond description. We've done everything we know how to do in providing a loving home for them, but it hasn't worked (authors' files).

Myth 4: Children are always appreciative   Most parents think of childrearing in terms of love, care, and nurturing—and also in terms of giving their children things. Parents may assume their children will appreciate their tender loving care and the material benefits, like clothes, computers, and cars, they bestow. That assumption may be wrong. Children often think parents are supposed to love them and give them things. They view material benefits as their birthright.

Myth 5: Parenting comes naturally   The lack of attention given to systematic parent education in our society (there are more driver education than parent education courses) reflects a belief that when one has a child, what the parent needs to know to take care of the child and rear the child will come naturally. Cooke (1991) challenged this belief by interviewing both novice and expert mothers of 6- to 10-month-old infants. The novice mothers had no previous children, no previous experience with children, and no formal education related to child development, childrearing, or related areas. The expert mothers had at least one older child, extensive experience with other children, and formal education in child development and childrearing.

Cooke observed clear differences between the thinking, knowledge, and behaviors of novice and expert mothers. In general, novices were less able to identify the cues provided by the child with regard to the child's goals and needs and were less knowledgeable about child development. For example, the novices did not know what behaviors it was appropriate to expect of a child at what age. Cooke recommended that new parents be exposed to parent education learning experiences, which provide skills in focusing on children's needs and goals, and become familiar with child development literature.

Myth 6: Family values are easy to instill   *Family values* is a concept most parents want to instill in their children. Cox (1992) identified what this term typically means:

1. Strong respect for other people; an appreciation for the differences that others bring.

2. The ability to discuss differences and find peaceful and cooperative means to resolve differences.

3. Sticking with something even when it becomes tough; working through difficult areas to build character, loyalty, and respect.

4. Making and keeping commitments.

5. Maintaining one's personal integrity at all times.

6. Being thoughtful toward others and providing a helping hand whenever possible.

7. Being aware of community needs and one's ability to provide some service to that community.

Principles of Effective Parenting

Most undergraduate students are aware that clear communication with and active listening to one's child are important parenting skills (Dickson & Dukes, 1992). Other principles of effective parenting include providing love and praise, discipline, and security, encouraging responsibility, and expressing confidence in the child. Parents might also want to consider supportively coparenting their children.

Give love and praise   Since children first depend on their parents for the development of their self-concept, it is crucial that parents communicate that they love their children and provide them with positive reinforcement. Instead of focusing only on correcting or reprimanding "bad" behavior, parents should frequently comment on and reinforce "good" behavior. Comments like "I like the way you shared your toys," "You asked so politely," "I am so proud of you for telling me the truth," and "You did such a good job cleaning your room" enhance a child's self-concept. Renowned family therapist Virginia Satir noted over twenty years ago the importance of having a positive sense of self-worth or self-esteem:

> [T]he crucial factor in what happens both inside people and between people is the picture of individual worth that people have of themselves. . . . Integrity, honesty, responsibility, compassion, love—all flow easily from the person who has high self-esteem or self-worth (1972, 22).

Set limits and discipline inappropriate behavior   While looking for opportunities to reinforce desired behavior, parents also must provide limits to children's behavior. This sometimes involves punishing negative behavior. Unless parents provide negative consequences for lying, stealing, and hitting, children can grow up to be dishonest, to steal, and to be inappropriately aggressive. Time out (removing the child from being with others to a place of isolation for one minute for each year of the child's age) has been shown to be an effective consequence for inappropriate behavior. Withdrawal of privileges (watching television, playing with friends) is also effective. Physical punishment is less effective in reducing negative behavior; it teaches the child to be aggressive and encourages negative emotional feelings toward the parents. When using time out or the withdrawal of privileges, parents should make it clear that they disapprove of the child's behavior, not the child.

Provide security   Predictable responses from parents, a familiar bedroom or playroom, and an established routine help to encourage a feeling of security in children. Security provides children with the needed self-assurance to venture beyond the family. If the outside world becomes too frightening or difficult, a child can return to the safety of the family for support. Knowing it is always possible to return to an accepting environment enables a child to become more involved gradually with the world beyond the family.

Encourage responsibility   Giving children increased responsibility encourages the autonomy and independence they need to be assertive and independent. Giving children more responsibility as they grow older can take the form of encouraging them to choose healthy snacks and letting them decide what to wear and when to return from playing with a friend (of course, the parents should praise appropriate choices). Children who are not given any control and responsibility for their own lives remain dependent on others. Successful parents can be defined in terms of their ability to rear children who can function as independent adults. One way to ensure such success is to give children increasing responsibility as they get older.

Express confidence   "One of the greatest mistakes a parent can make," confided one mother, "is to be anxious all the time about your child, because the child

NATIONAL DATA
Sixty-three percent of adults in a survey conducted by the Massachusetts Mutual Life Insurance Company "strongly agreed" that parents today are too lenient and permissive with their children (Carey & Stacey, 1995).

INSIGHT
Although providing negative consequences for inappropriate behavior is important, it is more important to notice and give attention to appropriate behavior. Rather than look for and punish lying, stealing, and aggressiveness, it is often more effective to look for and comment on telling the truth, being honest, and negotiating.

RECENT RESEARCH
Leve and Fagot (1997) compared 67 two-parent families, 32 single-mother families, and 13 single-father families and found that single-parent families reported more positive behavior from their children and reported using more problem-solving strategies.

INSIGHT

African-American parents sometimes encounter unique issues in providing a feeling of security for their children. Growing up in a racist society may engender feelings of insecurity if African-American children are not taught to view black as beautiful, to confront racism appropriately, and to be proud of their racial heritage. These and other concerns are addressed in the book *Raising Black Children* (Comer & Poussaint, 1993).

INSIGHT

Elkind notes that "the needs of children and adolescents for protection and security, guidance, limit setting, and monitoring are being weighted less heavily than are the needs of the adults in our society" (1995, 28). His observation suggests that we have gone too far in our quest for individualism and that an absence of familistic values will have its price in terms of children and adolescents who are not provided a context within which to reach their full potential. Elkind recommends a new family pattern "that provides a more nearly equal balance between the needs of children and youths and those of parents and adults" (1995, 28).

interprets this as your lack of confidence in his or her ability to function independently." Rather, this mother noted that it is best to convey to the child that you know that he or she will be all right and that you are not going to worry about the child because you have confidence in him or her. "The effect on the child," said this mother, "is a heightened sense of self-confidence." Another way to conceptualize this parental principle is to think of the self-fulfilling prophecy as a mechanism that facilitates self-confidence. If the parents show the child that they have confidence in him or her, the child begins to accept these social definitions as real and becomes more self-confident.

Coparent Gable and colleagues observed that "new parenthood calls for husbands and wives to work together, coordinate attitudes and beliefs, and to respect, and perhaps negotiate between, one another's unique ideas about raising children" (Gable, Belsky, & Crinic, 1995, 609). Gable and colleagues studied the degree to which parents coparent. They operationally defined coparenting as the following:

> Supportive coparenting occurs either when parents explicitly or implicitly agree with each other by voicing the same general message to the child, or when one parent directly asks the other for assistance with child care and the request is granted. Unsupportive coparenting takes place when one parent subtly, or not so subtly, undermines the other parent's efforts with the child; when one parent interrupts the ongoing interaction of the parent and child; or when a direct request for help with child care is denied (Gable, Crinic, & Belsky, 1994, 382).

A specific example of what is meant by a supportive event follows:

> Mom feeds Sam some food, pleasantly saying, "Bite, bite." Sam, however, does not eat the food this time. He begins to bang the tractors together. Mom seriously says, "Hey, I'm gonna take the tractor if you keep doing that." Dad follows this by sternly saying, "Hey, I'm not going to let you play with them if you keep crashing them together." Sam stops crashing the tractors together (Gable, Crinic, & Belsky, 1994, 386).

Sixty-nine Caucasian, maritally intact, middle- and working-class families raising sons across the second and third year of life were observed. An average of twenty coparenting events across two hours of family observation were recorded. Sixty percent of these events were identified as supportive; 20 percent to 25 percent were identified as unsupportive. The remaining 15 percent to 20 percent were both supportive and unsupportive. Over a fifteen- to twenty-one-month period, the average frequency of supportive exchanges remained unchanged, unsupportive decreased, and mixed increased (Gable, Belsky, & Crinic, 1995). Whether parents engaged in coparenting behaviors was related more to their personality traits (extroversion and interpersonal sensitivity) than to their age, education, or childrearing attitudes. In addition, the greater the stress in the family, the less likely spouses were to coparent.

Ehrensaft (1990) discussed "shared" parenting whereby both parents act as the "primary" parent. In effect, from the child's point of view, shared parenting means that mother and father are interchangeable. Whenever one parent is

Doing something together may help parents and children communicate during the children's teen years.

Children begin by loving their parents. After a time they judge them. Rarely, if ever, do they forgive them.
Oscar Wilde
Anglo-Irish playwright

there, the child feels that a complete parent is with the child. Traditionally, when the mother went to the store and left the children with the father, the children (and the parents) defined the situation as a "baby-sitting" arrangement until the "real" parent returned. Because an increasing number of families include both spouses working outside the home at different times, it becomes increasingly important that the parents and children view parenting as shared.

When Children Become Teenagers and Adults

The demands of parenthood change as children age. Parenting teenage and adult children presents challenges that differ from those in parenting infants and young children.

Children as teenagers The teenage years have been characterized as a time when adolescents defy authority, act rebellious, and search for their own identity. Teenagers today are no longer viewed as innocent, naive children. Elkind noted, "Our new, postmodern perception of adolescents is that of sophistication. We now look at adolescents as sophisticated in matters of sex, drugs, media, and computer technology" (1995, 27).

Conflicts between parents and teenagers often revolve around money and independence. As to money, 15- to 17-year olds spend $43 per week, in contrast to 9- to 11-year olds, who spend $4.80 per week (Stipp, 1993). An increase in parent-child conflict is not inevitable during the teenage years. In a study of mothers, fathers, and adolescents in eighty families, the researchers observed no increase in conflict during early adolescence (Galambos & Almeida, 1992). Even when conflict exists, it is less likely to be over substantive issues such as sex and drugs and more likely to be over such issues as chores and dress (Barber, 1994). The following suggestions can help to keep conflicts with teenagers at a low level.

1. Catch them doing what you like rather than criticizing them for what you don't like. Adolescents are like everyone else—they don't like to be criticized but do like to be noticed for what they do that is good.

2. Ignore some things. One adult said that when he was 13, he stayed up late, stole one of his father's cigarettes, and smoked it while watching television. Although he thought that his father was asleep, he was surprised by his father, who walked into the room where he was smoking. When his father saw his teenager smoking, he said nothing, turned around, and went back to bed. The father never spoke of the incident. The effect on the adolescent was to feel the tolerance for experimentation from his father that he wanted. After finishing the cigarette, he never smoked again.

3. Provide information rather than answers. When teens are confronted with a problem, try to avoid making a decision for them. Rather, it is helpful to provide information on which they may base a decision. What courses to take in high school and what college to apply for are decisions that might be made primarily by the adolescent. The role of the parent might best be that of providing information or helping the teenager obtain information.

4. **Be tolerant of high activity levels.** Some teenagers are constantly listening to loud music, going to each other's homes, and talking on the telephone for long periods of time. Parents often want to sit in their easy chairs and be quiet. Recognizing that it is not realistic to expect teenagers to be quiet and sedentary may be helpful in tolerating their disruptions.

5. **Engage in some activity with your teenagers.** Whether it is renting a video, eating a pizza, or taking a camping trip, it is important to structure some activities with your teenagers. Such activities permit a context in which to communicate with them.

Rueter and Conger (1995) emphasized that a supportive and communicative context helps to diffuse disagreements between parents and adolescents.

Sometimes teenagers present parents with challenges beyond which the parents feel able to cope. Family therapy may be helpful. A major focus of such therapy is to increase the emotional bond between the parents and the teenagers and to encourage consequences for expected behavior.

Some parents have joined TOUGHLOVE. This is a self-help organization of parents (none of whom have professional qualifications) who have difficulty controlling severe problem behaviors of teenagers such as staying away from home without explanation, drug abuse, stealing from family members, physical abuse of parents, and using obscene language to parents. The larger community, consisting of teachers, probation officers, social workers, therapists, and citizens, may also be involved in helping parents in TOUGHLOVE. For example, a child who takes drugs and has a history of lying about doing so may be taken to school by the parents, be watched carefully at school by the teacher, have weekly meetings with a caseworker, and be taken home by another member of the TOUGHLOVE group. The community pulls together to try to help the parents control their child's negative behavior. The emphasis is not on blaming anyone but on correcting the behavior problem. There are more than 1,500 chapters of TOUGHLOVE in the United States and other countries. Information about a chapter in your community can be obtained from the cofounders of TOUGHLOVE, David and Phyllis York, (P.O. Box 1069, Doylestown, PA 18901, (215) 348-7090.

Children as adults Zarit and Eggebeen emphasized that parenting continues well beyond the teenage years into adulthood.

> Parent-child relationships are a life-span issue. Rather than ceasing when children are launched from the family, these relationships endure with often complex patterns of interaction, support, and exchange that wax and wane around key transitions in the adult years. Indeed, family issues such as intergenerational conflict, mutual assistance, and inheritance have a timeless feel to them (1995, 119).

As a result of the difficulty of their offspring's getting and maintaining a job, the high probability that their children will have children of their own, and the possibility of their children's getting a divorce, parents continue to be important sources of economic and emotional support for their children. Parents are most likely to function in the role of advice givers, but they also provide money,

Assessing Motivation for Parenthood

Individuals have various motivations for wanting children. The following provides a way to measure some of these motivations.

Indicate the answer you feel is the best by placing a one (1) in front of it. Rank the remaining answers (2, 3, and 4) to show your order of preference.

1. Parents expect their children
3 F (2) to fulfill the purpose of life.
2 I (3) to strengthen the family.
1 A (1) to be healthy and happy.
4 N (4) to follow in their footsteps.

2. Men want children because
4 N (4) they would like to prove their sexual adequacy.
2 F (1) it is a natural instinct.
3 I (3) they need them to enhance their social status.
1 A (2) they like children.

3. A mother expects her daughter
2 I (3) to give her companionship and affection.
3 F (2) to take the place in the world for which she is destined.
4 N (4) to be like herself.
1 A (1) to be happy and well.

4. Men want children because
3 I (2) children hold the marriage together.
1 A (4) they like to care and provide for children.
2 F (1) it is a function of the mature adult.
4 N (3) they want to perpetuate themselves.

5. A father expects his son
1 A (2) to be happy and well.
3 F (3) to take the place in the world for which he is destined.
2 I (1) to give him companionship and affection.
4 N (4) to be like himself.

6. Women want children because
1 A (1) they like children.
3 I (3) they need them to enhance their social status.
4 N (4) they would like to prove their sexual adequacy.
2 F (2) it is a natural instinct.

7. Generally, people want children because
4 F (2) they are destined to reproduce.
1 A (4) they desire to help someone grow and develop.
3 N (3) they can create someone in their own image.
2 I (1) they provide companionship.

8. A father expects his daughter
3 N (4) to believe in him.
1 A (2) to be happy and well.
4 F (1) to take her place in the world.
2 I (3) to give him companionship and affection.

9. Women want children because
3 I (3) children hold the marriage together.
2 F (4) it is a function of the mature adult.
1 A (1) they like to care and provide for children.
4 N (2) they want to perpetuate themselves.

10. Women want children because
4 F (3) they are destined to reproduce.
1 A (2) they desire to help someone grow and develop.
2 I (1) they provide companionship.
3 N (4) they create someone in their own image.

11. Generally, people want children because
1 A (2) they like to care and provide for children.
4 N (3) they want to perpetuate themselves.
3 I (4) children hold the marriage together.
2 F (1) it is a function of the mature adult.

12. A mother expects her son
4 F (2) to take his place in the world.
2 I (3) to give her companionship and affection.
1 A (1) to be happy and well.
3 N (4) to believe in her.

13. Men want children because
2 I (1) they provide companionship.
3 N (2) they create someone in their own image.
4 F (4) they are destined to reproduce.
1 A (3) they desire to help someone grow and develop.

14. Generally, people want children because
2 F (1) it is a natural instinct.
1 A (2) they like children.
3 I (3) they need them to enhance their social status.
4 N (4) they would like to prove their sexual adequacy.

F – 41
I – 34
A – 14
N – 51

F – 29
I – 34
A – 28
N – 49

SCORING: The Child Study Inventory is composed of sentences related to motivations for parenthood. Each sentence is followed by four completion choices. Each choice following the relevant sentence can be categorized into one of the basic CSI motivational categories: Altruistic (A), Narcissistic (N), Fatalistic (F), Instrumental (I).

Scoring is accomplished for each of the four categories by summing the rankings of all completion choices in that category (e.g., all rankings of altruistic choices to obtain an altruistic motivation score). A low score indicates high preference for a given category.

For a more detailed subdivision of scores, the user may score "motivational" items (#2, 4, 6, 7, 9, 10, 11, 13, 14) and "expectancy" items (#1, 3, 5, 8) separately.

References

Counte, M. A., et al. Factor structure of Rabin's Child Study Inventory. *Journal of Personality Assessment,* 1979, *43,* 59–63.

Gordon, R. S. Assessing motivation for parenthood of adult adoptees and adoptive parents. The Wright Institute Studies of Psychiatry. Berkeley, California, 1988.

Rabin, A. I. Motivation for parenthood. *Journal of Projective Techniques & Personality Assessment,* 1965, *29,* 405–411.

Rabin, A. I., and R. J. Green. Assessing motivation for parenthood. *Journal of Psychology,* 1968, *69,* 39–46.

Source: "The Child Study Inventory" (An instrument to assess motivation for parenthood) Developed by A. I. Rabin (Professor Emeritus) and Robert J. Greene, Department of Psychology, Michigan State University, East Lansing, MI 48824-1117. Reprinted by permission of Dr. Rabin.

child care, and household assistance. Support is particularly forthcoming at the time when one's child has a child of his or her own.

All racial groups provide support for their offspring, with women being more involved in the role of "kinkeeping activities." Support tends to wane as the parents age. For example, parents' own health has "an impact on giving to children" (Zarit & Eggebeen, 1995, 126).

GLOSSARY

emergency contraception   Also referred to as postcoital contraception, the various types of morning-after pills that are used in three circumstances: when a woman has unprotected intercourse, when a contraceptive method fails (such as condom breakage), and when a woman is raped.

female condom   A condom that fits inside the woman's vagina to protect her from pregnancy, HIV infection, and other STDs.

hysterectomy   Removal of a woman's uterus.

infertility   The inability to achieve a pregnancy after at least one year of regular sexual relations without birth control, or the inability to carry a pregnancy to a live birth.

laparoscopy   A form of salpingectomy (tubal ligation) that involves a small incision through the woman's abdominal wall just below the navel.

mifepristone   Also known as RU-486, a synthetic steroid that effectively inhibits implantation of a fertilized egg.

oophorectomy   Removal of a woman's ovaries.

periodic abstinence   Refraining from sexual intercourse during the one to two weeks each month when the woman is thought to be fertile.

postpartum depression   Severe depression experienced by the mother following the birth of her baby; it is characterized by crying, loss of appetite, and difficulty in sleeping.

pronatalism   A social philosophy that values children and encourages couples to have them.

salpingectomy   Tubal ligation, or tying a woman's fallopian tubes, to prevent pregnancy.

TOUGHLOVE   A self-help organization of parents (none of whom profess to have professional qualifications other than experience) who have difficulty controlling severe problem behaviors of their teenage children.

transition to parenthood   The period of time from the beginning of pregnancy through the first few months after the birth of a baby.

vasectomy   Male sterilization involving cutting out small portions of the vas deferens.

SUMMARY

The decision whether to become a parent is one of the most important decisions you will ever make. Unlike marriage, parenthood is a role from which there is no easy withdrawal. Individuals may try out marriage by living together, but there is no such trial run for would-be parents.

Do You Want to Have Children?

The decision to become a parent is encouraged (sometimes unconsciously) by family, friends, religion, government, and cultural observances. The reasons people give for having children include personal fulfillment and identity and the desire for a close affiliative relationship.

Some couples opt for the childfree lifestyle. Reasons women give for wanting to be childfree are more personal freedom, greater time and intimacy with their spouses, and career demands. Husbands are generally less committed than their wives to remaining childfree. Couples who want but cannot have children are devastated. About 15 percent of couples are infertile.

Contraception

The primary methods of birth control are contraception and sterilization. With contraception, the risk of becoming pregnant can be reduced to practically zero, depending on the method selected and how systematically it is used. Contraception includes birth control pills, which prevent ovulation; the IUD, which prevents implantation of the fertilized egg; condoms and diaphragms, which are barrier methods, vaginal spermicides, periodic abstinence, and emergency contraception. These methods vary in effectiveness and safety.

Sterilization is a surgical procedure that prevents fertilization, usually by blocking the passage of eggs or sperm through the fallopian tubes or vas deferens, respectively. The procedure for female sterilization is called salpingectomy, or tubal ligation. Laparoscopy is another method of tubal ligation. The most frequent form of male sterilization is vasectomy.

Transition to Parenthood

Parenthood marks a major change in a couple's relationship as they orient their attention from each other to their baby. Planning helps to ease the transition. The transition to parenthood occurs during the period from the beginning of pregnancy through the first few months after the birth of the baby. The transition is usually more profound for the mother, whose hormonal system is altered, who may experience postpartum depression, and who is confronted with reordering her priorities. Most women tend to place family considerations above career considerations.

Fathers are crucial to the development of their children. Children who have involved fathers make better grades, report higher life satisfaction, and have more stable marriages as adults.

Parenting

The active participation of fathers in the lives of their children is one of the greatest predictors of positive outcomes for children. While over 90 percent of a sample of college students reported that they wanted to be actively involved in the lives of their children, only half reported that their fathers had been active. In general, fathers tend to become more active as their children age. When U.S. fathers are involved with their children, it is more often with their sons. In most cases, wives and mothers are very influential in whether fathers have a close relationship with their children. Research findings regarding how children affect marital happiness are inconsistent. However, having children is associated with greater commitment and marital stability.

Most child specialists emphasize the importance of following various principles in rearing children. These principles include giving consistent love and praise, setting limits and disciplining inappropriate behavior, providing security, encouraging responsibility, and telegraphing confidence. Coparenting (sharing the responsibility with one's mate) also has decided benefits for the child. The teen years may be particularly challenging for parents, who should strive to keep the communication channels open during these years. Parents may also have particular challenges when their children become adults.

REFERENCES

Ahlburg, Dennis A., and Carol J. De Vita. 1992. New realities of the American family. *Population Bulletin* 47: 2–44.

Alexander, N. J. 1995. Future contraceptives. *Scientific American.* September, 136–41.

American Council on Education and University of California. 1997. *The American freshman: National norms for fall, 1997.* Los Angeles: Los Angeles Higher Education Research Institute.

Aquilino, William S. 1997. From adolescent to young adult: A prospective study of parent-child relations during the transition to adulthood. *Journal of Marriage and the Family* 59: 670–86.

Barber, B. K. 1994. Cultural, family, and personal contexts of parent-adolescent conflict. *Journal of Marriage and the Family* 56: 375–86.

Beets, G. C. N., A. C. Liefroer, and J. D. Jong Gierveld. 1997. Combining employment and parenthood: A longitudinal study of intentions of Dutch young adults. *Population Research and Policy Review,* 16: 457–74.

Bird, C. E. 1997. Gender differences in the social and economic burdens of parenting and psychological distress. *Journal of Marriage and the Family,* 59, 809–23.

Blankenhorn, David. 1995. *Fatherless America: Confronting our most urgent social problem.* New York: Basic Books.

Boyd, R. L. 1989. Minority status and childlessness. *Sociological Inquiry* 59: 331–42.

Buxton, Michael S., and James E. Deal. 1997. Infant effects on mothers' and fathers' adult development during the transition to parenthood. Paper presented at the Annual Conference of the National Council on Family Relations, Crystal City, Va., November.

Carey, A. R., and J. Stacey. 1995. Love is not tough enough? *USA Today*. 27 November, D1.

Casper, L. M. 1997. My daddy takes care of me! Fathers as care providers. U. S. Bureau of the Census, Washington, D.C. 20233. USA. Current Population Report No. P70–59.

Cohen, Sharon. 1995. Suits cite unexpected Norplant side effects. *Charlotte Observer*. 1 October, 19A.

Comer, J. P., and A. F. Poussaint. 1993. *Raising black children*. New York: NAL-Dutton.

Cooke, B. 1991. Thinking and knowledge underlying expertise in parenting: Comparisons between expert and novice mothers. *Family Relations* 40: 3–13.

Cowan, C. P., and P. A. Cowan. 1992. Is there love after baby? *Psychology Today*. July/August, 25, 58–63.

Cox, E. 1992. Strengthening our values. *CalFam*. Fall, 1–19.

Demo, David H., and Alan C. Acock. 1996. Singlehood, marriage, and remarriage. The effects of family structure and family relationships on mother's well-being. *Journal of Family Issues* 17: 388–407.

Dickson, L. F., and R. L. Dukes. 1992. The effects of gender and role context on perceptions of parental effectiveness. *Free Inquiry in Creative Sociology* 20: 11–24.

Dzindolet, M., Xiaolin Zie, and William Meredith. 1996. Marriage and family life attitude: Comparison of Chinese and American Students. Paper presented at the Annual Meeting of the National Council on Family Relations, Kansas City, Missouri, 9 November. Used by permission of Dr. William Meredith, Department of Consumer and Family Science, University of Nebraska-Lincoln.

Ehrensaft, D. 1990. *Parenting together: Men and women sharing the care of their children*. Urbana, Ill: University of Illinois Press.

Elkind, David. 1995. The family in the postmodern world. *Phi Kappa Phi Journal* 75: 24–28.

Forste, Renata, Koray Tanfer, and Lucky Tedrow. 1995. Sterilization among currently married men in the United States, 1991. *Family Planning Perspectives* 27: 100–107.

Foust, A. 1997. An investigation of voluntary childlessness and the marital satisfaction of couples over the life cycle. Paper presented at the annual conference of the National Council on Family Relations, Crystal City, VA, November.

Gable, S., J. Belsky, and K. Crinic. 1995. Coparenting during the child's second year: A descriptive account. *Journal of Marriage and the Family* 57: 609–16.

Gable, S., K. Crinic, and J. Belsky. 1994. Coparenting within the family system: Influences on children's development. *Family Relations* 43: 380–86.

Galambos, N. L., and D. M. Almeida. 1992. Does parent-adolescent conflict increase in early adolescence? *Journal of Marriage and the Family* 54: 737–47.

Giordano, P. C., S. A. Cernkovich, and A. DeMaris. 1993. The family and peer relations of black adolescents. *Journal of Marriage and the Family* 55: 277–87.

Glenn, N. D. 1991. Quantitative research on marital quality in the 1980s: A critical review. In *Contemporary families*, edited by Alan Booth. Minneapolis: National Council on Family Relations, 28–41.

Gold, M. A., A. Schein, and S. M. Coupey. 1997. Emergency contraception: A national survey of adolescent health experts. *Family Planning Perspectives* 29: 15–19.

Gollub, Erica L., Zena Stein, and Wafaa El-Sadr. 1995. Short-term acceptability of the female condom among staff and patients at a New York City hospital. *Family Planning Perspectives* 27: 155–58.

Grant, L., L. A. Simpson, Z. L. Rong, and H. Peters-Golden. 1990. Gender, parenthood, and work hours of physicians. *Journal of Marriage and the Family* 52: 39–49.

Gregersen, E., and B. Gregersen. 1990. The female condom: A pilot study of the acceptability of a new female barrier method. *Acta Obstetricia et Gynecologica Scandinavica* 69: 73.

Gribin, K. S. 1994. What do adolescents worry about? A quantitative study. *Progress: Family Systems Research and Therapy* 3: 121–35.

Groat, H. Theodore, Peggy C. Giordano, Stephen A. Cernkovich, and M. D. Pugh. 1997. Attitudes toward childrearing among young parents. *Journal of Marriage and the Family* 59: 568–81.

Haspels, A. 1994. Emergency contraception: A review. *Contraception* 50: 101–8.

Jones, Charles L., Lorne Tepperman, and Susannah J. Wilson. 1995. *The futures of the family.* Englewood Cliffs, N.J.: Prentice Hall.

Joshi, Anupama. 1997. Parents resolving conflicts with children: Personal strategy or an interpersonal process. Paper presented at the Annual Conference of the National Council on Family Relations, Crystal City, Va., November.

Kaufman, G. 1997. Men's attitudes toward parenthood. *Population Research and Policy Review* 16: 435–46.

Kelley, J., and N. D. DeGraaf. 1997. National context, parental socialization, and religious belief: The results from 15 nations. *American Sociological Review* 62: 639-59.

Kluwer, Esther S. 1997. Marital change across the transition to parenthood: Conflict about the division of labor. Paper presented at the 59th Annual Conference of the National Council on Family Relations, Crystal City, Va., November.

Knox, D. (with Kermit Leggett). 1998. *The divorced dad's survival book: How to stay connected with your kids.* New York: Insight Books.

Knox, D., and C. Schacht. 1997. Parental concerns on Isla Mujures, Mexico. Unpublished data.

Kozol, Jonathan. 1995. *Amazing Grace: The lives of children and the conscience of a nation.* New York: Crown.

Kraus, M. A., and E. S. Redman. 1986. Postpartum depression: An interactional view. *Journal of Marital and Family Therapy* 12: 63–74.

Laban, E., and M. Gwako. 1997. Conjugal power in rural Kenya families: Its influence on women's decisions about family size and family planning practices. *Sex Roles: A Journal of Research* 36: 127-47.

Lambert, J. D., M. Cwik, and K. Bogenschneider. 1995. Government's role in promoting positive father involvement. Paper presented at the 57th Annual Conference of the National Council on Family Relations. Portland, Ore.: November.

Lamborn, S. D., N. S. Mounts, L. Steinberg, and S. M. Dornbusch. 1991. Patterns of competence and adjustment among adolescents from authoritative, authoritarian, indulgent, and neglectful families. *Child Development* 62: 1049–65.

LaRossa, R. 1988. Fatherhood and social change. *Family Relations* 37: 451–57.

Lavee, Yoav, S. Sharlin, and R. Katz. 1996. The effect of parenting stress on marital quality: An integrated mother-father model. *Journal of Family Issues* 17: 114–35.

LeMasters, E. E., and J. DeFrain. 1989. *Parents in contemporary America: A sympathetic view.* 5th ed. Belmont, Calif.: Wadsworth.

Leve, L. D., and B. I. Fagot. 1997. Gender-role socialization and discipline processes in one- and two-parent families. *Sex Roles: A Journal of Research,* 36: 1–21.

Lindberg, L. D., F. L. Sonenstein, L. Ku, and G. Levine. 1997. Young men's experience with condom breakage. *Family Planning Perspectives* 29: 128–31.

Longman, P. J. 1998. The cost of children. *US News and World Report,* March 30, 51–56.

MacDermid, S. M., T. L. Huston, and S. M. McHale. 1990. Changes in marriage associated with the transition to parenthood: Individual differences as a function of sex-role attitudes and changes in the division of household labor. *Journal of Marriage and the Family* 52: 475–86.

Mischel, H. N., and R. Fuhr. 1988. Maternal employment: Its psychological effects on children and their families. In *Feminism, children and the new families,* edited by S. M. Dornbusch and M. H. Strober. New York: Guilford Press, 194–95.

Morris, J., and D. Schneider. 1992. Health risk behavior: A comparison of five campuses. *College Student Journal* 26: 390–98.

Mueller, K. A., and J. D. Yoder. 1997. Gendered norms for family size, employment, and occupation. Are there personal costs for violating them? *Sex Roles: A Journal of Research* 36: 207–20.

Neville, B., and R. D. Parke. 1997. Waiting for paternity: Interpersonal and contextual implications of the timing of fatherhood. *Sex Roles* 36: 45–59.

O'Hare, W. P., and J. C. Felt. 1991. Asian Americans: America's fastest growing minority group. *Population Trends and Public Policy*. February.

O'Hare, W. P., K. M. Pollard, T. L. Mann, and M. M. Kent. 1991. African Americans in the 1990s. *Population Bulletin* 46: 1–40.

Popenoe, D. 1996. *Life without father*. New York: Free Press.

Reimann, R. 1997. Does biology matter? Lesbian couples' transition to parenthood and their division of labor. *Qualitative Sociology* 20: 153–85.

Reiss, David. 1995. Genetic influence on family systems: Implications for development. *Journal of Marriage and the Family* 57: 543–60.

Ringheim, Karin. 1995. Evidence for the acceptability of an injectable hormonal method for men. *Family Planning Perspectives* 27: 123–28.

Rosenberg, M. J., and E. L. Gollub. 1992. Commentary: Methods women can use that may prevent sexually transmitted disease, including HIV. *American Journal of Public Health* 82: 1473–83.

Rosenfeld, J. 1997. Postcoital contraception and abortion. In *Women's health in primary care*, edited by J. Rosenfeld. Baltimore: Williams & Wilkins, 315–29.

Ross, Catherine E., and Marieke Van Willigen. 1996. Gender, parenthood, and anger. *Journal of Marriage and the Family* 58: 572–82.

Rueter, M. A., and R. D. Conger. 1995. Antecedents of parent-adolescent disagreements. *Journal of Marriage and the Family* 57: 435–48.

Satir, V. 1972. *Peoplemaking*. Palo Alto, Calif: Science and Behavior Books.

Schoen, R., Y. J. Kim, C. A. Nathanson, J. Fields, and N. M. Astone. 1997. Why do Americans want children? *Population and Development Review* 23: 333–58.

Schwiesow, D. R., and Marcia Staimer. 1994. Charting the cost of kids. *USA Today*. 15 December, A1.

Sly, D. F., D. Quadagno, D. F. Harrison, I. W. Eberstein, and K. Riehman. 1997. The association between substance use, condom use and sexual risk among low-income women. *Family Planning Perspectives* 29: 132–36.

Sly, D. F., D. Quadagno, D. F. Harrison, I. W. Eberstein, K. Riehman, and M. Bailey. 1997. Factors associated with female condom use. *Family Planning Perspectives* 29: 181–84.

Stanley, S. M., and H. J. Markman. 1992. Assessing commitment in personal relationships. *Journal of Marriage and the Family* 54: 595–608.

Statistical Abstract of the United States: 1997. 116th ed. Washington, D.C.: U.S. Bureau of the Census.

Stifel, E. N., and J. Anderson. 1997. Contraception. In *Women's health in primary care*, edited by J. Rosenfeld. Baltimore: Williams & Wilkins, 289–313.

Stipp, Horst. 1993. New ways to reach children. *American Demographics*. August, 50–56.

Swerdloff, Ronald. 1995. Male contraception. *Hard Copy* (NBC). 7 August.

Thomas, Karen. 1997. Minding kids in the electronic age. *USA Today*. 20 November, d1.

Thomson, E. 1997. Couple childbearing desires, intentions, and births. *Demography* 34: 343–54.

Trussell, J., et al. 1995. The economic value of contraception: A comparison of 15 methods. *American Journal of Public Health* 85: 494–503.

Tucker, M. Belinda. 1997. Economic contributions to marital satisfaction and commitment. Paper presented at the Annual Convention of the American Psychological Association, Chicago, 18 August.

Van Balen, F., J. Verdurmen, and E. Ketting. 1997. Choices and motivations of infertile couples. *Patient Education and Counseling* 31, no. 1: 19–27.

Van Dyke, David J., Kris K. Ramassini, and Hallie P. Duke. 1997. Effects of children's behaviors and parenting satisfaction. Paper presented at the Annual Conference of the National Council on Family Relations, Crystal City, Va., November.

Von Hertzen, H., and P. F. A. Van Look. 1996. Research on new methods of emergency contraception. *Family Planning Perspectives* 28: 52–57, 88.

Weaver, S. M., E. Clifford, D. M. Hay, and J. Robinson. 1997. Psychosocial adjustment to unsuccessful IVF and GIFT treatment. *Patient Education and Counseling* 13, no. 1: 7–18.

Winton, Chester A. 1995. *Frameworks for studying families*. Guilford, Conn: Dushkin Publishing Group.

Zarit, S. H., and D. J. Eggebeen. Parent-child relationships in adulthood and old age. In *Handbook of parenting*, Vol. 1, edited by Marc H. Bornstein. Mahwah, N. J.: Lawrence Erlaum, 119–40.

CHAPTER 9 Work, Leisure, and the Family

Corporate America is still quite toxic to human well-being.

Joyce Keen, Physician

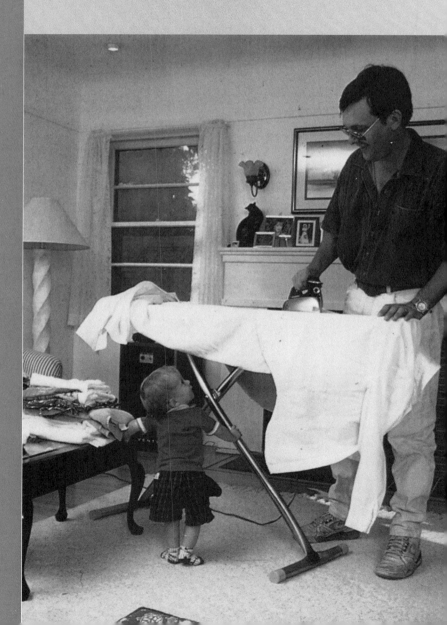

*Our materialistic society
encourages the belief that "the
bigger the diamond, the greater
the love."*

"**I** love to work. If you take a week's vacation, you're a week behind when you get back," says Jay Leno of *The Tonight Show*. Though his enthusiasm for work is probably not shared by most of us, we must nevertheless work. According to a Louis Harris & Associates and Intercep Research Division study, the average number of hours spent working (excluding commute time) each week is fifty-one (Boeck & Lynn, 1996). College students also work for pay. Sixty-three percent of all first-year college students reported that they were involved in some work for pay each week (American Council on Education and University of California, 1997). The effect of work and leisure on individuals, relationships, and children is the focus of this chapter. We begin by considering money—the reason most people work.

Social Meanings of Money

Economists view money as the medium of exchange for goods and services. Sociologists, particularly those who have a symbolic interaction perspective, emphasize the social meanings of money, including self-esteem, power, security, love, and conflict.

Self-Esteem

Money affects self-esteem because in our society human worth, particularly for men, is often equated with financial achievement. A young husband and father mused:

> My two closest friends are making a lot of money in their jobs. It makes me feel bad when I know that I can't provide for my family the way they provide for theirs. I'm a failure (authors' files).

Of course, the self-esteem of women is also influenced by money. Employed wives often report enhanced self-esteem as a result of increased economic independence.

Power in Relationships

Money is a central issue in relationships because of its association with power, control, and dominance (Riza et al., 1992). Blumstein and Schwartz found that "the greater the husband's income, the greater his decision-making and leadership power relative to his wife's. Likewise, the greater the wife's income, the greater her relative power on these two dimensions" (1991, 273).

Sharing the power over how money is spent is associated with increased marital satisfaction. Schwartz and Jackson (1989) found that wives whose influence in deciding how much cash to keep on hand, paying bills, and keeping track of expenditures was equal to or greater than their husbands' reported being much happier in their marriage. When the husband dominated the economic decisions, wives reported being much more unhappy in their marriage.

Money also provides women and men with the power to be independent. The higher a spouse's income, the more power that spouse has to leave the relationship. Indeed, some economically dependent unhappy wives may seek employment so that they can afford to leave.

Money translates into power not only between spouses but also in other family relationships. Parents use money to influence their children's decisions and behavior. For example, parents may threaten to withdraw their financial support from a son or daughter whom they discover is cohabitating. And some divorced parents complain that their ex-spouses use money to buy the affections of the children or to influence the children's preference concerning physical custody arrangements.

Security

Anna Nicole Smith was in her 20s when she married J. Howard Marshall, who was in his late 80s. Both before and after Marshall's death (at age 90), Smith was accused of marrying him for his money. Marshall had amassed a fortune—$725 million at its peak. Although only Smith knows her motivations, money represents security for many people. Oscar Wilde once said, "When I was young, I used to think that money was the most important thing in life; now that I am older, I know it is."

Buying life insurance expresses the desire to provide a secure economic future for loved ones. "If something happens to me," one wife said, "my husband and children will need more than the sympathy they'll get at my funeral. They'll need money." Without money, there is no security—either present or future. Money also secures us against ill health. Because medical care often depends on the ability to pay for it, our health is directly related to our financial resources. Money buys visits to the physician as well as food for a balanced diet.

Love

To some individuals, money also means love. While admiring the engagement ring of her friend, a woman said, "What a big diamond! He must really love you." The assumption is that big diamond equals high price equals deep love.

Similar assumptions are often made when gifts are given or received. People tend to spend more money on presents for the people they love, believing that the value of the gift symbolizes the depth of their love. People receiving gifts may make the same assumption. "She must love me more than I thought," mused one man. "I gave her a CD for Christmas, but she gave me a CD player. I felt embarrassed." His feeling of embarrassment may be based on the idea that the woman loves him more than he loves her because she paid more for her gift to him than he did for his gift to her.

Conflict

Money can also be a source of conflict in relationships. Seventy-seven percent of 15,000 spouses reported that money was the greatest problem in their marriage—outranking sex, in-laws, and infidelity (Schwartz & Jackson, 1989). Couples argue about what to spend money on (new car? vacation? pay off credit card?) and how much money to spend. One couple in marriage therapy

Money is power, and you ought to be reasonably ambitious to have it.
Russell Conwell
Acres of Diamonds

The only thing that counts is that stuff you take to the bank—that filthy buck that everybody sneers at but everybody slugs to get.
Joan Crawford
The Damned Don't Cry

It is better to have a permanent income than to be fascinating.
Oscar Wilde
Playwright

I never cared much for diamonds and pearls . . . cause honestly, honey, they just cost money.
Kitty Kalen
Little Things Mean a Lot

It was always the money.
Betty Davis, Actress
explaining why she divorced four times

reported that they argued over whether to buy orange juice that was fresh-squeezed or juice from concentrate (authors' files). As noted earlier, conflicts over money in a relationship often signify conflict over power in the relationship.

Family conflicts over money sometimes surface when a parent dies. One older brother noted that his father's death resulted in all of his siblings' arguing over who was due how much. "It has splintered the otherwise close family completely," he remarked.

Dual-Earner Marriages

Two-earner marriages have become more common. Over 60 percent of all wives are employed outside the home (*Statistical Abstract of the United States: 1997*, Table 632). The stereotypical family consisting of the husband who earns the income and the wife who stays at home with two or more children describes only 20 percent of all married couples (Ahlburg & De Vita, 1992). Although most couples need two incomes to survive, others who have two incomes can afford expensive homes, cars, vacations, and greater opportunities for their children. As previously noted, the cultural emphasis on equality for women is an important factor influencing the choice of an increasing percentage of wives to be employed. More married women seek careers and economic independence, goals traditionally reserved for men.

Because women still bear most of the child-care and other household responsibilities, women in dual-earner marriages are more likely than men to want to be employed part-time rather than full-time. If this is not possible, many women prefer to work only a portion of the year. The teaching profession allows employees to work only about ten months and to have two months in the summer free. Women are also twice as likely to be self-employed as to be organizationally employed (13% versus 6%) (Tuttle, 1996).

Some dual-earner marriages are dual-career in that each spouse has a career. Three types of dual-career marriages are those in which the husband's career takes precedence, the wife's career takes precedence, or both careers are regarded equally. These career types are sometimes symbolized as HIS/her, his/HER, and HIS/HER. When couples hold traditional gender role attitudes, the husband's career is likely to take precedence. This situation translates into the wife's being willing to relocate and to disrupt her career for the advancement of her husband's career (Bielby & Bielby, 1992).

For couples who do not hold traditional gender role attitudes, the wife's career may take precedence. In such marriages, the husband is willing to relocate and to disrupt his career for his wife's. Such a pattern is also likely to occur when the wife earns considerably more money than her husband.

Studies disagree on the happiness of spouses in marriages where the wife's career is given priority over the husband's. In a small sample of marriages in which the wife's career was given priority over the husband's (his/HER), almost all of the husbands (95 percent) and most of the wives (77 percent) felt that their marriage was "very happy" or "somewhat happy" (Atkinson & Boles, 1984). However, judging from interviews with twenty dual-career couples and a review of the literature, Silberstein noted that many

My work helps me have an
identity, which helps me feel like
I have a place, which makes me
feel like I have a purpose.
Mary Chapin Carpenter
Country music singer

INSIGHT

Symbolic interactionists emphasize that whether a woman's employment increases her power in the relationship depends on the meaning her husband attaches to her employment. "A woman married to a man who views her employment as a threat rather than as a gift for which he should reciprocate will derive less power from her employment" (Pyke, 1994, 75). The definition of the wife's income is also important. Though the primary reason that wives work is financial—to help pay bills and buy extras—the wives also report that they work to keep busy and to use their training. In a study of 186 dual-worker families, both husbands and wives tended not to define a wife's working role as one of coprovider (Spade, 1994).

RECENT RESEARCH

Tangri and Jenkins (1997) studied 117 women in regard to their future marital roles and career goals in 1967 and again in 1981. Those who expected marital conflict in 1967 asserted their career intentions, postponed having children, and had fewer children in 1981. Those with supportive husbands had less marital conflict.

women feel that "to surpass their husbands in occupational success would provoke marital unease as well as intrapersonal dissonance with expectations of both husbands and wives" (1992, 160).

Effects of a Dual-Earner Lifestyle

What are the effects on women, men, their marriage, and their children for couples involved in a dual-earner family?

Effects on Women

Employment for wives is typically associated with enhanced psychological well-being (Crosby, 1993). Employed wives report higher self-esteem and greater feelings of independence, as well as increased social interaction with a wider network of individuals. Employed wives also report more power in their relationships.

Benefits from employment may also surface when the children leave home. Adelmann et al. (1989) found that employed women adjust more easily than full-time homemakers to their children's leaving home. When the parenting role subsides, employed women still have meaningful work roles on which to focus their energy.

On the negative side, employed wives, as well as employed single mothers, often experience what sociologists call *role overload*, or *role strain*—not having the time or energy to meet the demands of one's role responsibilities. Because women have traditionally been responsible for most of the housework and child care, employed women come home from work to what Hochschild (1989) calls the second shift: the housework and child care that employed women do when they return home from their jobs.

> As a result, women tend to talk more intently about being overtired, sick, and "emotionally drained." Many women I could not tear away from the topic of sleep. They talked about how much they could "get by on" . . . six and a half, seven, seven and a half, less, more. . . . Some apologized for how much sleep they needed. . . . They talked about how to avoid fully waking up when a child called them at night, and how to get back to sleep. These women talked about sleep the way a hungry person talks about food (Hochschild, 1989, 9).

Employed women with children also excrete greater amounts of cortisol, associated with heart attacks, than women without children (Luecken et al., 1997).

OTHER CULTURES Who does the housework and child care is also influenced by culture. Martini (1997) studied parents in 120 families from four cultural groups about their dual-worker lifestyle. Caucasian-American households reflected the "second shift strategy," in which working mothers did the bulk of housework and child care. Filipino-American/Japanese-American households reflected the "help from extended family" pattern. Hawaiian-American families emphasized prioritizing quality time regardless of work demands. ●

NATIONAL DATA

Data from the National Survey of Families and Households on 656 dual-earner marriages revealed that wives performed about 70 percent of household tasks (Perry-Jenkins & Folk, 1994).

Another stressful aspect of employment for employed mothers, either in dual-earner or single-parent families, is *role conflict*—being confronted with incompatible role obligations. In a study of 135 married female professionals, over 75 percent of the women reported experiencing frequent conflict between work and family responsibilities. The average occurrence of role conflict was "two or three times a week" (Emmons et al., 1990). The most frequently reported role conflict situations were having to rush their children in the morning so that the women would not be late for work, having to leave work early because of their children, and having to work during times usually reserved for the family.

Role conflict may also create feelings of guilt. In a survey of over 3,000 employed mothers, about 40 percent reported that they felt guilty about not spending enough time with their children and not being home when their children got out of school (Rubenstein, 1991). Other situations that tended to produce role conflict included going to work when one's child was sick and missing school events because of work obligations. Other sources of guilt were also mentioned. Twelve percent of the wives reported feeling guilty over not spending enough time with their husband.

In general, working women with families experience more role overload and role conflict than men. A team of researchers analyzed national data from 1,989 full-time employees with families and concluded:

> The data show quite clearly that mothers experience more role overload, interference from work to family, and interference from family to work than do men. This suggests that combining paid employment with responsibilities at home is more problematic for women than it is for men. Women experience more problems because their family and combined work and family demands are higher (Duxbury et al., 1994, 464).

I have yet to hear a man ask for advice on how to combine marriage and a career.
Gloria Steinem
Ms. Magazine

OTHER CULTURES The "second shift" (women working during the day and coming home to work some more), is true not only in the United States but also in other parts of the world. For example, in a sample of 7,790 adults in the paid labor force in socialist Yugoslavia, women were primarily responsible for housework. "Neither education, occupation, urbanization, nor participation in the informal economy has a significant effect in reducing this" (Massey et al., 1995, 359). Skinner and Meredith (1996) also found that among the ninety-one Chinese couples in their study, women were engaged in 66 percent of the household labor. ●

Some women tire of paid work altogether and opt to return to being full-time moms. F.E.M.A.L.E. (Formerly Employed Moms at the Leading Edge) is an organization for women who have given up promising careers to be home with their children.

While women have sacrificed their careers for their families, men have sacrificed their families for their careers.
Marieke Van Willigen
Sociologist

Effects on Men

Men might also benefit from being in a dual-earner marriage. Benefits for the husband include being relieved of the sole responsibility for the financial support of the family and having more freedom to change jobs or go to school. Men

Researchers suggest that women who are most likely to benefit from employment are those who want to be employed and who enjoy their work (Moen, 1992; Spitze, 1991). Conversely, "women who dislike their work and have little control over it suffer the most conflict over their roles as wife, mother, and worker" (Rubenstein, 1991, 55). Studies also show that "unlike with men, part-time employment is associated with higher levels of well-being among women, since it presumably permits a better coordination of work and family responsibilities" (Moen, 1992, 50). Women who are self-employed also report higher levels of marital happiness. Not only do they work fewer hours, but they have more job flexibility (Tuttle, 1996).

Part of the income that working mothers earn is spent on work-related expenses. Estimates of such expenses are between 25 percent and 50 percent of a woman's take-home pay, with child care the largest expense of those who have young children (Israelsen, 1991). The cost at urban child-care centers may be $750 a month or higher (Shellenbarger, 1993). The cost of a day-care center in general ranges from $140 to $800 a month (Caminiti, 1992). Other expenses include taxes, transportation, and lunch money. In addition, when both spouses work, they often spend a larger portion of their family income on housecleaning services, eating out, and buying convenience foods such as frozen dinners.

also benefit by having spouses with whom to share the daily rewards and stresses of employment. And to the degree that women find satisfaction in their work roles, men benefit by having happier partners. Finally, men benefit from dual-earner marriages by increasing the potential to form closer bonds with their children through active child care.

Not all husbands want their wives to be employed. National longitudinal data analyzed by Blee and Tichameyer (1995) revealed that African-American husbands are more accepting than white husbands. Part of such acceptance may be because a greater proportion of African-American husbands were reared in homes in which their mothers worked outside the home. Regardless of their ideology, when men actually participate in house and child care, there are enormous benefits. A team of researchers (Hawkins et al., 1994) observed that a small increase in the actual time husbands devote to housework and child care results in large increases in the wives' satisfaction with domestic work arrangements. Increasingly, men are taking paternity leaves or becoming stay-at-home dads. While such dads are clearly the minority, their visibility is increasing. There are over 100 subscribers to the newsletter *At-Home Dad* (Armour, 1998).

Effects on Marriages

The effects of employment on marital satisfaction have been studied extensively. Most studies show either a positive effect for both marital partners or no effect (Spitze, 1991). However, Greenstein (1995) emphasized the importance of gender ideology on the marital stability of employed wives. He studied data on a national longitudinal sample of 3,284 women and found that full-time employed women with nontraditional gender ideology views were more likely to divorce than women who worked the same number of hours but had more traditional views of roles. Nontraditional views would include agreement with women pursuing careers. Traditional views would include agreement that "a woman place is in the home, not the office or shop" (Greenstein, 1995, 34). Hence, women who were employed but ideologically deferred to traditional roles were more likely to have stayed married.

Married couples report numerous benefits from dual-income lifestyles. Having two earners in a marriage allows couples to achieve a higher family income, although for many couples, two incomes are necessary to meet the basic financial needs of the family.

When both spouses earn an income, they tend to experience more joint decision making and more consensus on decisions (Godwin & Scanzoni, 1989). However, Moen noted that "couple decisions—whether to move, when to have a child—become more complicated when two jobs are involved" (1992, 69).

On the negative side, Moen noted that the marriage suffers in "terms of the availability and flexibility of time" (1992, 68). In a study of twenty dual-career couples, many women reported being stressed and exhausted, which resulted in little time together, lack of quality time, and less sex (Silberstein, 1992).

Researchers have linked the rise in women's employment with rising divorce rates (Spitze, 1991). Explanations given are that the stress of a dual-career

RECENT RESEARCH

Sefton (1998) calculated that the current value of a stay-at-home mother in terms of what a dual-income family might spend to cover all services that she provides (domestic cleaning, laundry, meal planning, shopping, providing transportation/errands, etc.) is $36,000 per year.

INSIGHT

After coming home from a stressful day at work, some individuals socially withdraw from interaction with others in the home. Family members may perceive such withdrawal in negative terms as they feel ignored or rejected. However, such withdrawal "may not only facilitate mood repair . . . and replenishment of energy, but it may also protect family members from workers' displaced aggression" (Crouter & Manke, 1994, 120). Nevertheless, Tuttle (1996) studied a national sample of male and female workers and found that as work hours increased, marital happiness declined.

When he brings home the bacon, she fries it. When she brings home the bacon, too, they eat out.

Natasha Josefowitz
Management consultant

lifestyle may contribute to divorce, that couples are better able to financially afford divorce, and that women may have sought work to become financially independent so they could leave.

Several factors may influence whether a dual-earner lifestyle has negative or positive effects on a marital relationship. One factor is the degree to which spouses experience job satisfaction. In a study of 180 marriages in which both spouses (mean age 35) were employed full-time, the greater the job satisfaction, the happier the marriage. Conversely, the lower the job satisfaction, the less happy the marriage. There were no gender differences in how the job affected the marriage (Barnett et al., 1994).

Another factor that affects marital quality among dual-earner couples is the amount of support the spouses give each other for working and the degree to which both spouses are satisfied with the division of housework and child care in the marriage (Vannoy & Philliber, 1992). Husbands are more likely to share housework and child care if they have egalitarian gender role attitudes (Pyke & Coltrane, 1996). In addition, the less the discrepancy in income and in employed hours between husbands and wives, the more comparable the amount of time they spend in housework and child care. This was a finding by Skinner and Meredith in their study of ninety-one Chinese couples (1996).

When the Wife Earns More Money

Four in ten wives earn more money than their husbands. Cultural norms dictate that the man is supposed to earn more money than his partner and that something is wrong with him if he doesn't. On the other hand, some men want an ambitious, economically independent woman who makes a high income. If she earns more than he does, he regards it as a plus. Such a perspective is particularly likely when the man wants to pursue a creative though not necessarily lucrative endeavor (e.g., oil painting).

Casamassima (1995) notes first that not equating individual worth with financial contribution is essential to making the higher income of the wife a nonissue. Second, each spouse must view the other as making an equal contribution to the relationship. Whether it is picking up the kids at day care or earning money at a job, both activities are necessary for the unit to function. Third, Casamassima emphasizes the need for couples to talk about their marital roles. "It's essential for the two of them to develop the same vision of the partnership and to discuss problems as they arise" (1995, 66).

Hendershott (1995) observed that among couples who relocate for job-related reasons, a small percentage do so to facilitate the wife's being able to make more money or to help her upwardly mobile career. The fact that few couples move in response to the wife's work emphasizes gender role ideology, which traditionally values the husband's career as more important than the wife's. "Even though more women have infiltrated management and the executive office suite, they are still likely to subscribe to traditional beliefs about the male provider role" (Hendershott, 1995, 29). Hence, since both spouses feel that their marriage will suffer if they move because of the wife's work, few do.

The United States has no national standards for child care.

Sandra Scarr

Mother Care/Other Care

Effects on Children

According to social historian Stephanie Coontz (1995), the current cultural concern about whether the mother works outside the home and its effects on the child was nonexistent in colonial America.

> The dominant family values of colonial days left no room for sentimentalizing childhood. Colonial mothers, for example, spent far less time doing child care than do modern working women, typically delegating this task to servants or older siblings (p. 11).

The Personal Application at the end of this chapter concerns whether or not maternal employment has negative consequences for children.

Nevertheless, parents today want to know how children are affected by maternal employment. Spitze (1991) reviewed the research and concluded that there are no direct effects, positive or negative, of maternal employment on children. Thornton (1992) also found no substantial or consistent effects of maternal employment on gender role attitudes, divorce attitudes, and premarital sexual behavior attitudes of adolescents. Research shows that children of employed mothers develop just as well emotionally, intellectually, and socially as children whose mothers stay at home (Berg, 1986). Recent research confirms no direct harmful effects of maternal employment on children (Pett, Vaughn-Cole & Wampold, 1994). One of the only clear effects is that children of employed mothers tend to do more housework and share more housework with the mother (Bryant & Zick, 1996). They may also experience less supervision. Jenson and Jenson (1997) compared 354 adolescents (seventh, ninth, eleventh grades) whose fathers and mothers (still in first marriage) worked full-time with 211 adolescents (same grades) where only the father worked full-time (these parents were also in their first marriage). Adolescents from homes in which both parents worked full-time were significantly more likely to report having "made out" and "touching another person's sex organ" and to report more arguing between family members.

Some efforts are being made to provide structural support for parents with infants. Support for Families is a new program in Albemarle, North Carolina, that pays new mothers or fathers $250 a month to stay at home during an infant's first year of life (Nowell, 1995). Flexible hours, child-care programs at the work site, and maternal/paternal leaves are also important.

Day Care

The murder conviction (and subsequent reduction to manslaughter) of child-care worker Louise Woodward for the death of 8-month-old Matthew Eappen in 1997 sensitized our nation to the issue of who is taking care of children when the parents work.

NATIONAL DATA   Of the over 11 million United States children under the age of 6 whose mothers are employed full-time outside the home, 39 percent are cared for by organized day-care facilities (*Statistical Abstract of the United States: 1997*, Table 612). The nationwide median monthly cost for day care is $388 per month (Carey & Stacey, 1996).

Muller (1995) examined data from over 13,000 adolescents and their parents and emphasized the need for a social structure that supports the development of children.

The results of this research suggest that aspects of the parent-child relationship that are important for adolescent development would benefit if the institutions in our society were structured in a way that did not force parents to choose between involvement in important social relationships with their child and working outside the home. It appears presently that opportunities to balance activities may be more available to families with more resources such as income, education, and two natural parents. These results suggest that better availability of supervised after school policies that adapt to the needs of parents' work schedules would be beneficial (1995, 97).

One effect of dual-earner marriages is increasing day care for infants.

For children aged 2 and under, the preferred day-care arrangements by parents are (in order of preference) as follows: the parent takes care of the child in the home (68%); the child is taken to the home of someone else other than a family member (13%); the child is taken care of by a member of the extended family, usually a grandmother (9%); and the child is taken care of by a sitter, friend, or neighbor (5%). Only 5% at this age are taken to a day-care center (Caruso, 1992). These data are based on responses from 476 parents of 2-year-olds who live in four Connecticut communities.

Employed parents are concerned that their children get good-quality care. Friends are most often consulted to recommend a day-care facility. Priorities in day-care selection include health and safety issues, caregiver quality, and the child's social and educational development (Bradbard et al., 1992).

State licensing regulations of day-care centers are supposed to provide a "floor of quality" for day-care programs. However, these regulations often fall short of recommended standards (Teleki et al., 1992). In a survey of 400 day-care centers in California, Colorado, North Carolina, and Connecticut, a team of researchers rated only 14 percent as good or excellent (Helburn et al., 1995). In these centers, the children's health and safety needs were met, the children received warmth and support from their caretakers, and learning was encouraged. Twelve percent of the centers were rated as minimal—the children were ignored, their health and safety were compromised, and no learning was encouraged. The remaining 74 percent received a rating just above the minimal, suggesting that the health/safety needs, nurturing needs, and educational needs were barely met. Care for infants was particularly lacking. Of 225 infant or toddler rooms observed, 40 percent were rated less than minimal with regard to hygiene and safety.

Some day-care centers are deadly. A *U.S. News World Report* query of all fifty states and the District of Columbia revealed seventy-six deaths in 1996. Causes included drownings, falls, being struck by autos, and sudden infant death syndrome (Pope, 1997).

Jay Belsky (1995), recognized as a leading expert on day-care centers over the past twenty-five years, makes the following observations:

1. Day care works best when children receive "care from individuals who will remain with them for a relatively long period of time (low staff turnover), who are knowledgeable about child development, and who provide care that is sensitive and responsive to their individualized needs" (p. 36).

2. Since few day-care centers are characterized by the above description, children with "early, extensive, and continuous day care experience" (that is, children whose care begins in the first year or so of life, on a full- or near full-time basis—more than twenty to thirty hours per week—and continues through their entry into public school) are the ones who seem most susceptible to negative outcomes. Such outcomes include insecure attachment bonds to parents, aggressiveness toward agemates, and disobedience toward adults.

3. Federal mandates for quality day-care standards must be established and tax provisions must be made for more economic support of families rearing children to increase their childrearing options.

In the meantime, parents should be cautious in choosing day-care facilities for their children. To help them, the National Association for the Education of Young Children (NAEYC) has recommended criteria for accrediting day-care centers. Such recommendations include having a 1:3 to 1:4 adult-child ratio for children from 0 to 24 months, with from six to twelve children per group. For 2-year-olds, the ratio is 1:4 to 1:6, with from eight to twelve children per group (Helburn & Howes, 1996, 67).

Balancing the Demands of Work and Family

Juggling one's external demands (whether school or paid employment) with one's relationships so as to achieve a sense of accomplishment and satisfaction in each area is no easy feat. This section examines an array of strategies individuals use to cope with the stress of role overload and role conflict, including (1) the superperson strategy, (2) cognitive restructuring, (3) delegation of responsibility, (4) planning and time management, and (5) shift work. The section concludes with a brief discussion of the stress of caring for an elderly parent.

Superperson Strategy

The superperson strategy involves working as hard and as efficiently as possible to meet the demands of work and family. The person who uses the superperson strategy often skips lunch and cuts back on sleep and leisure in order to have more time available for work.

I can work half as hard, make half as much money and have twice as much fun.

Robert Urich

Actor

One of the symptoms of approaching nervous breakdown is that one's work is terribly important, and that to take a holiday would bring all kinds of disaster.

Bertrand Russell

Philosopher

Hochschild noted that the term *superwoman* or *supermom* is a cultural label that allows the woman to regard herself as "unusually efficient, organized energetic, bright, and confident" (1989, 32). However, Hochschild also pointed out that this is a "cultural cover up" for an overworked and frustrated woman. Such a label also reveals that much of one's stress is self-induced. Type A personalities thrive in the role of the workaholic.

Cognitive Restructuring

Another strategy used by some women and men experiencing role overload and role conflict is cognitive restructuring, which involves choosing to view the situation in positive terms. In a study of dual-career couples, some spouses coped with the stress by "believing that our family life is better because both of us are employed," "believing there are more advantages than disadvantages to our lifestyle," and "believing that my career has made me a better wife/husband than I otherwise would be" (Schnittger & Bird, 1990, 201). Other examples of cognitive restructuring include "I told myself that it wouldn't be the end of the world if I didn't get all my work done on time" and "I tried to recognize that I'd have to lower my standards about such things as how clean the house is and how elaborate the meals are" (Emmons et al., 1990).

Other employed mothers "emotionally downsize life" by convincing themselves that leaving children alone makes them self-sufficient, birthday parties organized by professionals are fun, and Hallmark cards are as good as a kiss (Hochschild, 1997).

Delegation of Responsibility

INSIGHT

Since dual-earner couples have more income than one-earner couples, one might ask to what degree they use this money to hire help. Two researchers observed that the extra income is not used to employ people to help with housework. Rather "employed wives may simply reduce their leisure time (or their sleep), or other family members may be doing more of the housework, or household standards may be lowered" (Rubin & Riney, 1994, 134).

A third way couples manage the demands of work and family is to delegate responsibility to others for performing certain tasks. Because women tend to bear most of the responsibility for child care and housework, they may choose to ask their partners to contribute more or to take responsibility for these tasks. Parents may also involve their children more in household tasks, which not only relieves the parents but also benefits children by requiring them to learn domestic skills and the value of contributing to the family. If it is financially possible, the couple may hire someone to clean the house.

Another form of delegating responsibility involves the decision to reduce one's current responsibilities and not take on additional ones. For example, women and men may give up volunteer work or avoid agreeing to additional volunteer responsibilities or commitments. In the realm of paid work, women and men can choose not to become involved in professional activities beyond what is required.

OTHER CULTURES Martini (1995) compared forty-five Japanese-American and thirty-two Caucasian-American working mothers and found that 50 percent of the former in contrast to 15 percent of the latter reported help from extended family as the most important factor helping them to cope with the demands of work and family. The Japanese mothers also talked of being "at ease and able to concentrate on their job" when they knew their children were being cared for by extended family members (Martini, 1995, 114). ●

Planning and Time Management

Hard work never killed anybody, but why take a chance?
Edgar Bergen
Ventriloquist

Still another strategy for coping with the demands of work and family is the use of planning and time management. This involves setting priorities and making lists of what needs to be done each day. Time planning also involves allocating time for activities regarded as important and not letting other pressures interfere with those activities. In addition, time planning involves trying to anticipate stressful periods and planning ahead for them.

The recent negative publicity on day-care centers has reaffirmed for shift-worker parents the value of one parent's being present at all times to take care of their children. This is a major advantage of shift work.

Shift work also has its downside. Shift workers may find it difficult to adapt to a changing sleep pattern and may experience sleep deprivation and fatigue, which may interfere with their domestic role responsibilities. In addition, shift work may have a negative effect on a couple's marriage because they may rarely see each other. One is coming home from a night of work while the other is just getting ready to go to work. When that person gets off work, the other person goes to work. There is little time to share when both are rested. Another factor that may weigh heavily on a couple is the responsibility for taking care of an elderly parent.

Caring for an Elderly Parent

Balancing work and family demands is especially challenging for those who are caring for an elderly parent. Adults who must care for their own children as well as their elderly parents are members of the "sandwich generation." As with other domestic responsibilities, women tend to do more caregiving to elderly parents than do men. Women and men of the sandwich generation sometimes take time off from work or end their careers to take care of elderly parents. Reasons for an increase in the number of individuals caught in the sandwich generation include the following:

1. *Longevity.* The over-85 age group, the segment of the population most in need of care, is the fastest-growing segment of our population.

2. *Chronic disease.* In the past, diseases took the elderly quickly. Today, prevalent diseases such as arthritis and Alzheimer's do not offer an immediate death sentence, but rather a lifetime imprisonment.

3. *Fewer children to help.* The current generation of elderly have fewer children than the elderly in previous generations. Hence, the number of adult children to look after them is smaller. An only child has no one with whom to share the care of elderly parents.

4. *Commitment to parental care.* Contrary to the myth that adult children in the United States abrogate responsibility for taking care of their elderly parents, children institutionalize their parents only as a last resort.

Women more often than men care for their elderly parents.

RECENT RESEARCH

Starrels et al. (1997) examined data provided by 1,585 employees who cared for a parent or parent-in-law aged 60 or older. Caring for those elderly who were cognitively impaired was particularly stressful.

Caring for a dependent, aging parent requires a great deal of effort, sacrifice, and decision making on the part of several million adults in the United States who cope with the situation daily. The emotional toll is heavy. Guilt, resentment, and anger are the most commonly reported feelings. The guilt comes from having promised the parents that they would be cared for when they became old and frail. Paying off the promise often entails more than the children ever expected. Or the offspring may feel guilty that they resent disrupting their own lives to care for their parents. Or they may feel guilty that they are angry about the frustration their parents are causing them. "I must be an awful person to begrudge taking my mother supper," said one daughter. "But I feel that my life is consumed by the demands she makes on me, and I have no time for myself."

Because of the effect of work on families, government and corporate "work-family" policies and programs have developed. The social policy analysis of this chapter focuses on this issue.

RECENT RESEARCH

Menec and Chipperfield (1997) studied a sample of 1,053 elderly women and men and found that exercising and leisure activities were predictive of better perceived health and greater life satisfaction.

Leisure and the Family

Importance of Leisure

Leisure refers to the use of time to engage in freely chosen activity perceived as enjoyable and satisfying. When a national sample of university students was asked to identify various leisure activities they enjoyed for six or more hours in the last

Government and Corporate Work-Family Policies and Programs

In 1993, President Clinton signed into law the Family and Medical Leave Act, which requires all companies of twenty-five or more employees to provide each worker with up to twelve weeks of unpaid leave for reasons of family illness, birth, or adoption of a child. Nevertheless, the United States still lags behind other countries in providing paid time off for new parents. For instance, Germany provides fourteen weeks off with 100 percent salary (Caminiti, 1992).

Aside from government-mandated work-family policies, corporations and employers have begun to initiate policies and programs that address the family concerns of employees. It is good business to do so (worker retention, worker satisfaction). These include on- or near-site child care, flexible family leave policies, on- or near-site elder care, flexible scheduling opportunities, telecommuting opportunities, after-school care programs, and summer school programs for employee's children. Employees at SAS Institute, a computer software company in Cary, North Carolina, can eat meals with family members and guests at the company cafeteria. The older children from the company preschool can join their parents for lunch (Russo, 1993). Pepsi-Co, Inc., in Purchase, New York, has an on-site concierge to do personal chores for its 800 employees (Lopez, 1993). In general, though companies that provide help for parents and their children are the exception rather than the rule.

Work-family policies benefit both employees and their families and the corporations they work for. Hewlett (1992) suggests:

> Companies need not be farsighted or altruistic to have [family support] policies; all they need to do is consult their bottom line. Family supports are fast becoming win-win propositions; good for the working parent, good for the child, good for the company (p. 26).

For example, Corning Glass Works in upstate New York found that its turnover rate for female employees was twice as high as that for male employees. This high turnover rate was costly—replacing each lost worker cost $40,000 (for search costs, on-the-job training costs, and the like). Corning conducted a survey and discovered that "family stress—particularly child-care problems—was the main reason so many women quit their jobs" (Hewlett, 1992, 27). Corning decided to implement a family support package that included parenting leave, on-site child care, part-time work options, job sharing, and a parent resource center. The company's chairman, James P. Houghton, said that Corning's efforts "go way beyond simple justice; it's a matter of good business sense in a changing world . . . it's a matter of survival" (cited in Hewlett, 1992, 27).

Fran Rodgers, president of Work/Family Directions explains the need for work-family policies:

> For over 20 years we at Work/Family Directions have asked employees in all industries what it would take for them to contribute more at work. Every study found the same thing: They need aid with their dependent care, more flexibility and control over the hours and conditions of work, and a corporate culture in which they are not punished because they have families. These are fundamental needs of our society and of every worker. (Galinsky et al., 1993, 51)

Corporate adoption of a work-family ethic also benefits the larger community. Corporations that are sensitive to the family responsibilities and needs of employees often make contributions to community agencies that serve families. For example, the Mutual of New York Life Insurance Company funds community day-care programs, training programs for day-care employees, and child care for teenage mothers so they can continue school or receive job training (Stekas, 1993). The Communications Workers of America and the International Brotherhood of Electrical Workers have a Dependent Care Development Fund designed to improve the dependent care programs in the communities where employees live and work (Green, 1993).

REFERENCES

Caminiti, S. 1992. "Who's minding America's kids? *Fortune.* 10 August, 50–53.

Galinsky, E. J., E. Riesbeck, F. S. Rodgers, and F. A. Wohl. 1993. Business economics and the work-family response. In *Work-family needs: Redefining the business case.* New York: The Conference Board, 51–54.

Green, M. 1993. NYNEX's dependent care development fund. In *Work-family needs: Redefining the business case.* New York: The Conference Board, 45–46.

Hewlett, S. A. 1992. *When the bough breaks: The cost of neglecting our children.* New York: HarperCollins.

Lopez, J. A. 1993. Undivided attention: How PepsiCo gets work out of people. *Wall Street Journal.* 1 April, 221:1.

Russo, D. 1993. Do you employ women? How about common sense? In *Work-family needs: Redefining the business case.* New York: The Conference Board, 23–24.

Stekas, L. Corporate contributions. 1993. *Work-family needs: Redefining the business case.* New York: The Conference Board, 43–44.

week, 77 percent noted "socializing with friends," 50 percent "exercising or sports" and 30 percent "partying" (American Council on Education and University of California, 1997). In regard to the larger population, according to a Louis Harris & Associates and Intercep Research Division study, the average number of hours spent in leisure each week is 19 (Boeck & Lynn, 1996). Leisure is becoming more important to people. In a nationwide poll commissioned by the Merck Family Fund, 800 respondents identified the three things they needed to make their lives more satisfying: spending more time with family and friends, reducing stress, and being more involved in their community (Schor, 1995). Being available for more leisure involves the tradeoff of taking a lower-paying job or working fewer hours. Over one quarter (28%) reported that in the past five years they had voluntarily made changes in their lifestyle that resulted in making less money. Women, those with children, and those under 40 were more likely to do so.

Increasingly, individuals dream of and seek contexts in which they can completely relax.

While some companies have become aware of the link between work and family and have implemented policies such as paternity leave, companies are less likely to promote workers who actually use these policies.

Dana Friedman
Head, Corporate Solutions

I break out in a cold sweat if I don't have access to my e-mail.

Edward Melia
Manager of a computing firm

We need to reorganize work to make it more compatible with family life.

Stephanie Coontz
Social historian

Leisure fulfills important functions in our individual and interpersonal lives (Henderson et al., 1989; Schor, 1991). Leisure activities may relieve work-related stress and pressure; facilitate social interaction and family togetherness; foster self-expression, personal growth, and skill development; and enhance overall social, physical, and emotional well-being.

Barriers to Leisure

Many individuals feel they do not have enough leisure. What factors have contributed to the "leisure shortage" that many women and men complain about? Barriers to leisure include the rising demands of the workplace, materialistic values, traditional gender roles, the "commodification of leisure," and treating leisure as work.

Demands of the workplace A major barrier to leisure has been the rising demands of the workplace. Many jobs demand that employees work overtime; other jobs pay so little that an employee must take a second job or work two shifts to make ends meet.

> The 5:00 Dads of the 1950s and 1960s (those who were home for dinner and an evening with the family) are becoming an "endangered species." Thirty percent of men with children under fourteen report working fifty or more hours a week. And many of these 8:00 or 9:00 Dads aren't around on the weekends either. Thirty percent of them work Saturdays and/or Sundays (Schor, 1991, 21).

Studies estimate that the total working time of employed mothers (including housework) averages between sixty-five and eighty-nine hours per week (Schor, 1991). Very little time is left over after the responsibilities of work and home have been met, and women and men have very little energy to participate in leisure activities after working nine to sixteen hours a day.

Adults are not the only ones working long hours. Schor (1991) reported that by 1990, 53.7 percent of U.S. teens were in the workforce; this figure is nearly ten points higher than it had been in 1965. While some lower-income families cannot survive economically without the income generated by teenage children, many middle-class teenagers work for consumeristic reasons—expensive clothes, the latest CD sound system, and a car. Participating in the workforce may teach teenagers the value of work and money and the importance of education to avoid being stuck in low-wage jobs. However, working too many hours may interfere with one's education. "Teachers report that students are falling asleep in class, getting lower grades, and cannot pursue after-school activities" (Schor, 1991, 27).

Materialistic values The media bombard us daily with advertisements that increase our aspirations to have the things that money can buy. So we work long and hard hours to achieve a certain standard of living. Once we have achieved that standard of living, we may choose to scale back the time devoted to work and devote more time to leisure and our family. However, once we attain the

Adult women and men in the United States spend an average of fifteen hours a week watching television (Robinson, 1990). According to Robinson's estimates, 38 percent of the total leisure time of women and men is spent watching TV. This may be partly because people are too tired to engage in more active leisure activities. However, when a national sample of women and men were asked, "If there were one more hour in the day, how would you spend it?" the most common response was "participate in active sports"; only 1 percent said they would watch more TV.

Most of the luxuries, and many of the so-called comforts of life are not only dispensable, but hindrances to the elevation of mankind.
Henry David Thoreau
Author

I have no money, no resources, no hopes. I am the happiest man alive.
Henry Miller
Author

RECENT RESEARCH

Dempsey (1997) focused on the attempts of wives in 128 Australian marriages to get their husbands to do more housework and found only three or four who had experienced any lasting success. Dempsey concluded that structural and cultural forces result in superior power to the husbands which weakened the negotiation leverage of the wives.

goods that we desire, we often find we want more. Thus, many couples get caught up in a vicious cycle of working long hours to achieve a certain standard of living, only to find that their standard of living rises again.

Traditional gender roles   Traditional gender roles may also create barriers to leisure in the lives of women and men. Whether employed or unemployed, women bear most of the responsibility for childrearing, housework, and general care of the family. The result is that women have little time for leisure. Cultural norms not only influence *who* does household labor (women versus men), but also the nature of household labor itself.

Women also tend to spend their leisure time engaged in hobbies related to household tasks, such as preserving and canning fruits and vegetables, cooking, and sewing (Shelton, 1992). In addition, women's leisure is often combined with household chores (Henderson et al., 1989). For example, women often watch TV while they iron clothes, talk on the phone with a friend while cooking in the kitchen, socialize with neighbors while looking after the children, listen to music while cleaning house, or clip coupons and make a grocery list while riding a stationary bicycle. In addition, women's leisure activities are often constrained by the leisure needs of the family. For example, a woman may spend an afternoon at the pool but may do so primarily to provide her children a fun day at the pool. Similarly, a woman may walk in the park primarily so that her children may enjoy the playground or feed the ducks at the pond. (It is also true that some fathers schedule their leisure activities in reference to their children).

Family recreational activities often represent a great deal of work for women—more so than for men. For example, a family picnic involves shopping for and packing the food and cleaning up the picnic site afterwards—tasks that are traditionally viewed as women's work. Going on a family vacation involves many tasks, also done primarily by women, including packing the suitcases, canceling the newspaper delivery for the week, packing snacks for the car ride, making arrangements with the neighbors to feed the cat and get the mail, and unpacking the suitcases after returning from vacation. In addition, women may also do more of the emotional work of trying to ensure that all family members have a good time. It is not uncommon to hear women say that after taking a family vacation, they need another vacation.

Holidays provide another example of how gender roles constrain women's leisure. After the Thanksgiving holidays, many women are exhausted by the hours they spent cooking to achieve the perfect Thanksgiving meal. Christmas holidays also represent a great deal of work for women. Traditionally, women in our society who celebrate Christmas have learned it is their primary responsibility to do the Christmas shopping, mail the Christmas presents and cards, put up decorations, and bake Christmas cookies. Many women spend considerable time and energy to ensure that Christmas is a positive experience for the family. But many drive themselves to exhaustion during the process.

Women's leisure is also constrained by safety concerns. Women are less likely to use parks at certain hours or to participate in outdoor activities if they are fearful of their safety (Henderson, 1990).

Men's leisure is also affected by gender role expectations that equate a man's self-worth with his income and put men in the role of primary breadwinner. One man said:

The following Personal Application Scale allows you to assess your beliefs about the effect a mother's employment has on the development of her children.

Consequences of Maternal Employment Scale

DIRECTIONS Using the following scale, please mark a number on the blank next to each statement to indicate how strongly you agree or disagree with it.

1	2	3	4	5	6
Disagree Very Strongly	Disagree Strongly	Disagree Slightly	Agree Slightly	Agree Strongly	Agree Very Strongly

__14__ (1.) Children are less likely to form a warm and secure relationship with a mother who is working full-time outside the home.

__16__ 2. Children whose mothers work are more independent and able to do things for themselves.

__12__ (3.) Working mothers are more likely to have children with psychological problems than mothers who do not work outside the home.

__12__ (4.) Teenagers get into less trouble with the law if their mothers do not work full-time outside the home.

__15__ 5. For young children, working mothers are good role models for leading busy and productive lives.

__14__ 6. Boys whose mothers work are more likely to develop respect for women.

__12__ (7.) Young children learn more if their mothers stay at home with them.

__45__ 8. Children whose mothers work learn valuable lessons about other people they can rely on.

__24__ 9. Girls whose mothers work full-time outside the home develop stronger motivation to do well in school.

__26__ 10. Daughters of working mothers are better prepared to combine work and motherhood if they choose to do both.

__44__ (11.) Children whose mothers work are more likely to be left alone and exposed to dangerous situations.

__46__ 12. Children whose mothers work are more likely to pitch in and do tasks around the house.

__43__ (13.) Children do better in school if their mothers are not working full-time outside the home.

__45__ 14. Children whose mothers work full-time outside the home develop more regard for women's intelligence and competence.

__41__ (15.) Children of working mothers are less well-nourished and don't eat the way they should.

__45__ 16. Children whose mothers work are more likely to understand and appreciate the value of a dollar.

__42__ (17.) Children whose mothers work suffer because their mothers are not there when they need them.

__42__ (18.) Children of working mothers grow up to be less competent parents than other children because they have not had adequate parental role models.

__45__ 19. Sons of working mothers are better prepared to cooperate with a wife who wants both to work and have children.

__41__ (20.) Children of mothers who work develop lower self-esteem because they think they are not worth devoting attention to.

__36__ 21. Children whose mothers work are more likely to learn the importance of teamwork and cooperation among family members.

4 2 (22). Children of working mothers are more likely than other children to experiment with alcohol, other drugs, and sex at an early age.

4 6 23. Children whose mothers work develop less stereotyped views about men's and women's roles.

4 5 24. Children whose mothers work full-time outside the home are more adaptable: they cope better with the unexpected and with changes in plans.

Costs – 32
Benefits – 38

Costs – 25
Benefits – 68

SCORING INSTRUCTIONS: Items 1, 3, 4, 7, 11, 13, 15, 17, 18, 20, and 22 refer to "costs" of maternal employment for children and yield a Costs Subscale score. High scores on the Costs Subscale reflect strong beliefs that maternal employment is costly to children. Items 2, 5, 6, 8, 9, 10, 12, 14, 16, 19, 21, 23, and 24 refer to "benefits" of maternal employment for children and yield a Benefits Subscale score. To obtain a Total Score, reverse score all items in the Benefits Subscale so that $1 = 6$, $2 = 5$, $3 = 4$, $4 = 3$, $5 = 2$, and $6 = 1$. The higher one's Total Score, the more one believes that maternal employment has negative consequences for children.

Source: E. Greenberger, W. A. Goldberg, T. J. Crawford, and J. Granger. 1988. Beliefs about the consequences of maternal employment for children. *Psychology of Women Quarterly* 12:58–59, Appendix A. Used by permission of Cambridge University Press.

Note: This Personal Application is included in this text to be thought-provoking. It is not intended to be used by students or instructors as a clinical evaluation device.

Housework may not be Everest, but it is an adventure that awaits any man who wants to forge ahead and meet the challenges of unexplored territory.
William Beer
Househusbands

Being a man means being willing to put all your waking hours into working to support your family. If you ask for time off, or if you turn down overtime, it means you're lazy or you're a wimp (cited in Schor, 1991, 149).

Leisure as a commodity   Many leisure activities cost money that families simply do not have in their budget. One father said, "I used to take the family to a movie every weekend. Not anymore. After you add the cost of popcorn and drinks, I just can't afford it" (authors' files). Schor described the *"commodification of leisure"* in our society:

> Private corporations have dominated the leisure "market" encouraging us to think of free time as a consumption opportunity. Vacations, hobbies, popular entertainment, eating out, and shopping itself are all costly forms of leisure. How many of us, if asked to describe an ideal weekend, would choose activities that cost nothing? How resourceful are we about doing things without spending money? (1991, 162)

If you watch a game, it's fun. If you play it, it's recreation. If you work at it, it's golf.
Bob Hope
Comedian

Leisure as work   Roberts emphasizes that leisure has become work that we use "as a means to other ends—stress reduction, therapy, fitness, and self-actualization" (1995, 36). He suggests that true playfulness involves a sense of forgetting time and being completely absorbed or focused," something that we no longer know how to do. One suggestion is to simplify your play objectives. For example, try taking a walk without your watch or without monitoring your heart rate. And don't take on an activity that is so ambitious, challenging, or expensive that you spend the entire time worrying about whether you can afford it or whether you're performing well enough, or having enough fun, to justify the time expended" (Roberts, 1995, 41). Some people are reluctant to take time off for a vacation. A nationwide poll of over 1,000 adults revealed that 11 percent of full-time workers who were entitled to a vacation did not take one. Another 32 percent took fewer days than earned. Sixty percent of those who did not use their vacation time said that work was more important than taking a vacation; 27 percent feared something would go wrong at work if they went on vacation (Grossman, 1995). This fear of not working enough is surprising in view of the fact that Americans get only 11.37 days a year in paid vacation, according to a study by Primark Decision Economics (Neuborne, 1997).

This chapter has emphasized the effect of work on a couple's relationship and the importance of making time for individual and joint leisure. Just as some billboards read, "The family that prays together, stays together," the same might also be said of couples who play together.

GLOSSARY

commodification of leisure   The perception of free time as a consumption opportunity whereby one expects to spend money (e.g., on vacations) to enjoy leisure.

leisure   The use of time to engage in freely chosen activity perceived as enjoyable and satisfying.

role conflict   Being confronted with incompatible role obligations (e.g., the wife is expected to work full-time and also to be the primary caretaker of children).

role overload   The convergence of several aspects of one's role resulting in having neither time nor energy to meet the demands of that role. For example, the role of wife may involve coping with the demands of employee, parent, and spouse.

superwoman (or supermom)   A cultural label that allows the mother who is experiencing role overload to regard herself as particularly efficient, energetic, and confident.

SUMMARY

The need to work to provide income to support one's family, with the concomitant loss of leisure time for one's self, spouse, and children, is a common issue with which most individuals and families cope.

Social Meanings of Money

Money is a source of self-esteem, power in relationships, security, love, and conflict. It may also be used to express love, as when couples buy each other gifts. Symbolic interactionists also emphasize that work, the principal means by which money is obtained, is a major source of personal social identity and stress.

Dual-Earner Marriages

Over 60 percent of wives are employed outside the home. Because women are more active in child care, they are more likely to be employed part-time. If this is not possible, many women prefer to work in occupations (such as teaching) that allow them to be available when their children are not in school.

Effects of a Dual-Earner Lifestyle

In dual-earner relationships, benefits for the woman include enhanced self-esteem, more power in the relationship, greater independence, and a wider set of social (work) relationships. Negatives for women include exhaustion as a result of role overload and frustration or guilt caused by role conflict. Women all over the world do more housework than men, regardless of their being employed outside the home.

Men benefit from their wives' employment by being relieved of full responsibility for the financial support of the family, having freedom to change jobs, and having partners with whom to share the concerns of the work world. Men who report being dissatisfied with their wives' employment interpret such employment as a reflection of their own inadequacy to support the family. Some men are also torn between their work and family responsibilities.

Not all husbands want their wives to be employed. African-American husbands are more approving than white husbands. Most African-American husbands were reared in homes in which their mothers worked outside the home. Husbands who participate in child care or housework markedly increase their wives' satisfaction with the domestic work arrangements.

Couples in dual-earner relationships tend to report positive effects. Not only do they have a higher income, but they also report having similar experiences and concerns to talk about. A major drawback of involvement in a dual-earner marriage is the lack of time together. Also, when the wife earns more money than the husband, which is true of 40 percent of wives, marital satisfaction is lower.

Children seem to experience no direct negative effects from both parents' being employed. Rather, children may enjoy having more money in the family (for cable TV, music lessons, karate training), and daughters may benefit from observing an employed mother as a role model. Disadvantages include less time with parents, a higher level of stress at home, and more restricted activities if parents are not available to transport children. Young children who are put in day-care centers for most of the day may be at risk. A survey of 400 day-care centers nationwide revealed that only 14 percent were rated "good" or "excellent" by a team of researchers. Forty percent of infant or toddler rooms were rated less than minimal with regard to hygiene and safety.

Balancing the Demands of Work and Family

Strategies used for balancing the demands of work and family include the superperson strategy, cognitive restructuring, delegation of responsibility, planning and time management, and shift work. Government and corporations have begun to respond to the family concerns of employees by implementing work-family policies and programs.

Leisure and the Family

Spouses and couples are increasingly beginning to value their leisure time. Leisure helps to relieve stress, facilitate social interaction and family togetherness, and foster personal growth and skill development. However, leisure may also create conflict over how to use leisure time.

Barriers to having enough leisure time include incessant demands of the workplace, materialistic values (which require work to earn the money to buy "things"), traditional gender roles (women do most of the work during vacations and holidays), commodification of leisure (the idea that you need to spend money to have a good time), and treating leisure as work.

REFERENCES

Adelmann, P. K., T. C. Antonucci, S. E. Crohan, and L. M. Coleman. 1989. Empty nest, cohort, and employment in the well-being of midlife women. *Sex Roles* 22:173–89.

Ahlburg, D. A., and C. J. De Vita. 1992. New realities of the American family. *Population Bulletin* 47, no. 2: 2–44.

American Council on Education and University of California. 1997. *The American freshman: National norms for fall, 1997.* Los Angeles: Los Angeles Higher Education Research Institute.

Armour, S. 1998. Dad is job one: Paternity leaves increasingly popular. *USA Today,* February 23, p. A1.

Atkinson, M. P., and J. Boles. 1984. WASP (Wives as Senior Partners). *Journal of Marriage and the Family* 46: 861–70.

Barnett, Rosalind C., Robert T. Brennan, and Nancy L. Marshall. 1994. *Journal of Family Studies* 15: 229–52.

Belsky, J. 1995. A nation still at risk. *Phi Kappa Phi Journal* 75: 36–38.

Berg, B. 1986. *The crisis of the working mother.* New York: Summit Books.

Bielby, W. T., and D. D. Bielby. 1992. I will follow him: Family ties, gender-role beliefs, and reluctance to relocate for a better job. *American Journal of Sociology* 97: 1241–68.

Blair, S. L. 1993. Employment, family, and perceptions of marital quality among husbands and wives. *Journal of Family Issues* 14: 189–212.

Blee, Kathleen M., and Ann R. Tichameyer. 1995. Racial differences in men's attitudes about women's gender roles. *Journal of Marriage and the Family* 57: 21–30.

Blumstein, P., and P. Schwartz. 1991. Money and ideology: Their impact on power and the division of household labor. In *Gender, family, and economy,* edited by R. L. Blumberg. Newbury Park, Calif.: Sage, 261–88.

Boeck, S., and Genevieve Lynn. 1996. Balancing work and play. *USA Today,* 26 December.

Bradbard, M. R., C. A. Readdick, R. C. Endsley, and E. G. Brown. 1992. How and why parents select day care for their school age children: A study of three communities. Paper presented at the 54th Annual Conference of the National Council on Family Relations, Orlando, Florida, November.

Bryant, W. K., and C. D. Zick. 1996. An examination of parent-child shared time. *Journal of Marriage and the Family* 58: 227–37.

Caminiti, S. 1992. Who's minding America's kids? *Fortune.* 10 August, 50–53.

Carey, Anne R., and J. Stacey. 1996. Lowest cost areas for daycare. *USA Today.* 19 March, D1.

Caruso, G. L. 1992. Patterns of maternal employment and childcare for a sample of two-year-olds. *Journal of Family Issues* 13: 297–311.

Casamassima, Christy. 1995. Battle of the bucks. *Psychology Today* 28: 42 passim.

Coontz, Stephanie. 1995. The way we weren't: The myth and reality of the "traditional" family. *Phi Kappa Phi Journal* 75: 11–14.

Crosby, F. E. 1993. *Juggling: The unexpected advantages of balancing career and home for women and their families.* New York: Free Press.

Crouter, Ann C., and Beth Manke. 1994. The changing American workplace: Implications for individuals and families. *Family Relations* 43: 120–23.

Dempsey, K. 1997. Trying to get husbands to do more work at home. *Australian and New Zealand Journal of Sociology* 33: 216–55.

Duxbury, Linda, Christopher Higgins, and Catherine Lee. 1994. Work-family conflict: A comparison by gender, family type, and perceived control. *Journal of Family Issues* 15: 449–66.

Emmons, C., M. Biernat, L. B. Tiedje, E. L. Lang, and C. B. Wortman. 1990. Stress, support, and coping among women professionals with preschool children. In *Stress between work and family,* edited by J. Eckenrode and S. Gore. New York: Plenum Press, 61–93.

Godwin, D. D., and J. Scanzoni. 1989. Couple consensus during marital joint decision making: A context, process, outcome model. *Journal of Marriage and the Family* 51: 943–56.

Greenstein, Theodore N. 1995. Gender ideology, marital disruption, and the employment of married women. *Journal of Marriage and the Family* 57: 31–42.

Grossman, Cathy L. 1995. For many, vacation is no vacation. *USA Today.* 14 July, 1D.

Hawkins, Alan J., Tomi-ann Roberts, Shawn L. Christiansen, and Christina M. Marshall. 1994. An evaluation of a program to help dual-earner couples share the second shift. *Family Relations* 43: 213–320.

Helburn, Suzanne, Mary L. Culkin, John Morris, Naci Moran, Carollee Howes, Leslie Phillipsen, Donna Bryant, Richard Clifford, Debby Cryer, Ellen Peisner-Feinberg, Margaret Burchinal, Sharon Lynn Kagan, and Jean Rustici. 1995. Cost, quality, and child outcomes in child care centers: Key findings and recommendations. *Young Children* 50: 40–44.

Helburn, Suzanne W., and Caroline Howes. 1996. Day care cost and quality. *Future of Children* 6, no. 2: 62–82.

Hendershott, Anne. 1995. A moving story for spouses and other wage-earners. *Psychology Today* 28: 28–30.

Henderson, K. K. 1990. The meaning of leisure for women: An integrative review of research. *Journal of Leisure Research* 22: 228–43.

Henderson, K. A., M. D. Bialeschki, S. M. Shaw, and V. J. Freysinger. 1989. *A leisure of one's own: A feminist perspective on women's leisure*. State College, Pa.: Venture Publishing.

Hewlett, Sylvia Ann. 1992. *When the bough breaks: The cost of neglecting our children*. New York: HarperCollins.

Hochschild, A. 1989. *The second shift*. New York: Viking.

Hochschild, A. 1997. *The time bind*. New York: Metropolitan.

Holman, T. B., and M. Jacquart. 1988. Leisure-activity patterns and marital satisfaction: A further test. *Journal of Marriage and the Family* 50: 69–77.

Israelsen, C. L. 1991. Family resource management. In *Family research: A sixty-year review*, edited by S. J. Bahr. New York: Lexington Books, 1, 171–234.

Jenson, G. Eric, and G. O. Jenson. 1997. The effects of dual career employment on adolescent sexual behaviors, school experiences, and family relationships. Paper presented at the Annual Conference of the National Council on Family Relations, Crystal City, Va., November.

Liskowsky, David R. 1992. Biological rhythms and shift work. *Journal of the American Medical Association*. 268: 3047.

Lopez, J. A. 1993. Undivided attention: How PepsiCo gets work out of people. *Wall Street Journal*. 1 April, 221: 1.

Luecken, L. J., E. C. Suarez, C. M. Kuhn, J. C. Barefoot, J. A. Blumenthal, L. C. Siegler, and R. B. Williams. 1997. Stress in employed women: Impact of marital status and children at home on neurohormone output and home strain. *Psychosomatic Medicine* 59, no. 4: 352–59.

Lye, D. N., and T. J. Biblarz. 1993. The effects of attitudes toward family life and gender roles on marital satisfaction. *Journal of Family Issues* 14: 157–58.

Martini, Mary. 1997. Balancing work and family in four cultural groups. Paper presented at the Annual Conference of the National Council on Family Relations, Crystal City, Va., November.

Martini, Mary. 1995. Balancing work and family in Hawaii: Strategies of parents in two cultural groups. *Family Perspective* 29: 103–27.

Massey, Garth, Karen Hahn, and Dusko Sekulic. 1995. Women, men, and the "second shift" in socialist Yugoslavia. *Gender and Society* 9: 359–79.

Menec, V. H., and J. G. Chipperfield. 1997. Remaining active in later life: The role of locus of control in seniors' leisure activity participation, health, and life satisfaction. *Journal of Aging and Health*. 9: 105–25.

Moen, Phyllis. 1992. *Women's two roles: A contemporary dilemma*. New York: Auburn House.

Muller, Chandra. 1995. Maternal employment, parent involvement, and mathematics achievement among adolescents. *Journal of Marriage and the Family*. 85–100.

Neuborne, Ellen. 1997. Firms today less willing to pay for play. *USA Today*. 12 March, B1.

Nowell, Paul. 1995. Stay home pay. *Daily Reflector*. 26 August, A-1.

Perry-Jenkins, Maureen, and Karen Folk. 1994. Class, couples, and conflict: Effects of the division of labor on assessments of marriage in dual-earner families. *Journal of Marriage and the Family* 56: 65–180.

Pett, Marjorie A., Beth Vaughn-Cole, and Bruce E. Wampold. 1994. Maternal employment and perceived stress. *Family Relations* 43: 151–58.

Pope, Victoria. 1997. Day-care dangers. *U.S. News & World Report*. 4 August, 31–37.

Pyke, Karen. 1994. Women's employment as a gift or burden? Marital power across marriage, divorce, and remarriage. *Gender and Society* 8: 73–91.

Pyke, Karen, and Scott Coltrane. 1996. Entitlement, obligation, and gratitude in family work. *Journal of Family Issues* 17: 60–82.

Riza, W. R., R. N. Singh, and V. T. Davis. 1992. Differences among males and females in their perception of spousal abuse. *Free Inquiry in Creative Sociology* 20: 19–24.

Roberts, Paul. 1995. Goofing off. *Psychology Today.* July/August, 34–41.

Robinson, John P. 1990. The leisure pie. *American Demographics* 12, no. 9: 39.

Rosenblatt, P. C., S. L. Titus, A. Nevaldine, and M. R. Cunningham. 1979. Marital system differences and summer-long vacations: Togetherness-apartness and tension. *American Journal of Family Therapy* 7: 77–84.

Rubenstein, C. 1991. Guilty or not guilty. *Working Mother.* May, 53–56.

Rubin, Rose M., and Bobye J. Riney. 1994. *Working wives and dual-earner families.* Westport, Conn.: Praeger.

Schnittger, M. H., and G. W. Bird. 1990. Coping among dual-career men and women across the family life cycle. *Family Relations* 39: 199–205.

Schor, Juliet. 1991. *The overworked American: The unexpected decline of leisure.* New York: Basic Books.

Schor, Juliet. 1995. Why (and how) more people are dropping out of the rat race. *Working Woman.* August, 14.

Schwartz, P., and D. Jackson. 1989. How to have a model marriage. *New Woman.* February, 66–74.

Sefton, B. W. 1998. The market value of the stay-at-home mother. *Mothering* 86: 26–29.

Shellenbarger, Sue. 1993. Longer commutes force parents to make tough choices where to leave the kids. *Wall Street Journal.* 18 August, B1.

Shelton, Beth Anne. 1992. *Women, men, and time: Gender differences in paid work, housework, and leisure.* New York: Greenwood.

Silberstein, L. R. 1992. *Dual career marriage: A system in transition.* Hillsdale, N.J.: Lawrence Erlbaum.

Skinner, K. B., and William Meredith. 1996. The division of labor and child care in urban Chinese families. Paper presented at the Annual Conference of the National Council on Family Relations, Kansas City, Missouri, 9 November.

Spade, Joan Z. 1994. Wives' and husbands' perceptions of why wives work. *Gender and Society* 8: 170–88.

Spitze, G. 1991. Women's employment and family relations: A review. In *Contemporary families: Looking forward, looking back,* edited by A. Booth. Minneapolis: National Council on Family Relations, 381–404.

Starrels, M. E., B. Ingersoll-Dayton, D. W. Dowler, and M. B. Neal. 1997. The stress of caring for a parent: Effects of the elder's impairment on an employed adult child. *Journal of Marriage and the Family* 59: 860–72.

Statistical Abstract of the United States: 1995. 115th ed. Washington, D.C.: U.S. Bureau of the Census.

Statistical Abstract of the United States: 1997. 117th ed. Washington, D.C.: U.S. Bureau of the Census.

Stohs, Joanne H. 1995. Predictors of conflict over the household division of labor among women employed full-time. *Sex Roles* 33: 257–75.

Straus, M., R. Gelles, and S. Steinmetz. 1980. *Behind closed doors.* New York: Doubleday.

Tangri, S. S., and S. R. Jenkins. 1997. Why expecting conflict is good. *Sex Roles* 36: 725–46.

Teleki, J. K., C. W. Snow, and J. Reguero de Atiles. 1992. A comparative study of 1981 and 1991 childcare center licensing regulations in the United States. Paper presented at the Annual Conference of the National Association for the Education of Young Children, New Orleans, November. Used by permission.

Thornton, A. 1992. The influence of the parental family on the attitudes and behavior of children. In *The Changing American Family,* edited by S. J. South and S. E. Tolnay. Boulder, Colo.: Westview, 247–66.

Tuttle, Robert. 1996. Marital happiness among self-employed and organizationally employed workers. Paper presented at the Annual Conference of the National Council on Family Relations, Kansas City, Missouri, 9 November. Used by permission.

Vannoy, D., and W. W. Philliber. 1992. Wife's employment and quality of marriage. *Journal of Marriage and the Family* 54: 387–98.

CHAPTER 10 Abuse in Relationships

My husband is going to kill me.

Nicole Simpson, calling 911 about O. J. Simpson

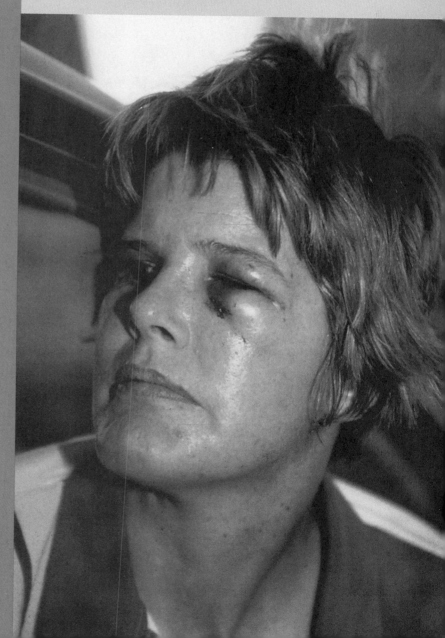

"I cared so much about him," said country-and-western singer Tanya Tucker of a previous partner, "and I saw what things can happen, even with people who love each other very much, and how angry and violent love can be" (Tucker, 1997). Her words reflect the reality of abuse in an intimate love relationship which can be initiated by either partner.

As social beings, we become involved with others partly for love, intimacy, and companionship. Because we regard relationships as a major source of our happiness and fulfillment, we are sometimes shocked to learn of or to experience abuse (and sometimes murder) between lovers and spouses. In this chapter we examine abusive relationships—factors that cause violence/abuse, the effects of abuse, and the cycle of abuse. Our focus is not only dating, cohabiting, and married couples but abuse in reference to children, siblings, and elderly parents. Violence/abuse in relationships has always occurred; its visibility has been recent.

Definitions of Violence, Abuse, and Neglect

Violence may be defined as the intentional infliction of physical harm by either partner on the other. Examples of physical violence/abuse include pushing, throwing something at the partner, slapping, hitting, and forcing sex on the partner. Though use of a weapon does occur, O'Keefe (1997) found that less than 2 percent of the physically aggressive behaviors in dating relationships reported by 939 high school students involved a knife or gun.

While physical violence can be injurious, *emotional abuse* (also known as *verbal abuse* or *symbolic aggression*) may be even more damaging to the individual (O'Hearn & Davis, 1997). Though such abuse does not involve physical harm, it is designed to reduce the victim's status and increase the victim's vulnerability, thereby giving the abuser more control. Examples of emotional abuse by either partner include

threatening to leave the partner

controlling who the partner socializes with

cutting off the partner from his or her family

controlling the money of the partner to ensure dependence

refusing to talk to the partner

accusing the partner of infidelity

telling the partner that he or she is repulsive

threatening to harm self if the partner leaves relationship

threatening to harm the partner

calling the partner obese, stupid, or ugly

telling the partner she or he is pathetic and no one else would want him or her

turning one's face when the partner attempts a kiss

refusing to touch the partner

threatening to harm the partner's relatives

telling the partner that he or she is a terrible lover

threatening to have an affair with someone else

. . . he didn't speak to her until six weeks later when she discovered he had taken all her credit cards from her wallet.

Charlotte Fedders
Shattered Dreams

threatening to take one's children away

insulting the partner in front of others

restricting the partner's use of the car

criticizing the partner's child care, food preparation, house care, or job performance

threatening to have the partner committed to a mental institution

telling the partner he or she is crazy

demanding the partner do as he or she is told

anything that demeans the partner

INSIGHT

Emotional abuse often precedes physical aggression in episodes between intimate partners (Murty & Roebuck, 1992). One way to avoid physical violence in a relationship is for partners to withdraw from interaction that involves escalating verbal aggression.

A common motivational thread of these behaviors is the feeling of ownership and the desire to exercise power and control over the partner (Dutton & Starzomski, 1997). Partners who question the way they are being treated are told that they are insane and that they deserve such treatment.

Abuse can also involve neglect. So much attention is typically paid to physical violence that the impact of neglect is often minimized. Neglect most often occurs in reference to children or dependent adults, whose caretakers "neglect" to provide them with adequate food, medical treatment, and personal hygiene. Finally, abuse in relationships can involve using force or coercion to engage in sexual activity with a child or nonconsenting adult. Having defined the various parameters of violence and abuse, we now examine these phenomena in dating relationships.

RECENT RESEARCH

Krishnan et al. (1997) surveyed 242 Anglo, Hispanic, and Native American women who had left an abusive relationship. Results indicated that violence started early in relationships, was frequent, and severe enough to be disclosed to others.

RECENT RESEARCH

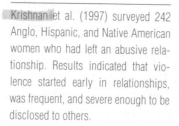

Bowman and Morgan (1998) assessed the abusive behaviors of 209 undergraduates and found that women more often than men reported being verbally or physically aggressive. In addition, bisexuals and lesbians reported higher rates than heterosexuals of verbal and/or physical aggressiveness toward their partners. However, gay men reported lower rates than heterosexuals of verbal and/or physical aggressiveness.

Abuse (Physical or Emotional) in Dating Relationships

Many spouses who have been abused by their partners have noted that during dating and courtship, their partners never mistreated them. Indeed, in many cases, abuse in relationships does not occur until after the couple is married or living together. However, violent and abusive behavior may also be experienced in dating relationships. The Personal Application, Abusive Behavior Inventory at the end of this chapter allows you to assess the level of abuse in your current relationship.

During courtship, women and men usually try to make the most positive impression possible on their dating partners. Nevertheless, violence in dating relationships is not uncommon. It is estimated that 15 percent to 20 percent of high school students have at least one violent interaction with a dating partner (Riggs & Caulfield, 1997). Such estimates may be low. In a study of 939 high school students involved in dating relationships, 43 percent of the women and 39 percent of the men reported that they had inflicted some form of physical aggression on their partners at least once (O'Keefe, 1997). Table 10.1 shows the percentage of men and women reporting various forms of physical aggression.

These data reflect that when dating violence occurs, it is often mutual. Gray and Foshee (1997) also studied violence in relationships among 185 adolescents from the sixth through twelfth grades and found that 60 percent of the

Table 10.1

	Men (N = 385)	Women (N = 554)
Forms of Physical Aggression Reported by High School Students Who Reported Ever Inflicting Violence on Someone They Were Dating		
Form of Aggression		
Pushed, grabbed, or shoved him or her	27.5%	32.0%
Threw something at him or her	16.0%	27.4%
Slapped him or her	12.0%	25.6%
Kicked, bit, or hit with a fist	9.6%	19.3%
Hit or tried to hit with something	9.6%	16.8%
Beat up	3.4%	3.1%
Threatened with a knife or gun	1.3%	2.5%
Used a knife or gun	.5%	1.3%
Forced sexual activities on partner	12.7%	3.0%

Source: Adapted from Maura O'Keefe, 1997. Predictors of dating violence among high school students. *Journal of Interpersonal Violence* 12, no. 4, 546–68. Table is on page 556. Adapted with the permission of the publisher (Sage Publications, Inc., 2455 Teller Road, Thousand Oaks, CA 91320) and the author.

Jim knotted one hand in Molly's hair and pounded her head against the dashboard.
Angela Browne
When Battered Women Kill

students reporting violence said that it was mutual. The longer these individuals had dated, the more involved their relationship, and the more exclusive their relationship, the greater their violence potential. Also, 96.1 percent of the individuals involved in mutually violent relationships reported that their relationships were "very good," and 63 percent were still dating the partner (p. 136).

Estimates of violence in dating relationships among college students are even higher than among high school students and range from 20 percent to 50 percent (Riggs & Caulfield, 1997). In one study of college dating couples, one-third reported violence in their relationships in the previous twelve months (Hanley & O'Neill, 1997). Also, among these couples, the more emotionally committed the partners, the greater the likelihood that they had experienced violence in their relationship. The researchers suggested that with increasing involvement in a relationship, the potential to disagree over various issues (including the level of commitment) also increases, and such disagreements may become heated and violent (Hanley & O'Neill, 1997).

OTHER CULTURES Cross-cultural data on 944 21-year-olds in New Zealand revealed that women were four times as likely as men to be victims of a physical assault (hit, punched, kicked, etc.) by their partners (Langley et al., 1997). The researchers hypothesized that men may be less willing to report that they were assaulted as well as less likely to define physical aggression as serious and worthy of reporting. ●

Alcohol influences the likelihood of being raped by an acquaintance.

Furthermore, individuals in emotionally and physically abusive relationships may be more vulnerable to being stalked when the relationship ends. *Stalking* is defined as the willful, repeated, and malicious following or harassment of another person. Coleman (1997) studied data from 144 undergraduate college women and found that those who had been in romantic but abusive relationships were more likely to have been stalked than those who had not been in such abusive relationships. Such stalking is usually designed either to seek revenge or to win the partner back.

Twenty-two percent of 872 women and 13 percent of 527 men, aged 19 to 22 (both college and noncollege), in Michigan reported that they had been pressured into having intercourse. When asked specifically if they had been raped, 13 percent of the women and 1 percent of the men reported that they had been raped. Hence, a much larger percentage of individuals reported having experienced pressure to have sex than reported having been raped (Zweig et al., 1997). Rape rates also differed between college and noncollege students, with the latter more likely to have been raped. Ten percent of the college women, in contrast to 16 percent of the noncollege women, reported having been raped. Differences between college and noncollege men were not significant.

There are different definitions of rape. Criminal law distinguishes between forcible rape and statutory rape. *Forcible rape* includes three elements: (1) vaginal penetration (2) by force or threat of force (vs. being persuaded or cajoled) (3) without consent of the victim. Hence, while some sexual assaults may not be considered "rape" per se, they are still considered abuse and are punishable by law. *Statutory rape* involves sexual intercourse without use of force with a person below the legal age of consent. *Marital rape*, now recognized in all states, is forcible rape by one's spouse.

The word *rape* often evokes images of a stranger jumping out of the bushes or a dark alley to attack an unsuspecting victim. However, most rapes are perpetrated not by strangers but by persons who have a relationship with the victim.

OTHER CULTURES While sex between a 21-year-old and a 14-year-old is considered statutory rape in the United States, in Mexico it is perfectly acceptable. Pedro Sotelo, who was living with his pregnant girlfriend in Houston, Texas, said, "I didn't think I was doing anything wrong. It's our culture in Mexico; young love is the blood of life. When there is love, there is no problem" (Sanchez, 1996). But Houston authorities jailed Pedro and placed him under $200,000 bond. If convicted, he could face 99 years in prison. ●

NATIONAL DATA   Only 15 percent of a national sample of rape victims reported that they had been raped by a stranger; 85 percent had been raped by someone they knew (Koss et al., 1988).

Among the rape victims studied by Koss et al. (1988), 35 percent were attacked by their boyfriend or lover; 29 percent by a friend, coworker, or neighbor; 25 percent by a casual date; and 11 percent by their husband. In a random sample of sorority women at Purdue University, 63 percent reported that since attending college they had experienced a man's attempting to force sexual

Rohypnol (called the "date rape drug") is a drug some men slip into the drinks of women they are about to rape. Although the drug is illegal in the United States, it causes profound, prolonged sedation, a feeling of well-being, and short-term memory loss. Universities have begun to warn students not to take drinks from guys they do not know.

It was a real kinship we had together. But when his mood changed, he could be violent and vicious.

Lenore Walker
The Battered Woman

intercourse (get on top of her, attempt to insert penis) against their will; 95 percent of the women reported that they knew their attacker (Copenhaver & Grauerholz, 1991).

These data reflect the fact that most rape is *acquaintance rape,* which may be defined as "nonconsensual sex between adults who know each other" (Bechhofer & Parrot, 1991, 12). One type of acquaintance rape is *date rape,* which refers to nonconsensual sex between people who are dating or on a date.

Date rapes are not necessarily planned. Bechhofer and Parrot (1991) noted:

> He plans the evening with the intent of sex, but if the date does not progress as planned and his date does not comply, he becomes angry and takes what he feels is his right—sex. Afterward, the victim feels raped, while the assailant believes that he has done nothing wrong. He may even ask the woman out on another date (Bechhofer & Parrot, 1991, 11).

Some date rapes, however, occur in predictable contexts. Three members of a fraternity at the University of North Carolina at Chapel Hill issued a letter to potential pledges implying that their fraternity parties create the context in which date rapes may occur:

> The forecast calls for a 99 percent chance of getting sex from one of the many new beautiful sorority pledges as they stumble around the dance floor in a drunken stupor bordering on the brink of alcohol poisoning (O'Brien, 1995, 1B).

Other date rapes occur when a boyfriend decides to rape his girlfriend. The following describes such an experience as recalled by a woman who was raped by her boyfriend on a date:

> Last spring, I met this guy and a relationship started, which was great. One year later, he raped me. The term was almost over and we would not be able to spend much time together during the summer. Therefore, we planned to go out to eat and spend some time together.
>
> After dinner we drove to a park. I did not mind or suspect anything for we had done this many times. Then he asked me into the back seat. I got into the back seat with him because I trusted him and he said he wanted to be close to me as we talked. He began talking. He told me that he was tired of always pleasing me and not getting a reward. Therefore, he was going to "make love to me" whether I wanted to or not. I thought he was joking so I asked him to stop playing. He told me he was serious and after looking at him closely, I knew he was serious. I began to plead with him not to have sex with me. He did not listen. He began to tear my clothes off and confine me so that I could not move. All this time I was fighting him. At one time I managed to open the door, but he threw me back into the seat, hit me, then he got on me and raped me. After he was satisfied, he stopped, told me to get dressed and stop crying. He said he was sorry it had to happen that way.
>
> He brought me back to the dorm and expected me to kiss him good night. He didn't think he had done anything wrong. Before this happened, I loved this man very much, but afterward I felt great hatred for him.

My life has not been the same since that night. I do not trust men as I once did, nor do I feel completely comfortable when I'm with my present boyfriend. He wants to know why I back off when he tries to be intimate with me. However, right now I can't tell him, because he knows the guy who raped me (authors' files).

While date rapes do occur, it is important to keep in mind, that friends, coworkers, and neighbors also rape and that these are trusted "acquaintances" who do not fit into the category of "date."

Shapiro and Schwarz (1997) studied the effects of date rape on forty-one college women. They reported higher levels of anxiety, depression, anger/irritability, and sexual dysfunctions than college women who had not experienced date rape. They also had lower levels of sexual self-esteem.

OTHER CULTURES Choquet et al. (1997) studied the effects of rape on sixty-one French adolescents and compared them with a group of adolescents who had not been raped. Sleep difficulties, depression, violent behavior, stealing, and somatic complaints were associated with those adolescents who had been raped. ●

Abuse in Cohabitation and Marriage Relationships

Abuse occurs not only in dating relationships but also in cohabitation and marriage relationships.

Abuse in Cohabitation Relationships

RECENT RESEARCH

Magdol et al. (1998) compared cohabitors and daters among 21-year-olds and found that cohabitors were significantly more likely to report abusive behaviors.

NATIONAL DATA   In a national sample of 5,768 couples, 35 percent of the cohabiting couples reported that one of the partners had been physically assaulted by the other partner during the previous year (Stets & Straus, 1989).

This information may be particularly accurate in that data from couples have been shown to be more reflective of the actual violence in a relationship than reports from only one partner (Bohannon et al., 1995).

Not only is the rate of physical violence highest among unmarried cohabiting couples, but also the severity of violence is greater among cohabiting couples than among dating and marital partners (Stets & Straus, 1989). In living-together relationships that experience violence, the precipitating disagreement is likely to be about jealousy or sex (Makepeace, 1989).

Forced Sex in Cohabitation Relationships

Almost 30 percent (29%) of a national sample of cohabiting females reported that they had at some time been forced by a man to do something sexual

(Michael et al., 1994, 224). Though not all of these females were forced to have penile-vaginal or penile-anal intercourse by the man who was living with them, some involvement with the partner is not unusual. The following National Data chart shows the relationship of a sample of women who had been forced by the man to do something sexual.

NATIONAL DATA   Relationship of Women Respondents to Man who Forced the Woman to Do Something Sexual

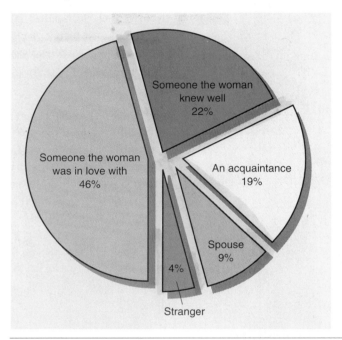

Source: From *Sex in America* by Michael et al. Copyright © 1994 by CSG Enterprise, Inc., Edward O. Laumann, Robert T. Michael, and Gina Kolata. By permission of Little, Brown and Company, p. 225.

Abuse in Marriage Relationships

Johnson identified two forms of violence against women. Common couple violence is an "intermittent response to the occasional conflicts of everyday life, motivated by a need to control in the specific situation, but not by a more general need to be in charge of the relationship" (1995, 286). (Either the man or the woman may become frustrated and angry and hit the other.)

NATIONAL DATA   Researchers estimate that common couple violence occurs an average of once every two months (Johnson, 1995, 287).

My body was not unmarked on any limb.
Brett Butler
Comedian

More sinister violence against women is that which has its roots in patriarchal terrorism "adopted with a vengeance by men who feel that they must control 'their' women by any means necessary" (Johnson, 1995, 286). Men who say "I married you so I own you" may beat their wives not only to control them but also because of a need to display that control.

Violence or a threat of it is not the only means by which men control women. Priscilla Presley noted that Elvis, at age 24, selected her as his steady girlfriend (age 14) and later as his wife because he could control and train her to do what he wanted. Noting that Elvis was rarely without his entourage, Priscilla said that no woman his age would put up with his demands to always have his friends around. She also noted that he controlled her by implied threat of replacement—if she did not do as he wished, because he had so many women available to him, he would simply replace her. In addition, the relationship was not without violence. Priscilla reported occasions in which Elvis would lose his temper and strike her (Walters, 1995).

When a child can get a gun more easily than a library card, we know we are heading in the wrong direction.

Lonnie Bristow

President

American Medical Association

NATIONAL DATA   Researchers estimate that patriarchal terrorism beatings by the husband occur an average of more than once a week (Johnson, 1995, 287).

"The central motivating factor behind the violence is a man's desire to exercise general control over 'his' woman" (Johnson, 1995, 287). Such control is the quintessential reflection of patriarchal thinking, where women are viewed as property and must be regularly beaten to "keep them in their place." The conflict theorist views such violence in the context of power.

> Violence is usually a vehicle to power. In the absence of authority—legitimized power—people can get others to do what they want them to do through coercion or force. Likewise, through violence or the threat of violence a person can gain dominance in a relationship (Winton, 1995, 99).

Rape in Marriage

Abuse in marital relationships may take the form of rape and sexual assault. In Russell's study of female San Francisco residents, 14 percent of married women revealed having been sexually assaulted by their husbands (1990). Ten percent of married women in a Boston survey reported that they had been raped by their husbands (Finkelhor & Yllo, 1988). Such rapes may have included not only intercourse but also other types of sexual activities in which the wife did not want to engage, most often fellatio and anal intercourse. One type of marital rape identified by the researchers is battering rape. Battering marital rapes occur in the context of a regular pattern of verbal and physical abuse. The husband yells at his wife, calls her names, slaps, shoves, and beats her. Such husbands are angry and belligerent and frequently are alcohol abusers. An example follows:

> One afternoon she came home from school, changed into a housecoat, and started toward the bathroom. He got up from the couch where he had been lying, grabbed her, and pushed her down on the floor. With her face pressed into a pillow and his hand clamped over her mouth, he proceeded to have anal intercourse with her. She screamed and struggled to no avail. Her injuries were painful and extensive. She had a torn muscle in her rectum, so that for three months she had to go to the bathroom standing up (Finkelhor & Yllo, 1988, 144–45).

Factors Contributing to Abuse in Relationships

Research suggests that numerous factors contribute to violence and abuse in intimate relationships. These factors include those that occur at the cultural, community, and individual and family levels.

Cultural Factors

In many ways, American culture tolerates and even promotes violence. Violence in the family stems from the acceptance of violence in our society as a legitimate

U.S. culture glorifies violent interpersonal aggression.

Dr. Murray A. Straus, Co-Director of the Family Research Laboratory at the University of New Hampshire, has conducted extensive research on child spanking.

RECENT RESEARCH

Forty-four percent of 807 mothers of children aged 6 to 9 in a national survey reported having spanked their children during the week prior to the study and that they spanked them an average of 2.1 times that week. The more spanking they engaged in, the higher the level of antisocial behavior of the children two years later (Straus et al., 1997).

means of enforcing compliance and solving conflicts at interpersonal, familial, national, and international levels (Viano, 1992). Violence and abuse in the family may be linked to cultural factors, such as violence in the media, acceptance of corporal punishment, gender inequality, and the view of women and children as property.

1. *Violence in the media.* According to a study conducted by MediaScope (an independent monitor) of 2,500 hours of television programming over a twenty-week period, almost 60 percent (57%) of the dramas, comedies, children's series, movies, and music videos contained violence. Such massive and relentless exposure to violence reveals that most Americans who watch television are on a steady diet of violence.

NATIONAL DATA   The results of seven national data sets show that 68 percent of adults (men more than women and more blacks than whites) agree that "it is sometimes necessary to discipline a child with a good hard spanking" (Straus & Mathur, 1996).

2. *Acceptance of corporal punishment.* Violence has become a part of our cultural heritage through the corporal punishment of children. Undergraduates testify to the fact that their parents used corporal punishment. Eighty-three percent of over 11,000 undergraduate students at the University of Iowa reported that they had experienced some form of physical punishment during their childhood (Knutson & Selner, 1994). In his book *Beating the Devil Out of Them*, Murray Straus (1994) noted that one of the effects of corporal punishment on children is that the child has an increased chance of being violent in his or her own life (p. 151). In the short run, children who are victims of corporal punishment display more antisocial behavior (Straus et al., 1997).

3. *Gender inequality.* Domestic violence and abuse may also stem from traditional gender roles. Traditional male gender roles have taught men to be

aggressive. Traditionally, men have also been taught that they are superior to women and that they may use their aggression toward women, believing that women need to be "put in their place." Traditional female gender roles have also taught women to be submissive to their male partners' control.

Anderson (1997) studied the demographic characteristics of a national sample of individuals involved in domestic violence and found that men who earn less money than their partners are more likely to be violent toward them. "Disenfranchised men then must rely on other social practices to construct a masculine image. Because it is so clearly associated with masculinity in American culture, violence is a social practice that enables men to express a masculine identity (p. 667).

4. *View of women and children as property.* Prior to the late nineteenth century, a married woman was considered the property of her husband. A husband had a legal right and marital obligation to discipline and control his wife through the use of physical force.

OTHER CULTURES Barker and Loewenstein (1997) investigated the attitudes of 127 low-income adolescent and young adult males in Rio de Janeiro, Brazil, and found that males viewed violence toward women as acceptable in many circumstances. ●

Community Factors

Community factors that contribute to violence and abuse in the family include social isolation, poverty, and inaccessible or unaffordable health-care, day-care, elder-care, and respite-care services and facilities.

Returning violence for violence multiplies violence, adding deeper darkness to a night already devoid of stars.
Martin Luther King, Jr.
Civil rights leader

1. *Social isolation.* Living in social isolation from extended family and community members increases a family's risk for abuse. Isolated families are removed from material benefits, caregiving assistance, and emotional support from extended family and community members. Also, parents who have little contact with others in the community do not have exposure to positive role models for effective parental behavior (Harrington & Dubowitz, 1993).

2. *Poverty.* Abuse in adult relationships occurs among all socioeconomic groups. However, Kaufman and Zigler point to a relationship between poverty and child abuse:

   Although most poor people do not maltreat their children, and poverty, per se, does not *cause* abuse and neglect, the correlates of poverty, including stress, drug abuse, and inadequate resources for food and medical care, increase the likelihood of maltreatment (1992, 284).

3. *Inaccessible or unaffordable community services.* Failure to provide medical care to children and elderly family members sometimes results from the lack of accessible or affordable health-care services in the community. Failure to provide supervision for children and adults may result from inaccessible

day-care and elder-care services. Without elder-care and respite-care facilities, families living in social isolation may not have any help with the stresses of caring for elderly family members and children.

Individual and Family Factors

Individual and family factors associated with domestic violence and abuse include psychopathology, biology, family history of violence, alcohol and other drug abuse, and fatherless homes.

1. *Psychopathology.* Some abusing spouses and parents have psychiatric conditions that predispose them to abusive behavior. Symptoms of psychiatric conditions that are related to violence and abuse include low frustration tolerance, emotional distress, and inappropriate expression of anger. Adults who sexually abuse children in their family may have developed a sexual fetish that necessitates the presence of a young child to provide sexual arousal.

   A number of personality characteristics have been associated with persons who are abusive in their intimate relationships. They include dependency, jealousy, need to control, unhappiness and dissatisfaction, anger and aggressiveness, quick involvement, blaming others for problems, and a Jekyll-and-Hyde personality (Vaselle-Augenstein & Ehrlich, 1992).

   a. *Dependency.* Therapists who work with batterers have observed that they are extremely dependent on their partners. Because the thought of being left by their partners induces panic and abandonment anxiety, batterers use physical aggression and threats of suicide to keep their partners with them.

   b. *Jealousy.* Along with dependence, batterers exhibit jealousy, possessiveness, and suspicion. An abusive husband may express his possessiveness by isolating his wife from others; he may insist she stay at home, not work, and not socialize with others. His extreme, irrational jealousy may lead him to accuse his wife of infidelity and beat her for her presumed affair.

   c. *Need to control.* Batterers are often described as individuals who have an excessive need to be in control. They do not let their partners make independent decisions, and they want to know everything their partners do. They like to be in charge of all aspects of family life, including finances and recreation. In abusive relationships, one partner's need for control takes precedence over the needs of the other partner, who is submissive to the controlling partner (Stets, 1992).

   d. *Unhappiness and dissatisfaction.* Abusive partners often report being unhappy and dissatisfied with their lives, both at home and at work. Many batterers have low self-esteem and high levels of anxiety, depression, and hostility. Extremely unhappy people often do not care how unhappy they make others and expect others to "make" them happy.

   e. *Anger and aggressiveness.* Batterers are often described as aggressive individuals with a history of interpersonal aggressive behavior. They

In a very deep-seated way, it is the fear of loss that drives someone to use intimidation to keep a partner in line.

Jacques Cook

Marriage therapist

In all the videotapes we made, never did we hear a batterer say anything like, "That's a good point."

Neil Jacobson

John Gottman

Psychologists

RECENT RESEARCH

Ronfeldt et al. (1998) analyzed the violence patterns of 156 male college students and found that the more they were dissatisfied with the power they had in their relationships, the more psychologically and physically abusive they were likely to be.

Even if I did do this, it would have to have been because I loved her very much, right?

O. J. Simpson

in *Esquire* on the murder of ex-wife Nicole Brown-Simpson

RECENT RESEARCH

Ascione (1997) studied 38 women seeking shelter at a safe house for battered women. Of the women reporting present or past pet ownership, 71 percent reported that their partner had threatened and/or actually hurt or killed one or more of their pets. Of the women who had children, 32 percent reported that one or more of their children had hurt or killed pet animals.

are also described as having poor impulse control in dealing with anger. Battered women report that episodes of violence are often triggered by minor events, such as a late meal or an unironed shirt.

f. *Quick involvement.* Because of feelings of insecurity, the potential batterer will move his partner quickly into a committed relationship. If the woman tries to break off the relationship, the man will often try to make her feel guilty for not giving him and the relationship a chance.

g. *Blaming others for problems.* The abuser takes little responsibility for his problems and blames everyone else. For example, when he makes a mistake, he will blame the woman for upsetting him and keeping him from concentrating on his work. He tells the woman she is at fault for almost everything that is wrong in the relationship.

h. *Jekyll-and-Hyde personality.* The abuser has sudden mood changes so that his partner is continually confused. One minute he is nice, the next minute angry and accusatory. Explosiveness and moodiness are typical.

i. *Isolation.* The abusive person will try to cut off the person from all family, friends, and activities. Ties with anyone are prohibited. Isolation may reach the point at which the abuser tries to stop the victim from going to school, church, or work.

2. *Biological factors.* Higher levels of testosterone have been associated with increased violence. Football players have used steroids (synthetic testosterone) to increase their aggression (as well as muscle development and strength). However, increased levels of testosterone alone do not produce violence but interact with other social and psychological factors (McKenry et al., 1995).

3. *Family history of abuse.* Most adults who abuse or resort to violence with their own children either were abused themselves or witnessed such abuse as children. However, a majority of those who were abused or witnessed abuse do not continue the pattern. Pagelow (1992) emphasized that a family history of violence is only one factor out of many that may be associated with a greater probability of adult violence. Lackey and Williams (1995) emphasized that regardless of the history of violence in one's family of origin, persons who are socially bonded/emotionally attached to their partners have a decreased probability of being violent with them.

4. *Alcohol and other drug use.* O'Keefe (1997) found that higher use of alcohol and other drugs was associated with violence in one's dating relationship. Whether alcohol acts to inhibit one's use of violence, acts to allow one to avoid responsibility for being violent, or increases one's aggression are still being debated. But the relationship between using alcohol and violence was clear in this sample of over 1,000 high school students in eight schools in the Los Angeles area.

5. *Fatherless homes.* Living in a home where the father is absent increases a child's risk for being abused. Numerous studies show that children are more likely to be sexually abused by a stepfather or a mother's boyfriend

than by their biological father (Blankenhorn, 1995). This is largely because stepfathers and mothers' boyfriends are not constrained by the cultural incest taboo that prohibits fathers from having sex with their children.

Sexual Abuse: Some Causes

A number of factors may help to explain why rape and sexual assault occur between dating partners, cohabiting partners, and spouses. These factors include gender role socialization, rape-tolerant attitudes, and low self-esteem.

Gender role socialization   In our society, some men are socialized to be sexually aggressive and to view women as objects for sexual gratification. To be sexually aggressive is, for some men, a part of the masculine role.

As noted earlier, women have long been viewed as the property of men. Forcing his wife to have sex was viewed as a husband's right. Married women who were raped by their husbands could not have their husband arrested because marital rape was not considered a crime. In 1978, only three states recognized marital rape as a crime. In 1993, North Carolina became the fiftieth state to recognize marital rape as a crime (National Clearinghouse on Marital and Date Rape, 1994).

Rape-tolerant attitudes   Rape-tolerant attitudes, also called "rape-supportive attitudes" and "rape myths," are associated with self-reported rapes (Frank, 1991). Rape-tolerant attitudes, which may be learned from family, friends, and mass media, "are the mechanism that people use to justify dismissing an incident of sexual assault from the category of 'real' rape" (Burt, 1991, 27). For example, 13 percent of all first-year college students throughout the United States *did not* agree with the statement, "Just because a man thinks that a woman has led him on, does not entitle him to have sex with her" (American Council on Education and University of California, 1997). Believing that a man is entitled to have sex with a woman who has led him on is an example of a rape-tolerant attitude. Such attitudes lead some individuals to initiate acts of forced sex and also serve as a subsequent justification for engaging in rape behavior.

In a study of rape-tolerant attitudes among college students, Gilmartin (1994) found that men are more likely than women to have rape-tolerant attitudes. For example, men are less likely than women to believe that when a woman says no to sexual advances, she really means it (see Table 10.2).

Low self-esteem   Sexual abuse has been linked to low self-esteem. In a study of sexual abuse among college dating partners, Pirog-Good (1992) suggested that women with low self-esteem may be more likely to force sexual behaviors on their partners because the act of sex gives them the feeling of being wanted, desired, loved, "and thus important" (p. 108). Hence, when they have sex, even though they may force it, it has the effect of increasing their positive feelings about themselves. Women who are victims of sex abuse may also have low self-esteem and may feel that they deserve the abuse.

Table 10.2

Percentage of College Men and Women Agreeing with Rape-Tolerant Statements		
Statement	**Men** **(N = 407)**	**Women** **(N = 422)**
A man sees sex as an achievement or notch in his belt.	47.9	67.8*
Deep down, a woman likes to be whistled at on the street.	54.8	38.9*
If a woman is heavily intoxicated, it is OK to have sex with her.	22.6	1.9*
Some women ask to be raped and enjoy it.	44.7	20.6*
Rape is often provoked by the victim.	30.5	14.7*
If a woman says "no" to having sex, she means "maybe" or even "yes."	36.9	21.1*

*Indicates a statistically significant gender difference.
Source: D. R. Holcomb, L. C. Holcomb, K. A. Sondag, and N. Williams. 1991. Attitudes about date rape: Gender differences among college students. *College Student Journal* 25: 434–39. Reprinted by permission of *College Student Journal.*

Effects of Abuse

Not surprisingly, marital violence is associated with unhappy marital relationships (Bowman, 1990). Physical and emotional abuse are no doubt a factor in many divorces. In addition to affecting the happiness and stability of relationships, abuse affects the physical and psychological well-being of the victim.

Effects of Abuse on Partner

The most obvious effect of physical abuse by an intimate partner is physical injury. Indeed, former Surgeon General Antonia Novello noted that battering is the single major cause of injury to women in the United States. As many as 35 percent of women who seek hospital emergency room services are suffering from injuries incurred by battering (Novello, 1992).

When the abuse is sexual, it may be more devastating than sexual abuse by a stranger. The primary effect is to destroy the woman's ability to trust men in intimate interpersonal relationships. In addition, the woman raped by her husband lives with her rapist and may be subjected to repeated assaults. Most women raped by their husbands are raped on multiple occasions (Finkelhor & Yllo, 1988).

Violence among intimate partners or ex-partners may also include unintentional death and intentional murder. Each day, four women in the United

States are killed by an abusing partner. The FBI reports that 30 percent of female homicide victims are killed by their husbands or boyfriends and 6 percent of male homicide victims are killed by their wives or girlfriends (North Carolina Coalition Against Domestic Violence, 1991).

Other less obvious effects of abuse by one's intimate partner include fear, feelings of helplessness, confusion, isolation, humiliation, anxiety, depression, stress-induced illness, symptoms of post-traumatic stress disorder, and suicide attempts.

Effects of Partner Abuse on Children

Abuse between adult partners affects children. About 40 percent of battered women are abused during their pregnancy, resulting in a high rate of miscarriage and birth defects (North Carolina Coalition Against Domestic Violence, 1991). The March of Dimes has concluded that the physical abuse of pregnant women causes more birth defects than all the diseases put together for which children are usually immunized (Gibbs, 1993).

Witnessing marital violence is related to emotional and behavioral problems in children. O'Keefe (1994) studied 185 children and their mothers who were residents at women's shelters and found that the children exhibited more externalizing (aggression, acting out) and internalizing (depression, anxiety) than children who did not come from maritally violent homes. "The finding that the amount of marital violence witnessed by the child is significantly related to child adjustment highlights the deleterious effects of family violence on children's adjustment and underscores the need for intervention with this vulnerable group" (O'Keefe, 1994, 412). Many children are resilient and not permanently harmed by abuse. But abuse does increase their potential for problems.

Cycle of Abuse

The cycle of abuse occurs when the person first is abused and the perpetrator then feels regret, asks for forgiveness, and starts acting nice. The victim, who perceives few options and feels guilty terminating the relationship with a partner who asks for forgiveness, does not call the police or file charges and stays in the relationship. There is usually a makeup or honeymoon period, during which the person feels good again about the partner. But tensions mount in the relationship, again resulting in stress, anger, and tension release in the form of violence. Such violence is followed by the familiar sense of regret and pleadings for forgiveness accompanied by being nice, etc.

The rest of this section discusses reasons why people stay in an abusive relationship and how to get out of such a relationship.

Why Individuals Stay in Abusive Relationships

One of the most frequently asked questions of people who are abused by their partners is, "Why don't you get out of the relationship?" Reasons why abused

RECENT RESEARCH

McNeal and Amato (1998) analyzed longitudinal data and found that parental reports of marital violence predicts offspring's reports of negative outcomes in early adulthood, including poorer parent-child relationships, lower psychological well-being, and more violence in their own relationships.

INSIGHT

Individuals in previously abusive relationships must be sensitive to mounting tension and reduce it or withdraw from it before it is expressed physically or emotionally. Some partners cue each other that "the stress level is rising" so that either or both can act to reduce it.

Community education programs on dating violence, marital violence, child abuse, elder abuse, etc., should also be expanded to a nationwide state-by-state awareness program. Such an emphasis would make it more acceptable to talk about abuse and to seek help for it.

Shelters for battered women are important. But individuals, communities, and states all have a responsibility to confront the problem of abusive relationships.

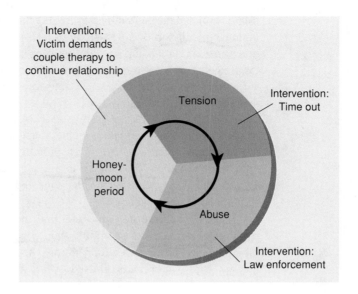

partners stay in the abusive relationship include love, emotional dependency, commitment to the relationship, hope, view of violence as legitimate, guilt, fear, economic dependency, and feeling stuck.

Love   Despite the physical and emotional pain of the abuse, an abused partner often feels love for the abuser. Love feelings may be maintained by the fact that abusive partners do not always behave in abusive ways; they may also act in positive and loving ways. In a study of abused women in dating relationships, those who were more likely to stay in the relationship had partners who engaged in a high frequency of positive behaviors (Kasian & Painter, 1992). The researchers hypothesized that "the presence of positive behaviors maintains a relationship regardless of the level of negative experiences" (p. 361).

Emotional dependency   Just as abusive partners are often emotionally dependent on the persons they abuse, so are abused partners emotionally dependent on their abusers. Such codependency was expressed by a woman who said:

> I know my boyfriend treats me badly, but I wouldn't know what to do without him. I need him. I would rather put up with the abuse than be alone without him (authors' files).

Some battered women also have very low self-esteem and feel that they would be incapable of attracting a new partner, so that staying in an abusive relationship is not such a terrible alternative.

Commitment to the relationship   Abused partners, especially those in marital relationships, may feel committed to the relationship. Some abused spouses stay in the relationship because they don't believe in divorce; they believe that marriage is a permanent relationship, no matter what the quality of the marriage.

Hope   Abused partners may stay in the relationship because they hope the relationship will improve. In a study of forty-four abused women who stayed in the abusive relationship, over 60 percent stated that "no matter how bad my abuse gets, I believe there is a chance that our relationship will get better" and "if my partner promised never to abuse me again, I would believe him" (Herbert, Silver, & Ellard, 1991, 320).

View of violence as legitimate   Some abused partners stay in the relationship because they accept violence as a legitimate part of intimate relationships. This may be the result of growing up in a home in which the parents abused each other, which may convey the message that marital violence is natural, inevitable, and to be expected.

Some abused partners may feel that the violence directed toward them is legitimate because it is their fault; they either caused the abuse or deserved it. Some abused partners feel that if only they were a better partner or a better person, they would not be abused.

Guilt   Other abused partners stay in the abusive relationship because leaving would produce guilt. They may feel guilty about breaking up a family, especially if children are involved. Some abusers threaten suicide if their partner leaves or use other verbal pleas to guilt-induce their partner into staying.

Fear   Abused partners may stay in a relationship because they fear that the abuser will become even more violent if they leave. Often, such fear is the result of threats made by the abuser. One woman in the authors' classes reported that she stayed with an abusive husband for eleven years because he threatened to "slit my throat if I tried to leave." When he became abusive toward her children, she took them and left the state.

Economic dependency   Some spouses stay in an abusive relationship because they are economically dependent on the partner. This is most often true of a wife who has limited economic resources. Leaving the husband would mean giving up not only his monthly income but his retirement benefits and health insurance as well. The woman referred to in the preceding paragraph also noted that her husband kept the checkbook and gave her money only as he wished. When she finally left, she went to a women's shelter, which provided food for her and her children.

Isolation   Battered women may be physically and socially isolated from family, friends, and community resources. Isolation may be even greater for victims who live in rural areas or for foreign-born wives.

A team of researchers studied battered women in women's shelters and identified the primary reasons they stayed in abusive relationships.

> Battered women stay because they rarely have escape routes related to educational or employment opportunities, relatives were critical of plans to leave the relationship, parenting responsibilities impeded escape, and abusive situations contributed to low self-esteem and negative emotions—

Many of the battered women had gone from their fathers' homes to their husbands' homes, and having to justify the money they needed was an adolescent way of life they were used to.
Lenore Walker
The Battered Woman

especially anxiety and depression. Such troubles and the dangerousness of the abuser probably undermined rebellious and self-enhancing actions (Forte et al, 1996, 69).

Disengaging from an Abusive Relationship

Disengaging from a relationship may be particularly difficult, since the abused woman may already have limited coping resources—poor health, limited material resources, and inadequate social supports (Carlson, 1997).

Dr. Karen Rosen and Dr. Sandra Stith (1992) interviewed six young women who had stayed in violent relationships from three months to five and a half years and were eventually able to free themselves from the relationships. Disengaging was a complex process that included the following factors.

Seeds of doubt   Despite feeling stuck, each of the women had seeds of doubt that she should remain in the abusive relationship. One woman who had endured the relationship for five years reported "that a small voice somewhere inside her would occasionally whisper, 'You don't deserve this. You don't deserve this.' Although she had stayed with her partner for all this time and felt that she had loved him for most of those years, she never really trusted him fully after the first time he hit her" (1992, 17).

The first time there is abuse, get couple therapy; the second time, separate; the third time, divorce.
Barry Lubetkin
Psychologist

Turning points   Turning points were events in the relationship that had a significant impact on the motivation of the abused partner to leave the relationship. One woman said that when her husband kicked her in the stomach when she was pregnant, she knew she had to leave. Another woman said that becoming engaged was a turning point for her. "I knew it was now or never if I was going to leave this abusive relationship," she said.

Taking control   The abused women took control of their lives. One woman recognized that she had to save herself and began putting distance between her and her abusive partner. Her story follows:

The one thing that is unforgivable is deliberate cruelty.
Blanche
A Streetcar Named Desire

> Two weeks before school started I decided, this is it. This whole relationship is driving me crazy. I've got to do something to get away from it. So in an attempt to get away from it, I decided to go away to school. So in a two-week span, I got accepted to a college that was a distance away, got registered, got an apartment and moved down there (Rosen & Stith, 1992, 22).

Cognitive shifts   A cognitive shift occurred when the woman shifted her focus from staying in the relationship to getting out of it. "The initial position that the relationship is a means to meet her needs is replaced with the position that the relationship must be dissolved in order to meet her needs" (p. 25). Sometimes the shifts were prompted by last-straw events. Being severely abused, being coldly rejected, and recognizing that the future was hopeless are examples. The experience of one woman reflected the process of finally deciding that something had to change.

Leaving an abused partner is a process. Even when abused partners leave their abusers, they often return to the relationship. Seven out of ten shelters in Alabama reported that more than half of their clients returned to their abusive partners (Johnson, Crowley & Sigler, 1992). The shelter workers in this study were asked to rate their perceptions of important reasons for the victim's decision to return to the abusive relationship on a scale from most important (10) to least important (1). The results are listed in Table 10.3.

I had no more hope that things were going to work out. . . . And I saw Matt as the root of almost all of my problems. And I think that's part of what pushed me over the edge to decide that I couldn't be with him anymore. It was like, on one hand I had Matt, on the other I had hope. It was like which do I need more to survive. I needed the hope, so I had to get rid of him (p. 26).

Leaving and moving on Actually disengaging from the partner was difficult and painful. Researchers Rosen and Stith summarized their study by emphasizing the enormous power abusive relationships hold over the people involved in them and the difficulty of breaking through the cultural mandate to "stand by your man." Rather, women in abusive relationships must "manage to come to their senses and stand by themselves instead" (1992, 31).

Once a decision is made to withdraw from an abusive relationship, often the woman not only has to call the police and have the man arrested but also has to take out a restraining order restricting the man from contacting her or coming within a certain distance of her. Women who fear that the abusers will harm them (or their children) sometimes leave town or the state. Other women "hide"

Table 10.3

Perceived Reasons for Returning to Abusive Relationship	
Reasons	**Mean**
Give the abuser one more chance	10.0
Lack of financial resources	9.1
Emotional dependency on the abuser	9.0
Lack of housing resources	8.7
Lack of job opportunities	7.7
Denial of cycle of violence	7.6
Lack of support or follow-through by the legal system	7.6
Lack of child-care resources	7.1
Lack of transportation	6.7
Fear that the abuser will find her and do her harm	6.7
Lack of support from other family members	6.6
Fear that the abuser will get custody of the children	5.8
Fear that the abuser will kidnap the children	5.8
Children miss the absent parent	5.6
Lack of professional counseling	5.1
Fear that the abuser will harm the children	4.6

Source: I. M. Johnson, J. Crowley, and R. T. Sigler. 1992. Agency response to domestic violence: Services provided to battered women. In *Intimate Violence: Interdisciplinary Perspectives*, edited by E. C. Viano. Washington, D.C.; Hemisphere Publishing Co., 191–202 (Table on p. 199). Used by permission.

It is estimated that only 1 in 7 incidents of child abuse is reported.

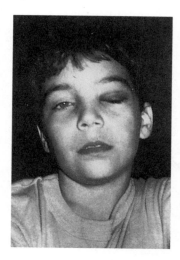

Some parental child abuse masquerades under the name of healthy competition for the Olympics. Some parents subject their children to gymnastics trainers who require 13-year-olds to train six to nine hours a day and who insist on a near-starvation diet to maintain a low body weight. According to the U.S. Olympic Committee, the average weight of the girl who qualified for competition in gymnastics in 1992 was 83 pounds. Young girls subjected to such strict criteria may also become bulimic, severely depressed, and suicidal (Sciere, 1995)

I was only 9 years old when I was raped by my 19-year-old cousin. He was the first of three family members to sexually molest me.
Oprah Winfrey
TV talk-show host

in the home of a friend or seek refuge in a women's shelter. In either case, disengagement from the abusive relationship is usually a very difficult decision and takes a great deal of courage. Calling 800-799-7233, the national domestic hot line, can be a beginning.

Child Abuse

Physical Abuse and Neglect

Child abuse can be defined as any interaction or lack of interaction between a child and his or her parents or caregiver that results in nonaccidental harm to the child's physical or psychological well-being. Typically, child abuse includes physical abuse, such as beating and burning; verbal abuse, such as insulting or demeaning the child; and neglect, such as failing to provide adequate food, hygiene, medical care, or adult supervision for the child.

NATIONAL DATA The number of children annually reported to protective services who are alleged to be abused is 2,959,237 (*Statistical Abstract of the United States: 1997*, Table 353).

Where physical abuse occurs, the cost to both the individual and society is high. The individual may die. Seventy percent of the abused children admitted to pediatric intensive care units at two hospitals died. The most frequent forms of death were skull fracture and internal bleeding. The medical bill for the acute care of a child abuse patient averaged $35,641 (Irazuzta et al., 1997).

The percentages of various types of child abuse in substantiated victim cases are illustrated in Figure 10.1. Notice that "neglect" is the largest category of abuse. Although our discussion will focus on physical abuse, it is important to keep in mind that children often are not fed, are not given medical treatment, or are left to fend for themselves.

The social policy of this chapter examines yet another child abuse issue—fetal abuse.

Intrafamilial Child Sexual Abuse

Another type of child abuse is child sexual abuse. Intrafamilial child sexual abuse (formerly referred to in professional literature as incest) refers to exploitive sexual contact or attempted sexual contact between relatives before the victim is 18. Sexual contact or attempted sexual contact includes intercourse, fondling of the breasts and genitals, and oral sex. Relatives include biologically related individuals but may also include stepparents and stepsiblings.

NATIONAL DATA In a national survey of adults concerning child sexual abuse, 27 percent of the women and 16 percent of the men reported being victims. These percentages refer to both intrafamilial child sexual abuse and sexual abuse by nonfamily members (Finkelhor et al., 1990).

Female children are more likely than male children to be sexually abused. A study of almost 4,000 intrafamilial child sexual abuse cases revealed that 85 percent of the victims were female and 15 percent were male (Solomon, 1992).

Figure 10.1
Child Abuse and Neglect Cases: 1995

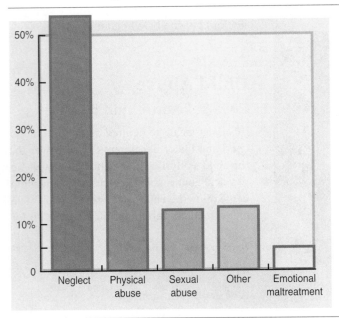

Source: *Statistical Abstract of the United States 1997*, 117th ed. Washington, D.C.: U.S. Bureau of the Census, 1997, Table 352.

OTHER CULTURES Sam and Kathy Kresnidi are an Albanian immigrant couple (now U.S. citizens) who had two children, Tim and Lima. One day in 1989 while Sam was attending his son's karate tournament, Sam was observed to be "molesting" his 4-year-old daughter Lima. The police were called, and Sam was arrested and taken to jail. Subsequently, his children were taken from him and adopted by a Christian family. The Kresnidis have been forbidden by the courts from any contact with their children (ABC, 1995).

Symbolic interactionists would emphasize that at issue were the meanings attached to a father's kissing and touching the genitals of his daughter. In Albanian culture, physical intimacy in the form of kissing and touching is normative and assigned no sexual connotations by either the parent or the child. In the United States, since such behaviors are assigned sexual meaning, they resulted in the children being permanently taken from their parents. Perhaps this is the downside of our cultural concern for child abuse. We may have become *too sensitive* to child abuse and may be inclined to report any act that could be perceived as abusive. This may lead to less affectionate parents as well as to overly permissive ones. ●

Intrafamilial child sexual abuse, particularly when the perpetrator is a parent, involves an abuse of power and authority. The following describes the experience of one woman who was forced to have sexual relations with her father:

I was around six years old when I was sexually abused by my father. He was not drinking at that time; therefore, he had a clear mind as to what he

Is Fetal Abuse Child Abuse?

Should women who abuse heroin, alcohol, crack cocaine, and other drugs harmful to the fetus be prosecuted? Though over 100,000 developing fetuses have been exposed to such drugs, the courts have been reluctant to prosecute (Farr, 1995). Arguments against such prosecution are based on the difficulty of defining when the fetus becomes a person whose rights have been violated, the lack of warning to women that drug abuse during pregnancy may be a prosecutable offense, the vagueness of what exactly constitutes "the crime," and the fact that fetal abuse is a lifestyle issue. Regarding the latter, many women who take drugs during their pregnancy live in poverty, which includes lack of prenatal care and nutritional deficiencies. Indeed, prenatal care, drug treatment, and general health services are least accessible for poor and minority women (Smith & Dabini, 1991). An additional problem is that sending mothers to jail or prison for drug abuse during pregnancy deters their coming into treatment during pregnancy.

The issue remains unresolved. Seven states continue to enact fetal abuse laws. For example, legislation in Ohio is seeking to include fetal drug abuse in the definition of child abuse and neglect. "Yet there is no evidence that in the long run, such policies are either socially effective or economically sound. Instead, the research seems to call for a policy direction that emphasizes the provision of health and drug education" (Farr, 1995, 242). Teratogens, substances that cross the placenta of the pregnant woman and cause harm to the fetus or baby, should be identified and avoided. Healthy development depends on lack of exposure to teratogens during pregnancy. While fetal abuse is still being debated, one case of similar abuse to an infant has reached the courts. In 1997, 24-year-old Amoret Powell of Tucson, Arizona, was charged with first-degree murder after her heroin-laced breast milk allegedly led to her 7-week-old daughter's death from oxygen deprivation (Powell, 1997).

REFERENCES

Farr, K. A. 1995. Fetal abuse and the criminalization of behavior during pregnancy. *Crime and Delinquency* 41: 235–45.

Powell, A. Milestones. 1997. *Time.* 11 August, 25.

Smith, G. B., and G. M. Dabini. 1991. Prenatal drug exposure: The constitutional implications of three governmental approaches. *Constitutional Law Journal* 2: 53–126.

was doing. On looking back, it seemed so well planned. For some reason, my father wanted me to go with him to the woods behind our house to help him saw wood. Once we got there, he looked around for a place to sit and wanted me to sit down with him. He said, "Susan, I want you to do something for Daddy. I want you to lie down, and we are going to play Mama and Daddy." Being a child, I said, "Okay," thinking it was going to be fun. I don't know what happened next and I can't remember if there was pain or whatever. I was threatened not to tell, and remembering how he beat my mother, I didn't want the same treatment. It happened a few more times. I remember not liking this at all. But what could I do? Until age 18, I was constantly on the run, hiding from him when I had to stay home alone with him, staying out of his way so he wouldn't touch me by hiding in the corn fields all day long, under the house, in the barns, and so on until my mother got back home, then getting punished by her for not doing the chores she had assigned to me that day. It was a miserable life, growing up in that environment (authors' files).

While the sex abuser in the example was the biological father, stepfathers are more often the perpetrators (Jensen & Doxey, 1996).

Although brother-sister incest taboos are nearly universal across cultures, there are exceptions. Royalty siblings in ancient Egypt, Hawaii, and the Incas of Peru were permitted to have intercourse for procreation to keep power invested in a small group. ●

Causes of and Contributing Factors to Child Abuse

A variety of factors may cause or contribute to child abuse, as discussed in the following subsections.

Parental psychopathology   Some abusing parents may have a psychiatric condition. Symptoms of such conditions that may predispose the parent to abuse children include low frustration tolerance, inappropriate expression of anger, and emotional distress. Some child-abusing parents are dependent on alcohol or other drugs, which may be associated with child abuse or neglect.

Unrealistic expectations   Abusing parents often have unrealistic expectations of their children's behavior. For example, a parent might view the crying of a baby as a deliberate attempt on the part of the child to irritate the parent. Or a parent may unrealistically expect a 1-year-old child to be toilet trained and may beat the child for soiling or wetting his or her pants. One mother, unaware that most children are not developmentally ready to walk until age 1, regarded her 8-month-old baby as "lazy" because he would not get up and walk (Blau et al., 1993). Changing parents' unrealistic expectations or irrational perceptions concerning their children's behavior is an important part of an overall plan to reduce child abuse (Martin & Walters, 1992).

A man who attended a lecture on child abuse approached the speakers after the lecture ended. With tears in his eyes, the man told the speakers, "You said that people who are abused grow up to be abusers. Well, I was an abused child. I don't want to get married and grow up to abuse my children, so I will not get married" (Gelles & Straus, 1988, 49).

Although parents who were abused as children are somewhat more likely than parents who were not abused to repeat the pattern, we would like to emphasize that the majority of parents who were abused do NOT abuse their own children. Indeed, many parents who were abused as children are dedicated to ensuring nonviolent parenting of their own children precisely because of their own experience of abuse.

History of abuse   Although the majority do not, some parents who were themselves physically or verbally abused or neglected may duplicate these patterns in their own families (Gelles & Conte, 1991).

The use of corporal punishment in schools and in homes may also teach children that physical punishment is acceptable.

NATIONAL DATA   It is estimated that more than 2 million children receive corporal punishment each year (Pinkney, 1992).

Problems of definition abound. One parent's "discipline" is another parent's love.

Displacement of aggression   One cartoon shows several panels consisting of a boss yelling at his employee, the employee yelling at his wife, the wife yelling at their child, and the child kicking the dog, who chases the cat up a tree. Washburne observed, "Women's abuse of children stems directly from their own oppression in society and within the family. . . . Some women displace their frustration and anger on their children, the family members who are less powerful than they" (1983, 291).

Social isolation   Unlike members of most societies of the world, many Americans rear their children in closed and isolated nuclear units. In extended kinship

The frustration and anger some parents displace onto their children may have its base in the larger society. Harrington and Dubowitz noted that "poverty, unemployment, lack of support for families and working parents, lack of health care, poor nutrition, substance abuse, and society's acceptance of violence are all related to child maltreatment" (1993, 262). Indeed, the strategy to eliminate child abuse must include social policies to attack each of these broader issues.

societies, other relatives are always present to help with the task of childrearing. Isolation means no relief from the parenting role as well as no supervision by others who might interfere in child-abusing situations.

In addition to the factors just mentioned, the following factors are associated with child abuse and neglect:

1. The pregnancy is premarital or unplanned, and the father or mother does not want the child.
2. Mother-infant attachment is lacking.
3. The child suffers from developmental disabilities or mental retardation.
4. Childrearing techniques are strict and harsh and include little positive reinforcement for the child.
5. The parents are unemployed.
6. Abuse between the husband and wife is present.
7. The children are adopted or are foster children.

As to the risk of a child's being abused by adoptive parents, adoption agencies have found that some people who adopt have unrealistic expectations for their children, and they lash out at their children when the children do not measure up to those expectations. One couple was told by the adoption agency, "We don't generally place children with teachers, social workers, or professionals because they tend to have very high expectations." The parents had to convince the placement worker that they were realistic in their expectations before they were allowed to adopt their son.

The Effects of Child Abuse

How does being abused—physically, verbally, or sexually—affect the victim as a child and later as an adult? In general, the effects are negative and vary according to the intensity and frequency of the abuse.

Reviews of research on the effects of child abuse suggest the following (Mullen et al., 1996; Gelles & Conte, 1991; Lloyd & Emery, 1993):

1. Abused children have a number of cognitive, social, and emotional deficits believed to be the result of both their experience of abuse and their development in a socially impoverished environment.
2. Abused children tend to exhibit aggression, low self-esteem, depression, and low academic achievement.
3. Children who experience more severe abuse suffer more from intellectual deficits, communication problems, and learning disabilities.
4. Adults who were physically abused as children may exhibit low self-esteem, depression, unhappiness, anxiety, an increased risk of alcohol abuse, and suicidal tendencies.
5. Physical injuries sustained by child abuse cause pain, disfigurement and scarring, physical disability, and even death.

Child sexual abuse may have serious, negative long-term consequences. Researchers on the effects of being sexually abused have found the following:

RECENT RESEARCH

Data from 3,128 girls in grades 8, 10, and 12 revealed that those who had been sexually abused were 2.3 times more likely to say they had had intercourse and 3.1 times more likely to say they had been pregnant than girls who had not been abused (Stock et al., 1997).

RECENT RESEARCH

Langehough et al. (1997) found that individuals abused in childhood reported higher levels of intrinsic spirituality and religious orientation than those not abused.

1. Among adolescent females, early forced sex is associated with lower self-esteem, higher levels of depression, antisocial behavior (e.g., running away from home, illegal drug use), and more sexual partners (Lanz, 1995). Women who were sexually abused as children also report a higher frequency of sexual problems (Mullen et al., 1996).

2. The most devastating effects of being sexually abused occur when the sexual abuse is forceful, is prolonged, involves intercourse, and when the abuse was perpetrated by a father or stepfather (Beitchman et al., 1992; Morrow & Sorrell, 1989).

3. Sexually abused girls are more likely to experience adolescent pregnancy (Stock et al., 1997).

4. Fear, guilt, shame, sleep disturbances, and eating disorders have been associated with child sexual abuse (Browne & Finkelhor, 1986).

5. Adult males who were sexually abused as children by their mothers revealed several problems, including difficulty establishing intimate relationships, depression, and substance abuse (Krug, 1989). Sexually abused males also tended to develop negative self-perceptions, anxiety disorder, sleep and eating disturbances, and sexual dysfunctions such as decreased sexual desire, rapid ejaculation, and difficulty with ejaculation (Elliott & Briere, 1992).

Parent/Sibling/Elder Abuse

As we have seen, intimate partners and children may be victims of relationship violence and abuse. Parents, siblings, and the elderly may also be abused by family members.

Parent Abuse

Some people assume that because parents are typically physically and socially more powerful than their children, they are immune from being abused by their children. But parents are often targets of their children's anger, hostility, and frustration. It is not uncommon for teenage and even younger children to physically and verbally lash out at their parents. In a national survey of family violence, almost 10 percent of the parents reported that they had been hit, bitten, or kicked at least once by their children (Gelles & Straus, 1988). The same researchers found that 3 percent of parents reported that they had been victimized at least once by a severe form of violence inflicted by a child aged 11 or older. Children have been known to push parents down stairs, set the house on fire while their parents are in it, and use weapons such as guns and knives to inflict serious injuries on or even kill a parent.

Sibling Abuse

Observe a family with two or more children and you will likely observe some amount of sibling abuse. Seventy-five percent of children with siblings report

INSIGHT

Heide (1992) reported that about 300 parents in the United States are killed each year by their children. However, according to one attorney who specializes in defending adolescents who have killed a parent, over 90 percent of youths who kill their parents have been abused by them. "In-depth portraits of such youths have frequently shown that they killed because they could no longer tolerate conditions at home" (Heide, 1992, 6).

having had at least one violent episode of conflict with their siblings during a year's time. An average of twenty-one violent acts take place between siblings in a family per year (Steinmetz, 1987).

Even in "well-adjusted" families, some degree of fighting among the children is expected. Most incidents of sibling violence consist of slaps, pushes, kicks, bites, and punches. However, serious and dangerous violent behavior between siblings occurs as well.

NATIONAL DATA Each year, an estimated 3 percent of children in the United States use a weapon on a brother or sister (Gelles & Straus, 1988).

Elder Abuse

The following are various forms of elder abuse (Johnson, 1991):

1. Physical abuse—inflicting injury or physical pain or sexual assault.
2. Neglect—failing to buy or give the elderly needed medicine, failing to take them to receive necessary medical care, or failing to provide adequate food, clean clothes, and a clean bed.
3. Psychological abuse—verbal abuse, deprivation of mental health services, harassment, and deception.
4. Social abuse—unreasonable confinement and isolation, lack of supervision, abandonment.
5. Legal abuse—improper or illegal use of the elder's resources.

NATIONAL DATA The National Aging Resource Center on Abuse reports that neglect is the most frequent type of domestic elder abuse (37.2%), followed by physical abuse (26.3%), financial/material exploitation (20%), and emotional abuse (11%) (U.S. House of Representatives, Select Committee on Aging, 1991).

It [the sweetheart swindle] is really about money.
Ray Allsdorf
Investigator

Another form of elder abuse is the *sweetheart swindle* (Sutherland, 1997). This involves young women who target elderly, frail, lonely men. After feigning love and promising care and companionship, a woman entices the trusting, lovestruck elderly man to sign over his bank accounts and property. Mark Moyer, a 77-year-old widower, said that Lisa Loeo took more than $200,000 of his assets after becoming his wife. She has since disappeared. This elder abuse is both emotional and financial. Since the swindlers get the consent of the elderly in writing to have access to the money/property, none of them have been convicted of a crime.

Still another type of elder abuse is "granny dumping." Adult children or grandchildren who feel burdened with the care of their elderly parent or grandparent drive the elder to the entrance of a hospital and leave him or her there with no identification. If the hospital cannot identify responsible relatives, it is required by state law to take care of the abandoned elder or transfer the person to a nursing-home facility, which is paid for by state funds. Relatives of the "dumped granny," hiding from financial responsibility, never visit or see "granny" again.

The scale to follow is designed to assess the amount of abuse occurring in a relationship. Circle the number that best represents your closest estimate of how often each of the behaviors happened in your relationship with your partner or former partner during the previous six months.

Abusive Behavior Inventory

1 Never
2 Rarely
3 Occasionally
4 Frequently
5 Very frequently

1. Called you a name and/or criticized you. 1 ②③ 4 5

2. Tried to keep you from doing something you wanted to do (e.g., going out with friends, going to meetings). 1 ②③ 4 5

3. Gave you angry stares or looks. 1 2 ③ 4 ⑤

4. Prevented you from having money for your own use. ① 2 3 4 5

5. Ended a discussion with you and made the decision himself/herself. 1 ② 3 4 5

6. Threatened to hit or throw something at you. ① 2 3 4 5

7. Pushed, grabbed, or shoved you. ① 2 3 4 5

8. Put down your family and friends. ① 2 3 ④ 5

9. Accused you of paying too much attention to someone or something else. 1 2 ③ ④ 5

10. Put you on an allowance. ① 2 3 4 5

11. Used your children to threaten you (e.g., told you that you would lose custody, said he/she would leave town with the children). ① 2 3 4 5

12. Became very upset with you because dinner, housework, or laundry was not done when he/she wanted it or done the way he/she thought it should be. ① 2 3 4 5

13. Said things to scare you (e.g., told you something "bad" would happen, threatened to commit suicide). 1 ② 3 4 5

14. Slapped, hit, or punched you. ① 2 3 4 5

15. Made you do something humiliating or degrading (e.g., begging for forgiveness, having to ask his/her permission to use the car or to do something). ① 2 3 4 5

16. Checked up on you (e.g., listened to your phone calls, checked the mileage on your car, called you repeatedly at work). ① 2 3 4 5

17. Drove recklessly when you were in the car. ① 2 ③ 4 5

18. Pressured you to have sex in a way you didn't like or want. ① 2 3 4 5

19. Refused to do housework or child care. ① 2 3 4 5

20. Threatened you with a knife, gun, or other weapon. ① 2 3 4 5

21. Spanked you. ① 2 3 4 5

22. Told you that you were a bad parent. ① 2 3 4 5

23. Stopped you or tried to stop you from going to work or school. ① 2 3 4 5

(Continued on following page)

24. Threw, hit, kicked, or smashed
    something.

25. Kicked you.

26. Physically forced you to have sex.

27. Threw you around.

28. Physically attacked the sexual parts of
    your body.

29. Choked or strangled you.

30. Used a knife, gun, or other weapon
    against you.

SCORING: Add the numbers you circled and divide the total by 30 to find your score. The higher your score, the more abusive your relationship.

The inventory was given to 100 men and 78 women equally divided into groups of abusers/abused and nonabusers/nonabused. The men were members of a chemical dependency treatment program in a veterans' hospital and the women were partners of these men. Abusing or abused men earned an average score of 1.8; abusing or abused women earned an average score of 2.3. Nonabusing/abused men and women earned scores of 1.3 and 1.6, respectively (Shepard & Campbell, 1992).

Source: Melanie F. Shepard and James A. Campbell. The abusive behavior inventory: A measure of psychological and physical abuse. *Journal of Interpersonal Violence,* September 1992, 7, no. 3, 291–305. Inventory is on pages 303–304. Used by permission of Sage Publications, 2455 Teller Road, Newbury Park, CA 91320.

As is true of all forms of domestic violence, reliable estimates of the prevalence of elder abuse are difficult to obtain.

NATIONAL DATA   Studies on the prevalence of elder abuse in the United States suggest that from 4 percent to 10 percent of the elderly population is abused (Johnson, 1991).

Though the elderly may be abused by their partners, more often the abuse is by their adult children. These caretakers may be under a great deal of stress and use alcohol or other drugs. They also tend to be white, middle-aged, and lower-middle or upper-lower-class (White, 1988). In some cases, parent abusers are "getting back" at their parents for maltreatment of them as children. In other cases, the children are frustrated with the burden of having to care for their elderly parents. Such frustration is likely to increase. As baby boomers age, they will drain already limited resources for the elderly, and their children will be forced to care for them with little governmental support.

GLOSSARY

acquaintance rape   Nonconsensual sex between adults who know each other.
date rape   Nonconsensual sex between two people who are dating or on a date.

emotional abuse The denigration of an individual with the purpose of reducing the victim's status and increasing the victim's vulnerability so that he or she can be more easily controlled by the abuser. Also known as verbal abuse or symbolic aggression.

forcible rape Rape that involves vaginal penetration, force or threat of force, and non-consent of the victim.

granny dumping A situation in which adult children or grandchildren who feel burdened with the care of their elderly parent or grandparent drive the elder to the entrance of a hospital and leave him or her there with no identification.

marital rape Forcible rape by one's spouse, now recognized as a crime in every state.

stalking The willful, repeated, and malicious following or harassment of another person.

statutory rape Sexual intercourse without the use of force with a person below the legal age of consent.

sweetheart swindle A swindle perpetrated by a young woman who feigns love, care, and companionship for an elderly man, who in return gives her access to his money and property. After she gets the money, she disappears.

violence The intentional infliction of physical harm by one individual toward another.

SUMMARY

Violence, physical injury, emotional abuse, and neglect sometimes occur in intimate relationships between dating, cohabiting, and marital partners.

Definitions of Violence, Abuse, and Neglect

Violence is the intentional use of physical force designed to hurt or injure another person. Abuse may also be emotional (denigrating a partner), neglect (withholding medical treatment or food from elderly parents), or sexual (rape or molestation).

Abuse in Dating Relationships

Estimates suggest that between 15 percent and 20 percent of dating relationships in high school and between 20 percent and 50 percent of dating relationships in college involve either physical or emotional abuse. Either the woman or the man may initiate the abuse, which often occurs in relationships defined by the partners as "good" and in which the partners are committed to each other. Abuse may not end when the relationship ends, but may continue in the form of stalking.

About 20 percent of college women and 10 percent of college men report that they have been pressured into having intercourse. Fewer report having been raped. Rape rates also differ between college and noncollege women, with the latter more likely to have been raped.

Abuse in Cohabitation and Marriage Relationships

Abuse, both sexual and nonsexual, occurs among cohabitants and marital partners. Marital abuse is rooted in patriarchal terrorism, an attitude under which men feel they must control "their" women by any means possible. Marital rape includes not only intercourse but oral and anal sex.

Factors Contributing to Abuse in Relationships

Violence in the media, gender roles (men are superior to and should control women), a family history of violence, alcohol or other drug use, personality characteristics (dependency, jealousy, need to control), and acceptance of corporal punishment are among the numerous factors that contribute to abuse in relationships.

Effects of Abuse

Effects of abuse include admission to the hospital for physical injury or death. Other less obvious effects of abuse by one's intimate partner include fear, feelings of helplessness, confusion, isolation, humiliation, anxiety, depression, stress-induced illness, symptoms of post-traumatic stress disorder, and suicide attempts.

Children are affected by partner abuse in that the abused woman may be pregnant at the time. Alternatively, children who witness abuse are more likely to be aggressive themselves and to experience depression and anxiety.

Cycle of Abuse

Persons who stay in abusive relationships report that they are in love, are committed to the partner/relationship, have hope the violence will stop, view the abuse as part of loving someone, feel guilty about abandoning the partner, and fear what the partner might do if they leave. Those who eventually leave an abusive relationship begin to see more negatives than positives in staying and find an opportunity to leave. Many return several times before finally breaking free.

Child Abuse

Child abuse may include physical, emotional, and sexual abuse, as well as neglect. Sexual abuse may be by family members or nonfamily members. The consequences of sexual abuse can be devastating to the child's self-esteem, health, and relationships. It is important for the child to understand that he or she is not responsible for such abuse.

Parent/Sibling/Elder Abuse

Children and teenagers sometimes abuse their parents verbally and/or physically. About 300 parents in the United States are killed each year by their children. Fighting among siblings may be considered as a form of abuse. Neglect is the most frequent type of domestic elder abuse, followed by physical abuse, financial exploitation, and emotional abuse.

REFERENCES

ABC (American Broadcasting Company). 1995. The death of a family. 19 August. *20/20*.

Anderson, Kristin L. 1997. Gender, status, and domestic violence: An integration of feminist and family violence approaches. *Journal of Marriage and the Family* 59: 655–69.

Andrews, B. 1997. Bodily shame in relation to abuse in childhood and bulimia: A preliminary investigation. *British Journal of Clinical Psychology* 36: 41–49.

American Council on Education and University of California. 1997. *The American freshman: National norms for fall, 1997*. Los Angeles: Los Angeles Higher Education Research Institute.

Ascione, F. R. 1998. Battered women's reports of their partners' and their children's cruelty to animals. *Journal of Emotional Abuse* 1: 119–33.

Barker, G., and I. Loewenstein. 1997. Where the boys are: Attitudes related to masculinity, fatherhood, and violence toward women among low-income adolescent and young adult males in Rio de Janerio, Brazil. *Youth and Society* 29: 166–96.

Bechhofer, L., and A. Parrot. 1991. What is acquaintance rape? In *Acquaintance rape: The hidden crime*, edited by A. Parrot and L. Bechhofer. New York: Wiley, 9–25.

Beitchman, J. H., K. J. Zuker, J. E. Hood, G. A. daCosta, D. Akman, and E. Cassavia. 1992. A review of the long-term effects of child sexual abuse. *Child Abuse and Neglect* 16: 101–19.

Blankenhorn, David. 1995. *Fatherless America: Confronting our most urgent social problem*. New York: Basic Books.

Blau, G. M., M. B. Dall, and L. M. Anderson. 1993. The assessment and treatment of violent families. In *Family violence: Prevention and treatment*, edited by Robert L. Hampton, Thomas P. Gullotta, Gerald R. Adams, Earl H. Potter III, and Roger P. Weissberg. Newbury Park, Calif.: Sage, 198–229.

Bohannon, Judy Rollins, David A. Dosser, and S. Eugene Lindley. 1995. Using couple data to determine domestic violence rates: An attempt to replicate previous work. *Violence and Victims* 10, 133–41.

Bowman, M. L. 1990. Measuring marital coping and its correlates. *Journal of Marriage and the Family* 52: 463–74.

Bowman, R. L., and H. M. Morgan. 1998. A comparison of rates of verbal and physical abuse on campus by gender and sexual orientation. *College Student Journal* 32: 43–52.

Browne, A., and D. Finkelhor. 1986. Initial and long-term effects: A review of the research. In *A sourcebook on child sexual abuse*, edited by D. Finkelhor. Newbury Park, Calif.: Sage, 143–79.

Burt, M. R. 1991. Rape myths and acquaintance rape. In *Acquaintance rape: The hidden crime*, edited by A. Parrot and L. Bechhofer. New York: Wiley, 26–40.

Carlson, Bonnie E. 1997. A stress and coping approach to intervention with abused women. *Family Relations* 46: 291–98.

Choquet, M., Jean-Michel Darves-Bornoz, Sylvie Ledoux, Robert Manfredi, and Christine Hassler. 1997. Self-reported health and behavioral problems among adolescent victims of rape in France: Results of a cross-sectional survey. *Child Abuse and Neglect* 21: 823–32.

Coleman, Frances L. 1997. Stalking behavior and the cycle of domestic violence. *Journal of Interpersonal Violence* 12: 420–32.

Copenhaver, S., and E. Grauerholz. 1996. Sexual victimization among sorority women: Exploring the link between sexual violence and institutional practices. *Sex Roles* 24: 31–41.

Dutton, Donald G., and Andrew J. Starzomski. 1977. Personality predictors of the Minnesata Power and Control Wheel. *Journal of Interpersonal Violence* 12: 70–82.

Elliott, D. M., and J. Briere. 1992. The sexually abused boy: Problems in manhood. *Medical Aspects of Human Sexuality* 26: 68–71.

Finkelhor, D., G. Hotaling, I. A. Lewis, and C. Smith. 1990. Sexual abuse in a national survey of adult men and women: Prevalence, characteristics, and risk factors. *Child Abuse and Neglect* 14: 19–28.

Finkelhor, D., and K. Yllo. 1988. Rape in marriage. In *Abuse and victimization across the life span*, edited by M. B. Straus. Baltimore: Johns Hopkins University Press, 140–52.

Forte, J. A., D. D. Franks, J. A. Forte, and D. Rigsby. 1996. Asymmetrical role-taking: Comparing battered and nonbattered women. *Social Work* 41: 59–73.

Frank, J. G. 1991. Risk factors for rape: Empirical confirmation and preventive implications. Poster session presented at the 99th Annual Convention of the American Psychological Association, San Francisco, 16 August.

Gelles, Richard J., and Jon R. Conte. 1991. Domestic violence and sexual abuse of children: A review of research in the eighties. In *Contemporary families: Looking forward, looking back*, edited by Alan Booth. Minneapolis: National Council on Family Relations, 327–40.

Gelles, R. J., and M. Straus. 1988. *Intimate violence*. New York: Simon and Schuster.

Gibbs, N. 1993. Till death do us part. *Time*. 18 January, 38–45.

Gilmartin, Pat. 1994. Gender differences in college students' perceptions about rape: The results of a quasi-experimental research design. *Free Inquiry in Creative Sociology* 22: 3–12.

Gray, Heather M., and Vangie Foshee. 1997. Adolescent dating violence: Differences between one-sided and mutually violent profiles. *Journal of Interpersonal Violence* 12: 126–41.

Hanley, M. Joan, and Patrick O'Neill. 1997. Violence and commitment: A study of dating couples. *Journal of Interpersonal Violence* 12: 685–703.

Harrington, Donna, and Howard Dubowitz. 1993. What can be done to prevent child maltreatment? In *Family violence: Prevention and treatment*, edited by Robert L. Hampton, Thomas P. Gullotta, Gerald R. Adams, Earl H. Potter III, and Roger P. Weissberg. Newbury Park, Calif.: Sage.

Heide, K. M. 1992. *Why kids kill parents: Child abuse and adolescent homicide*. Columbus, Ohio: Ohio State University Press.

Herbert, T. B., R. C. Silver, and J. H. Ellard. 1991. Coping with an abusive relationship: I. How and why do women stay? *Journal of Marriage and the Family* 53: 311–25.

Irazuzta, J. E., J. E. McJunkin, and J. Zhang. 1997. Outcome and cost of child abuse. *Child Abuse and Neglect* 21: 751–57.

Jensen, Larry, and Cynthia Doxey. 1996. Family, type, and denomination in reported sexual abuse. Poster session presented at the 58th Annual Conference of the National Council on Family Relations, Kansas City, Missouri. Used by permission.

Johnson, I. M., J. Crowley, and R. T. Sigler. 1992. Agency response to domestic violence: Services provided to battered women. In *Intimate violence: Interdisciplinary perspectives,* edited by E. C. Viano. Washington, D.C.: Hemisphere, 191–202.

Johnson, Michael P. 1995. Patriarchal terrorism and common couple violence: Two forms of violence against women. *Journal of Marriage and the Family* 57: 283–94.

Johnson, T. F. 1991. *Elder mistreatment: Deciding who is at risk.* New York: Greenwood Press.

Kasian, M., and S. L. Painter. 1992. Frequency and severity of psychological abuse in a dating population. *Journal of Interpersonal Violence* 7: 350–64.

Kaufman, Joan, and Edward Zigler. 1992. The prevention of child maltreatment: Programming, research, and policy. In *Prevention of child maltreatment: Developmental and ecological perspectives,* edited by Diane J. Willis, E. Wayne Holden, and Mindy Rosenberg. New York: Wiley, 269–95.

Knutson, John F., and Mary Beth Selner. 1994. Punitive childhood experiences reported by young adults over a 10-year period. *Child Abuse and Neglect* 18: 155–66.

Koss, M. P., T. E. Dinero, C. A. Seibel, and S. L. Cox. 1988. Stranger and acquaintance rape. *Psychology of Women Quarterly* 12: 1–24.

Krishnan, S. P., J. C. Hilbert, D. VanLeeuwen, and R. Kolia. 1997. Documenting domestic violence among ethnically diverse populations. Results from a preliminary study. *Family and Community Health* 20: 32–48.

Krug, Ronald S. 1989. Adult male report of childhood sexual abuse by mothers: Case description, motivations, and long-term consequences. *Child Abuse and Neglect* 13: 111–19.

Kurrie, S. E., P. M. Sadler, K. Lockwood, and I. D. Cameron. 1997. Elder abuse: Prevalence, intervention and outcomes in patients referred to four aged care assessment teams. *Medical Journal of Australia* 122.

Lackey, Chad, and Kirk R. Williams. 1995. Social bonding and the cessation of partner violence across generations. *Journal of Marriage and the Family* 57: 295–305.

Langehough, Steven O.,C. Walters, D. Knox, and M. Rowley. 1997. Spirituality and religiosity as factors in adolescents' risk for antisocial behaviors and use of resilient behaviors. Paper presented at the 59th Annual Conference of the National Council on Family Relations, Crystal City, Va., November.

Langley, J., J. Martin, and S. Nada-Raja. 1997. Physical assault among 21-year-olds by partner. *Journal of Interpersonal Violence* 12: 675–84.

Lanz, Jean B. 1995. Psychological, behavioral, and social characteristics associated with early forced sexual intercourse among pregnant adolescents. *Journal of Interpersonal Violence* 10: 188–200.

Lloyd, S. A., and B. C. Emery. Abuse in the family: An ecological, life-cycle perspective. In *Family relations: Challenges for the future,* edited by T. H. Brubaker. Newbury Park, Calif.: Sage, 129–52.

Magdol, L., T. E. Moffitt, A. Caspi, and P. A. Silva. 1998. Hitting without a license: Testing explanations for differences in partner abuse between young adult daters and cohabitors. *Journal of Marriage and the Family* 60: 41–55.

Makepeace, James. 1989. Dating, living together, and courtship violence. In *Violence in dating relationships,* edited by M. S. Pirog-Good and J. E. Stets. New York: Greenwood Press, 94–107.

Martin, M. J., and J. C. Walters. 1992. Child neglect: Developing strategies for prevention. *Family Perspective* 26, 305–14.

Michael, Robert T., John H. Gagnon, Edward O. Laumann, and Gina Kolata. 1994. *Sex in America: A definitive survey.* Boston: Little, Brown.

McKenry, Patrick C., Teresa W. Julian, and Stephen M. Gavazzi. 1995. Toward a biopsychosocial model of domestic violence. *Journal of Marriage and the Family* 57: 307–20.

McNeal, C., and P. R. Amato. 1998. Parents' marital violence: Long-term consequences for children. *Journal of Family Issues,* 19: 123–39.

Morrow, R. B., and G. T. Sorrell. 1989. Factors affecting self-esteem, depression, and negative behaviors in sexually abused female adolescents. *Journal of Marriage and the Family* 51: 677–86.

Mullen, P. E., J. L. Martin, J. C. Anderson, S. E. Romans, and G. P. Herbison. 1996. The long-term impact of the physical, emotional, and sexual abuse of children: A community study. *Child Abuse and Neglect* 20: 7–21.

Murty, K. S., and J. B. Roebuck. 1992. An analysis of crisis calls by battered women in the city of Atlanta. In *Intimate violence: Interdisciplinary perspectives,* edited by E. C. Viano. Washington, D.C.: Hemisphere, 61–70.

National Clearinghouse on Marital and Date Rape. 1994. Personal communication. Berkeley, Calif.

North Carolina Coalition against Domestic Violence. 1991. Domestic violence fact sheet. Spring. P.O. Box 51875, Durham, NC 27717-1875, 919/490-1467.

Novello, A. 1992. The domestic violence issue: Hear our voices. *American Medical News* 35, no. 12: 41–42.

O'Brien, Chris. 1995. Phi Gamma Delta offers public apology for lewd letter. *News and Observer,* 16 November, 1B.

O'Hearn, R. E., and K. E. Davis. 1997. Women's experience of giving and receiving emotional abuse: An attachment perspective. *Journal of Interpersonal Violence* 12: 375–91.

O'Keefe, Maura. 1994. Adjustment of children from maritally violent homes. *Families in Society: The Journal of Contemporary Human Services* 75: 403–15.

O'Keefe, Maura. 1997. Predictors of dating violence among high school students. *Journal of Interpersonal Violence* 12: 546–68.

Pagelow, M. D. 1992. Adult victims of domestic violence: Battered women. *Journal of Interpersonal Violence* 7: 87–120.

Pinkney, D. S. 1992. Sparing the rod: Activists physicians argue that school punishments reinforce our nation's culture of violence. *American Medical News* 35: 45–50.

Pipes, R. B., and K. LeBov-Keeler. 1997. Psychological abuse among college women in exclusive heterosexual dating relationships. *Sex Roles* 36: 585–603.

Pirog-Good, Maureen A. 1992. Sexual abuse in dating relationships. *Intimate violence: Interdisciplinary perspectives,* edited by E. C. Viano. Washington, D.C.: Hemisphere, 101–10.

Riggs, David S., and M. B. Caulfield. 1997. Expected consequences of male violence against their female dating partners. *Journal of Interpersonal Violence* 12: 229–40.

Ronfeldt, H. M., R. Kimnerling, and I. Arias. 1998. Satisfaction with relationship power and the perpetuation of dating violence. *Journal of Marriage and the Family* 60: 70–78.

Rosen, K. H., and S. M. Stith. 1992. The process of leaving abusive dating relationships. Paper presented at the 54th Annual Conference of the National Council on Family Relations, Orlando, Florida. Used by permission.

Russell, D. E. 1990. *Rape in marriage.* Bloomington, Ind.: Indiana University Press.

Sanchez, Sandra. 1996. Expectant couple caught in clash of two cultures. *USA Today.* 29 January, 1D.

Sappington, A. A., R. Pharr, A. Tunstall, and E. Rickert. 1997. Relationships among child abuse, date abuse, and psychological problems. *Journal of Clinical Psychology* 53: 319–29.

Sawyer, Robin G., E. D. Schulken, and P. J. Pinciaro. 1997. A sexual victimization in sorority women. *College Student Journal* 31: 387–95.

Sciere, Kathleen. 1995. The stars of the parallel bars. CBS, *Sixty Minutes.* 8 September.

Shapiro, B. L., and J. C. Schwarz. 1997. Date rape. The relationship to trauma symptoms and sexual self-esteem. *Journal of Interpersonal Violence* 12: 407–19.

Solomon, J. C. 1992. Child sexual abuse by family members: A radical feminist perspective. *Sex Roles* 27: 473–85.

Statistical Abstract of the United States: 1997. 117th ed. Washington, D.C.: U.S. Bureau of the Census.

Steinmetz, S. K. 1987. Family violence. In *Handbook of marriage and the family,* edited by M. B. Sussman and S. K. Steinmetz. New York: Plenum.

Stets, J. E. 1992. Interactive processes in dating aggression: A national study. *Journal of Marriage and the Family* 54: 165–77.

Stets, J. E., and M. A. Straus. 1989. The marriage as a hitting license: A comparison of assaults in dating, cohabiting, and married couples. In *Violence in dating relationships,* edited by M. A. Pirog-Good and J. E. Stets. New York: Greenwood Press, 33–52.

Stock, J. L., M. A. Bell, D. K. Boyer, and F. A. Connell. 1997. Adolescent pregnancy and sexual risk-taking among sexually abused girls. *Family Planning Perspectives* 29: 200–203.

Straus, Murray A. 1994. *Beating the devil out of them: Corporal punishment in American families.* New York: Lexington Books/MacMillan.

Straus, Murray A., and Anita K. Mathur. 1996. Social change and the trends in approval of corporal punishment by parents from 1968 to 1994. In *Family violence against children: A challenge for society,* edited by D. Frehsee, W. Horn, and K. D. Bussmann. New York: Walter de Gruyter, 91–105.

Straus, Murray A., David B. Sugarman, and Jean Giles-Sims. 1997. Spanking by parents and subsequent antisocial behavior of children. *Archives of Pediatric Adolescent Medicine* 151: 761–67.

Sutherland, Kelly. 1997. Let me call you sweetheart. *Dateline.* NBC (National Broadcasting Company), 13 November.

Tucker, Tanya. 1997. *Nice dreams: My life.* New York: Hyperion.

U.S. House of Representatives, Select Committee on Aging. 1991. *Elder abuse: What can be done?* (Hearing, May 15). Washington, D.C.: Government Printing Office.

Vaselle-Augenstein, R., and A. Ehrlich. 1992. Male batterers: Evidence for psychopathology. In *Intimate violence: Interdisciplinary perspectives,* edited by E. C. Viano. Washington, D.C.: Hemisphere, 139–54.

Viano, C. Emilio. 1992. Violence among intimates: Major issues and approaches. In *Intimate violence: Interdisciplinary perspectives,* edited by C. E. Viano. Washington, D.C.: Hemisphere, 3–12.

Walters, Barbara. 1995. Interviews of a lifetime: Priscilla Presley. Lifetime Television Network. 4 August.

Washburne, C. K. 1983. A feminist analysis of child abuse and neglect. In *The dark side of families: Current family violence research,* edited by D. Finkelhor, R. J. Gelles, G. T. Hotaling, and M. A. Straus. Beverly Hills: Sage, 289–92.

White, Melvin. 1988. Elder abuse. In *Aging and the family,* edited by S. J. Bahr and E. T. Peterson. Lexington, Mass.: Lexington Books, 261–71.

Winton, Chester A. 1995. *Frameworks for studying families.* Guilford, Conn.: Dushkin.

Zweig, J. M., B. L. Barber, and J. S. Eccles. 1997. Sexual coercion and well-being in young adulthood: Comparison by gender and college status. *Journal of Interpersonal Violence* 12: 291–308.

In our study of 50 successful marriages, every couple had lived through at least one major tragedy, and some had experienced several traumatic events. No one escaped the wolves at the door.

Judith Wallerstein, Sandra Blakeslee, The Good Marriage

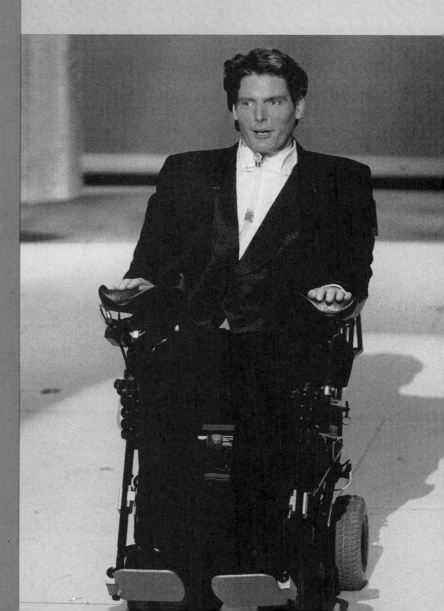

I n the spring of 1995, Christopher Reeve, star of three *Superman* movies, became paralyzed from the neck down by accidentally catapulting over his horse during a jumping competition. Although Reeve requires a respirator to breathe, he is in excellent spirits and sees "meaning and opportunity in the future" (ABC, 1995). Reeve's optimism is matched by the love, support, and encouragement from his wife, Dana, and two sons. His accident illustrates the suddenness with which a family crisis can occur, its impact on other family members, and its enormous expense ($400,000 annually in Reeve's case). It also illustrates the epitome of a family's positive response to a crisis and the importance of effective stress and crisis management.

Stress and Resilient Families

All families experience stress and crisis events. And many are resilient.

Definitions of Stress and Crisis Events

Stress is a reaction of the body to substantial or unusual demands made on it. Stress often involves tension, irritability, high blood pressure, and depression and may result from a wide range of both positive and negative events and situations. The effects of stress are profound.

> Stress is not something that just grips us and, with time or effort, then lets go. It changes us in the process. It alters our bodies—and our brains (Capri, 1996, 34).

RECENT RESEARCH

Thirty-seven percent of a national sample of university first-year women students reported that they were overwhelmed "with all I had to do." Twenty percent of first-year university men felt the same way (American Council on Education and University of California, 1997).

Managing crisis events is part of family living.

Stress is a process rather than a state. For example, a person will experience different levels of stress throughout a divorce—the stress involved in acknowledging that one's marriage is over, telling the children, leaving the home, getting the final decree, and becoming remarried will vary across time.

A *crisis* is a crucial situation that requires changes in normal patterns of behavior. A family crisis is a situation that upsets the normal functioning of the family and requires a new set of responses to the stressor. Sources of stress and crises can be external (e.g., flood, downsizing, military separation) or internal (e.g., alcoholism, infidelity, Alzheimer's disease). Stressors or crises may also be categorized as expected or unexpected. Examples of expected family stressors include the need to care for aging parents and the death of one's parents. Unexpected stressors include contracting HIV, miscarriage, and teen suicide. Both stress and crises are a normal part of family life and sometimes reflect a developmental sequence. Pregnancy, childbirth, job changes/loss, children leaving home, retirement, and widowhood are all stressful and predictable for most couples and families. There is also a cumulative effect of crisis events. Persons who report numerous crisis events also report higher levels of chronic stress (Turner & Lloyd, 1995).

Characteristics of Resilient Families

Resiliency refers to a family's ability to respond to a crisis in a positive way. Several characteristics associated with resilient families include having a joint cause or purpose, good problem-solving skills, the ability to delay gratification, and flexibility (Richardson & Hawkes, 1995). Two researchers (Kieren & Delehanty, 1995) illustrated the value of flexibility in their case study of a 25-year-old woman who had been thrown from a horse and suffered severe brain injury. When her physician noted that she would not likely recover, her family responded to the knowledge of this crisis by acknowledging "that she had become a 'new person' since her accident and that this new person required that family members relate to her differently" (p. 26). Hence, the family accepted the reality that one of their family members was permanently changed and adapted accordingly.

Family Stress Theories

Various theorists have suggested how individuals and families experience and respond to stressors.

ABC-X Model

Burr and Klein (1994) reviewed the ABC-X model of family stress, developed by Reuben Hill in the 1950s.

The model can be explained as follows:

A = stressor event
B = family's management strategies, coping skills

C = family's perception, definition of the situation
X = family's adaptation to the event

A is the stressor event, which interacts with B, which is the family's coping ability, or crisis-meeting resources. Both A and B interact with C, which is the family's appraisal or perception of the stressor event. X is the family's adaptation to the crisis. Thus, a family that experiences a major stressor (e.g., spinal cord-injured spouse or parent) but has great coping skills (e.g., love, communication, commitment) and perceives the event to be manageable will not experience an extensive crisis. But a family that experiences a less critical stressor event (e.g., child makes Ds in school) but has minimal coping skills (e.g., everyone blames everyone else) and perceives the event to be catastrophic will experience an extreme crisis. Hence, how a family experiences and responds to stress depends not only on the event but also on the family's coping resources and perceptions of the event.

Koos's Roller-Coaster Model

Tragedy is in the eye of the observer, and not in the heart of the sufferer.
Emerson
Natural History of Intellect

Other theorists have suggested other ways to conceptualize the developmental pattern of a crisis. In the mid-forties, Koos (1946) presented the "roller-coaster model" that visually displays the different stages many families go through in responding to a crisis (see Figure 11.1).

The horizontal line at the far left of the figure represents the precrisis level of family functioning. The stressor or crisis (e.g., unemployment) then throws the family into a period of disorganization (e.g., increased drinking, role disruption), which can last for a short or a long time, as illustrated by the angle of recovery. The level of reorganization depicts the family's return to its precrisis state. While some families return to the precrisis level of organization, others

What is to give light must endure burning.
Viktor Frankl
Nazi camp survivor

Figure 11.1
Koos's Roller-Coaster Model of Family Stress

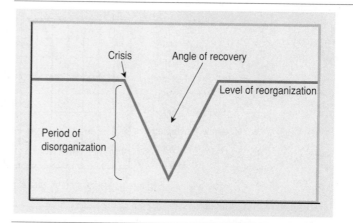

become worse, and still others improve. As to the latter, an alcoholic reported that his disease, though traumatic for him and his family, eventually strengthened his marriage and resulted in a closer relationship with his children. Some families become worse in some areas and improve in others.

Researchers and family therapists have commonly assumed that the rollercoaster pattern is universal in families experiencing a crisis. But recent research suggests that the roller-coaster model describes some, but not all, family responses to stress. Burr and Klein (1994) looked at the family functioning patterns of fifty-one families who had experienced a family crisis (such as bankruptcy, handicapped child, infertility, and troubled teen). The researchers asked eighty-two adults from these families to draw a graph illustrating how a particular ordeal affected their overall family functioning across time. The researchers explained to the participants that overall family functioning includes a variety of aspects such as marital satisfaction, family togetherness, communication, daily routines, and emotional climate. Figure 11.2 depicts the grid onto which participants drew these graphs. Analysis of the graphs, as well as interviews with the respondents, revealed five patterns in the way families respond to stress (see Figure 11.3). As noted in Table 11.1, only about half of the families in the Burr and Klein study reflect the roller-coaster model of family functioning.

Stress Management Strategies

Families and individuals confronted with a crisis use an array of management strategies (the B in the ABC-X model previously). Some stress-management strategies are helpful; others are not.

RECENT RESEARCH

Quatman (1997) identified perceptions of healthy family functioning in a sample of 130 parents and guardians of sixth graders. Characteristics of healthy families identified by the sample included emotional bondedness, commonness/mutuality, and communication.

If we do not handle stress properly as we age, a part of the brain shrinks and we lose short-term memory. This is not Alzheimer's disease but the effect of stress hormones.
Bruce Mcewen, MD

Figure 11.2
Overall Family Functioning Grid

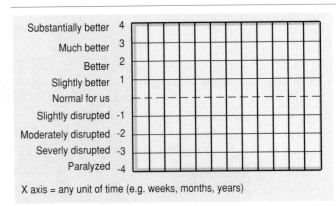

X axis = any unit of time (e.g. weeks, months, years)

Source: Wesley R. Burr, Shirley R. Klein and Associates. *Reexamining family stress: New theory and research,* p. 69. Thousand Oaks, Calif.: Copyright © 1994 by Sage Publications, Inc. Reprinted by permission of Sage Publications, Inc.

Figure 11.3
The Effects of Stress on Family Functioning: Five Patterns

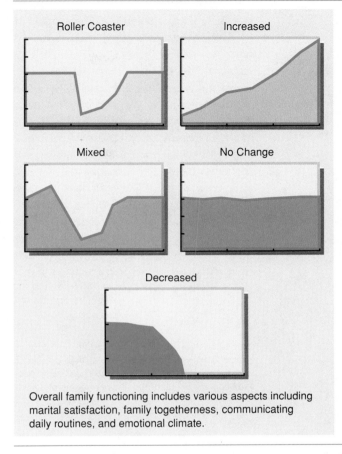

Overall family functioning includes various aspects including marital satisfaction, family togetherness, communicating daily routines, and emotional climate.

Source: Wesley R. Burr, Shirley R. Klein and Associates. *Reexamining family stress: New theory and research*, p. 159. Thousand Oaks, Calif.: Copyright © 1994 by Sage Publications, Inc. Reprinted by permission of Sage Publications, Inc.

Helpful Strategies

If your needs are not being met, drop some of your needs.
George Carlin
Comedian

Researchers Burr and Klein (1994) administered an eighty-item questionnaire to seventy-eight adults to assess how families experiencing various stressors such as bankruptcy, infertility, disabled child, and troubled teen used various coping strategies and how useful they found these strategies. Table 11.2 presents the rank order of ten helpful stress and management strategies and the percentage of respondents finding each strategy to be helpful.

The management behavior that the highest percentage reported as being helpful was changing basic values as a result of the crisis situation. After his accident, Christopher Reeve reported a complete change in his basic values. Previously, as an avid scuba diver, equestrian, and athlete, he had placed great value on his athleticism. Now, his limbs were dead weight. "You begin to see," he said, "that your body is not you and that your mind and spirit must take over" (ABC, 1995).

Table 11.1

		Percentage of Families in the Burr and Klein (1994) Study*
1. **Roller-Coaster Model:**	Stress results in decline in quality of family functioning followed by a recovery.	51
2. **Increased Model:**	Stress results in increase in quality of family functioning.	18
3. **No Change Model:**	Family functioning remains essentially unchanged throughout ordeal.	15
4. **Decreased Model:**	Stress results in a decrease in the quality of family functioning that is not followed by a recovery.	5
5. **Mixed Model:**	Stress results in increased quality of family functioning, followed by a decline and then, usually, a recovery.	11

*Based on data from eighty-two adults from fifty-one families.
Source: Wesley R. Burr, Shirley R. Klein and Associates. *Reexamining family stress: New theory and research,* pp. 70–78. Thousand Oaks, Calif.: Sage Publications, Copyright © 1994 by Sage Publications, Inc. Reprinted by permission of Sage Publications, Inc.

In responding to the crisis of bankruptcy, people may reevaluate the importance of money and conclude that relationships are more important. In coping with unemployment, people may reevaluate the amount of time they spend with family members versus the amount of time they have spent in the pursuit of money. Responding to crisis events by changing one's philosophy of life (item 10 in Table 11.3) is similar to changing one's basic values.

Items 2 (understanding), 4 (being sensitive), 6 (having common goals), 7 (listening), and 9 (cooperating) reflect relationship and communication items in which the partners convey to each other a sense of involvement with each other and a desire to support each other in the face of a crisis. Items 3 (talking with others), 5 (faith), and 8 (prayer) involve seeking help from others as well as from a higher spiritual support.

In another study, Wallerstein and Blakeslee (1995) identified five management strategies that spouses in fifty successful marriages found helpful in responding to a crisis.

1. *Developed realistic perspective.* After the initial shock of the crisis, the spouses get hold of themselves rather than "freezing into helplessness or rushing around in meaningless activity" (Wallerstein & Blakeslee, 1995, 123). They were careful to distinguish their fears of the worst-case sce-

Don't look forward to the day when you stop suffering. Because when it comes, you'll know you are dead.
Tennessee Williams
Playwright

Table 11.2

	Ten Helpful Stress and Crisis Management Strategies	
Rank	Item	Percentage Report Strategy Helpful
1	Changed basic values as a result of crisis	96
2	Tried to be understanding with each other	93
3	Tried to talk with others in a similar situation	93
4	Tried to be more sensitive to each other	92
5	Tried to have more faith in God	92
6	Tried to focus on common goals and values	91
7	Tried to listen to each other more	91
8	Tried to get help through prayer or spiritual assistance	90
9	Tried to cooperate more as a family	89
10	Changed basic philosophy of life as result of crisis	88

Source: Adapted from Wesley R. Burr, Shirley R. Klein and Associates. *Reexamining family stress: New theory and research.* Thousand Oaks, Calif.: Sage Publications, Copyright © 1994 by Sage Publications, Inc. Reprinted by permission of Sage Publications, Inc.

Table 11.3

Harmful Stress and Crisis Management Strategies		
Item	Percentage Reporting Strategy Harmful	Percentage Reporting Strategy Helpful
Kept feelings inside	75	14
Took out feelings on others	75	25
Denied, avoided, or ran away from problem	70	30
Sought to keep others from knowing how bad crisis was	70	14
Expressed affection less	67	33
Acted as if nothing had happened	60	33
Waited for problem to go away	59	33
Became more critical of others	51	19
Sought to forget entire matter	50	22

Source: Adapted from Wesley R. Burr, Shirley R. Klein and Associates. *Reexamining family stress: New theory and research.* Thousand Oaks, Calif.: Sage Publications, 1994, 173. Used by permission.

nario from what was likely to happen. For example, although the parents of a child who had cancer feared the worst, they also acknowledged that the child might benefit from treatment.

2. *Avoided blame.* The spouses did not blame each other for the crisis. Phrases such as "If you had done what I told you, you wouldn't have lost your job" or "If you had gotten x-rayed earlier, you would have had time to treat the cancer" were not spoken. Rather, the spouses were careful not only to avoid blaming each other but also to protect each other from self-blame and self-reproach.

3. *Sought opportunities for fun.* In spite of the tragedy the couples were coping with, they sought to enjoy life as best they could. They took the view of seizing opportunities for fun rather than spiraling downward into a deep depression. Whether taking trips, going out to eat, or renting a movie, they still sought to enjoy life. "They tried their best not to let the tragedy totally dominate their world" (Wallerstein & Blakeslee, 1995, 123).

4. *Kept destructive impulses in check.* Recognizing that fear can lead to anxiety and helpless rage, the spouses sought to avoid "driving away the people they needed and loved the most" (p. 123). Rather than exacerbate the crisis by responding to it with anger, drug/alcohol abuse, or emotional withdrawal, spouses made a great effort to "keep destructive tendencies from getting out of control and harming the marriage."

5. *Intervened early in a crisis.* When the spouses saw a crisis coming, they intervened early. Whether the crisis is depression or alcoholism or violence, early intervention helps to prevent the problem from mushrooming into a much larger issue.

One woman, whose husband started to drink and come home late because his job was in trouble, told him firmly, "I'm not going to tell our daughter that I don't know where her father is. Stop drinking." She put her demand in terms of the daughter he loved, and the man did indeed stop drinking (Wallerstein & Blakeslee, 1995, 123).

Exercise

Exercise is an effective stress reducer. In one study of 410 respondents, exercise was identified as the most helpful behavior they engaged in as a mood-regulating strategy (Thayer, Newman, & McClain, 1994). The Centers for Disease Control and Prevention (CDC) and the American College of Sports Medicine (ACSM) recommend that people aged 6 and older engage regularly, preferably daily, in light to moderate physical activity for at least thirty minutes at a time.

Sleep

Getting an adequate amount of sleep is also associated with low stress levels. Fourteen percent of a sample of adults who slept seven to eight hours each night reported feeling stress, in contrast to 43 percent who slept fewer than six hours each night (Kate, 1994). In spite of the need for sleep, most Americans feel sleep-deprived.

Life—a continuous process of getting used to things we hadn't expected.

Evan Esar
Esar's Comic Dictionary

If you can't control your body, you can't control your mind. Make an appointment with exercise and keep it.

Nolan Ryan
Baseball player

RECENT RESEARCH

A team of researchers (Finucane et al., 1997) examined the exercise patterns of 1788 individuals aged 70 years and older and found that, compared to those who did not exercise, the exercisers reported better health and their mortality rates were lower at two-year followup. Similarly, Williams (1997) studied 187 older women who became involved in a 12-month program of group exercise and found that anxiety reduction was associated with their exercising. Physiological and cognitive benefits were also realized.

Vigorous aerobic exercise is a natural antidote to stress.

INSIGHT

Beyond exercise and getting enough sleep, having a work environment with flexible hours and supportive supervisors helps to lower stress and reduces depression, somatic complaints, and blood cholesterol. Hence, structural changes in the workplace have an important influence on keeping stress under control (Thomas & Ganster, 1995). Other stress-relieving mechanisms are personal and include meditation, biofeedback, and relaxing postures. The latter suggests that sitting in certain positions has a pronounced effect on reducing stress (Capri, 1996).

Successful coping is also associated with the degree to which a person's life is integrated into a network of relationships that can serve as resources. "The simplest and most powerful measure of social support appears to be whether a person has an intimate, confiding relationship or not (typically with a spouse or lover; friends or relatives function equivalently but less powerfully)" (Thoits, 1995, 64).

NATIONAL DATA   Sixty-four percent of Americans in a Gallup Poll reported that they do not get enough sleep. Those who report lacking sleep the most are young adults with children. Those who get the most sleep are over the age of 64 (Saad, 1995).

Love

A love relationship also helps an individual cope with stress. Forty-four percent of 351 undergraduates responded "a lot" or "some" when asked how important is a love relationship in helping you cope with life? Five percent said "a great deal" (Langehough et al., 1997).

Multiple Roles

Another factor that helps individuals cope with stress is to be involved in multiple roles. Two researchers studied the psychological consequences of multiple roles of 461 mothers over the age of 55 who occupied an average of five roles—parent of a nonhandicapped child, spouse, relative, friend, and parent of an adult child with mental retardation (Hong & Seltzer, 1995). They found that the greater the number of roles, the lower the depression and the higher the psychological well-being. Starting a new job (adding even another role) also had positive psychological effects.

Humor

A sense of humor may be defined as the ability to laugh, smile, and find something amusing in a situation. A sense of humor is associated with a number of positive outcomes including stress reduction, physical health, mental well-being, and life satisfaction. In general, the greater the humor, the more satisfied people are with their lives. Humor is also associated with improving one's immune system (Frankenfield, 1996; Kamei et al., 1997).

Harmful Strategies

Some coping strategies are not only ineffective for resolving family problems, but also add to the family's stress by making the problem worse. Respondents in the Burr and Klein (1994) research identified several strategies they regarded as harmful to overall family functioning.

As indicated in Table 11.3, most respondents found these strategies to be harmful in that they "made things much worse."

Burr and Klein's research also suggests that there are gender differences between women and men in their perceptions of the usefulness of various coping strategies. Women were more likely than men to view such strategies as sharing concerns with relatives and friends, becoming more involved in religion, and expressing emotions as helpful. Men were more likely than women to view potentially harmful strategies such as using alcohol, keeping feelings inside, and keeping others from knowing how bad the situation was as helpful coping strategies.

Five Family Crises

Some of the more common crisis events faced by spouses and families include physical illness, infidelity, unemployment, alcohol/drug abuse, and death. We now examine each of these as it impacts the family.

Physical Illness and Disability

"In sickness and in health" is a promise most brides and grooms make to each other on their wedding day.

> Yet most of us rejoicing at a wedding have little idea about what life together with an illness or disability would mean. Chronic disorders can affect couples with devastating consequences. Couples must meet the challenge of maintaining a balanced mutual relationship while assuming roles of patient-caretaker (Rolland, 1994, 327).

Physical illness and disability are not uncommon. Almost a quarter (23 percent) of 1,393 adults aged 18 to 55 reported that they had had a major illness or accident that required them to spend a week or more in the hospital (Turner & Lloyd, 1995). Although short-term illness and disability often produce stress in the family, long-term illness and disability have profound and enduring effects on family members and family life. One woman described her family's experience (authors' files):

> After fourteen years of marriage and three lovely daughters, my husband was diagnosed with what is commonly known as Lou Gehrig's disease. It

is a disease of the muscles which is progressive, degenerative, and fatal. There is no cure. My husband went from an independent strong weight-lifter to a bony frail man who cannot feed himself and needs twenty-four-hour care. He has been in the bed for the last six years.

The effect on our lives, marriage, and family has been enormous. He has had to stop denying that he can't escape an early debilitating death. I have had to accept that our marriage is dramatically and permanently altered. Our daughters have had to accept that their daddy will not be here to walk them down the aisle on their wedding day.

In spite of the difficulty we have managed some wonderful times. We still go to basketball and football games and watch our girls perform in softball and gymnastics. We still watch movies and laugh and cry like other folks.

Our parents, friends, and church have been wonderful in their support. We feel blessed. Life is hard but we were never promised a rose garden and we are doing our best to cope. We are not looking forward to the end but are preparing for it.

Booth and Johnson (1994) studied a national sample of 1,298 spouses with regard to how the change in the health of one partner affected the quality of the marital relationship over a three-year period. They found that spouses of sick partners experienced a greater decline in marital quality than did the spouses who themselves became ill. "Changes in the financial circumstances, shifts in the division of labor, declines in marital interaction, and problematic behavior by the afflicted individual account for much of the health-marital quality relationship" (Booth & Johnson, 1994, 222).

In their study of fifty couples with successful marriages, Wallerstein and Blakeslee (1995) noted that a major developmental task is to confront the inevitable and unpredictable adversities of life in ways that enhance the relationship despite suffering. Rolland (1994) identified several issues that spouses might profit from discussing when one of them becomes seriously ill.

Nothing is terrible except fear itself.
Francis Bacon
English writer

Intimacy and threatened loss   Although the definitions of intimacy differ by social class, intimacy often includes emotional, intellectual, sexual, spiritual, and recreational aspects. The diagnosis of a serious condition can threaten the continuation of such intimacy, which may result in partners' either "pulling away from one another or clinging to each other in a fused way" (Rolland, 1994, 329).

The key to healthy coping and adaptation with chronic disorders depends largely upon a couple's willingness to address these basic issues. Couples adapt best when they learn to deal with these facts of life and use that consciousness in an empowering manner to live more fully rather than constrain their relationship. This means revising their intimacy to include rather than avoid issues of disability and threatened loss. This helps offset the dark side of chronic disorders (Rolland, 1994, 329).

Hence, rather than retreat into their respective thoughts of doom and gloom, spouses might benefit from talking about the disorder in a way that frames it in a positive light. Researchers have consistently shown that the choice to view something positively, as a challenge, is a major coping mechanism. For example, rather than viewing cancer as debilitating one partner, the partners can choose to regard it as a challenge to face adversity together.

OTHER CULTURES Symbolic interactionists emphasize that the labels that spouses assign to various maladies they experience and their definitions of those maladies influence their interaction. The same is true of afflictions of children in the family. Rounds, Weil, and Bishop observed considerable cultural diversity in the way disabilities are regarded. For example, "in traditional Hispanic and Asian Pacific culture, infant disability is associated with considerable social stigma and shame. The disability may represent 'wrongdoing' on the part of the family" (1994, 8). ●

Establishing healthy boundaries Couples coping with physical illness may need to express emotions and discuss concerns related to the illness. However, if these emotions and concerns permeate every discussion and activity in the marriage, then the illness becomes the focus of the relationship. Thus, couples are challenged to establish healthy boundaries between illness-related emotions and concerns and other aspects of the relationship. Rolland (1994) suggests ways of meeting this challenge.

> Specific areas of a home can be off-limits to illness-related functions. For instance, not discussing the illness in the bedroom can help preserve romance. When couples stop inviting friends to their home because it has become associated with illness, they can be encouraged to socialize in their home in order to preserve parts of their homelife that have been joyous and connect them to their wider social network (Rolland, 1994, 332).

Togetherness and separateness Couples must also balance the need to be together with the need to spend time away from each other. "Couples adapt best to chronic disorders when they can transform their understanding of 'we-ness' to include a new version of separateness that acknowledges different needs and realities" (Rolland, 1994, 336). Hence, while the partner with the malady may have fears of disability and death that may translate into the desire for more closeness, the other partner may "pull away in detachment that represents preparation for the final separation of death" (p. 336). The respective needs to become closer and to separate are normative, and an openness between the partners about their feelings/inclinations is preferable to silence.

Indeed, all couples, even those not experiencing a crisis, need to seek a balance between togetherness and separateness. A crisis event such as illness or disability may bring this need into focus.

We have been discussing the management of physical illness. But not all illnesses are physical. Some are mental. Everyone experiences problems in living from time to time, including relationship problems, criminal or domestic victimization, work-related problems, and low self-esteem. However, some conditions or mental disorders are more intense, persistent, and debilitating.

The toll of mental illness on a relationship can be immense. A major initial attraction of partners to each other includes intellectual and emotional qualities. When these are affected, the partner may feel that the mate has already died psychologically, since he or she is, literally, "not the same person I married."

There is no road but hath an end.

Montaigne

Essays

RECENT RESEARCH

Jacob and Johnson (1997) compared the effects of maternal and paternal depression on child functioning and found associations with child adjustment problems and more impaired parent-child communication. The study involved 50 depressed fathers, 41 depressed mothers, and 50 control families (all intact).

INSIGHT

"Gay and lesbian couples facing chronic disorders must deal with issues related to social stigma in addition to all the same issues heterosexual couples face. A chronic or life-threatening illness with its attendant need for health care often forces a hidden or extremely private relationship into the public for the first time at a moment of great vulnerability. Experiencing cold or distant professional healers at a time of need can be a particularly poignant rejection" (Rolland, 1994, 345).

NATIONAL DATA In a study of more than 8,000 respondents between the ages of 15 and 54, almost half (48 percent) reported having experienced at least one psychiatric disorder at some time in their lives, and nearly 30 percent reported experiencing at least one disorder in the past twelve months. The most common disorders were depression, alcohol dependence, and social phobias (Kessler et al., 1994).

In the rest of this chapter, we examine how spouses cope with infidelity, unemployment, drug abuse, and death. Each of these events can be viewed either as devastating and the end of meaning in one's life or as an opportunity and challenge to rise above. Review the opening quote for this chapter.

Infidelity

The public crisis Kathie Lee Gifford and Frank Gifford endured over his extramarital encounter with a flight attendant in 1997 reemphasized that such experiences do occur and that they often represent family crisis events. They also involve intense pain for the partner of the person involved in the affair. Princess Diana's biographer revealed the pain that she experienced when overhearing Prince Charles say on the phone to Camilla Parker Bowles, "I will always love you" (Morton, 1997). Another spouse expressed his pain at the discovery that his wife was having an affair:

> Forget sleeping for a while. Forget friendly conversations. Forget peace of mind. You will literally begin the process of grieving as if a death had occurred. And in reality a death has occurred. The total happiness with which you have embraced your spouse and your marriage is now dead. The innocence with which you have always pictured your spouse is dead. Trust is long gone and you now doubt every word said. You will consider yourself violated and your spouse has now become somewhat "unclean." You will have wide, sweeping mood changes within a five-minute period (author's files).

The term *extradyadic* includes sexual involvement with someone other than the person with whom one is involved in an emotional/sexual relationship. While the term *extramarital* refers to the attraction of a spouse to someone other than the mate, extradyadic refers to all pair-bonded individuals who are attracted to someone other than the partner.

The terms *cheating, unfaithfulness,* and *infidelity* reflect societal disapproval in reference to extramarital sexual involvements. While some couples have *open marriages* which allow for extradyadic involvements, most do not.

NATIONAL DATA In a national sample of adults in the United States, 57.1 percent of the men and 56.3 percent of the women agreed that extramarital sex is "always wrong" (Wiederman, 1997, 171).

OTHER CULTURES Societies differ as to how serious a violation of morality they regard adultery. In the United States, the penalty ranges from nothing to alimony. In countries that practice the Islamic religion (in Northern Africa and parts of the Middle East), adultery is considered a major sin and execution by stoning or beheading is considered appropriate. ●

"I can forgive, but I cannot forget," is another way of saying, "I cannot forgive."
H. W. Beecher

Nothing is more noble, nothing more venerable than fidelity. Faithfulness and truth are the most sacred excellences and endowments of the human mind.
Cicero

RECENT RESEARCH

Amato and Rogers (1997) interviewed a national sample of spouses (1,748) and identified the behaviors associated with subsequent divorce—sexual infidelity, jealousy, moodiness, not communicating, and anger (p. 622). Notice that all of these behaviors may occur together and in reference to infidelity.

INSIGHT

In spite of the difficulty infidelity imposes on a couple and their marriage, not one of the fifty spouses in the "successful" marriages studied by Wallerstein and Blakeslee (1995) said that they would automatically terminate their marriage if an affair were revealed.

NATIONAL DATA   In a national sample of adults in the United States, 22.7 percent of the men and 11.6 percent of the women reported ever having had extramarital sex. For the previous twelve months, 4.1 percent of the currently married men and 1.7 percent of the currently married women reported having had extramarital sex. When respondents under 40 years of age were compared, no differences between men and women were observed (Wiederman 1997, 170, 171).

Infidelities range from brief sexual encounters to full-blown romantic affairs. Infidelities that are brief and involve little or no emotional investment are referred to as brief sexual encounters. Although the partners may see each other again, more often than not, their sexual encounter is a one-night stand. Sexual involvement with prostitutes or with women who offer sex in massage parlors may also be viewed as a brief sexual encounter.

Other infidelities are characterized by intense emotional feelings and sexual involvements. The affair between Francesca Johnson and Robert Kincaid as personified in the novel and movie *The Bridges of Madison County* depicted a romantic affair. One condition of such love affairs is restriction. The *Bridges* couple were restricted by marriage and children (hers) and by time (he would be with her only four days). Such limited access makes the time that romantic lovers spend together very special. In addition, the lover in a romantic affair is not associated with the struggles of marriage—bills, child care, washing dishes, cleaning house, mowing the lawn—and so may experience the affair from a more romantic perspective. Romantic love affairs may be more devastating to an ongoing relationship than a sexual affair.

Becoming more common today is the computer affair. Although legally adultery does not exist unless two persons have physical sex, an on-line computer affair can be just as disruptive to a marriage or a couple's relationship. Computer friendships that move to feelings of intimacy, involve secrecy (one's partner does not know the level of involvement), include sexual tension (even though there is no overt sex), take the time, attention, energy, and affection away from one's partner. "There is really another person there, and that person can move you in various ways, emotionally and sexually" (Turkle, 1996). One New Jersey husband sued his wife for divorce, claiming that his wife committed infidelity during dozens of sexually explicit exchanges on America Online (Peterson, 1996).

Unemployment

Corporate America is downsizing. The result is massive layoffs and insecurity in the lives of American workers. Unemployment and/or the threatened loss of one's job is a major stressor for individuals, couples, and families.

NATIONAL DATA   Of 133,943,000 who are in the labor force, 5.4 percent, or 7,236,000, are unemployed (*Statistical Abstract of the United States: 1997*, Table 619).

The effects of unemployment may be more severe for men than for women. Our society expects men to be the primary breadwinners in their families and equates masculine self-worth and identity with job and income. Stress, depression, suicide, alcohol abuse, and lowered self-esteem are all associated with unemployment.

336   •   *Chapter II*

Turner (1995) noted that the damage to one's sense of self in response to being unemployed is associated with the probability of reemployment. On the basis of interviews with a national sample of unemployed individuals, those who lived in areas of low employment where there was little chance of being reemployed suffered the greatest depression. Those who had college educations were particularly vulnerable.

Even the divorced have their own support groups. Those who are laid off feel a sense of shame and have no such groups to turn to.
Paula Ramon
Sociologist

INSIGHT

Voydanoff (1991) noted that unemployment may influence major family choices. For example, couples experiencing unemployment may decide to postpone childbearing, move in with relatives, or have relatives or boarders join the household.

RECENT RESEARCH

Miller and Cervantes (1997) examined gender differences in a sample of 233 (83 women, 150 men) problem drinkers treated at the same clinic. Patterns of drinking and intoxication were similar. However, men drank more away from home and reported more drinking and driving. They were also more likely to accept a disease concept of alcoholism.

Unemployment affects not only individuals who lose their jobs but also their spouses and families.

> Wives and children of unemployed men are hit particularly hard because their economic well-being is often dependent on the employment of male breadwinners. . . . Wives experience stress not only from the need to live on reduced income but also because they are exposed to changes in their husbands' psychosocial functioning (Jones, 1992, 59).

Wives who earn more money than their husbands also report feeling a great deal of stress—they must cope with their husbands' feelings of inadequacy (Greaves et al., 1995, 66).

Liem and Liem (1990) studied 82 families in which the husbands had recently lost their jobs and found that the wives reported increases in hostility, depression, and anxiety. The degree to which they experienced these negative reactions depended on their perception of how much the unemployed partner contributed to the family in nonmonetary ways (e.g., housework and child care).

Unemployment has also been associated with declines in one's physical health (Ross & Mirowsky, 1995), increases in one's need to control the spouse (Stets, 1995), increases in family violence (Straus, Gelles, & Steinmetz, 1980), and divorce (Glyptis, 1989). However, financial setbacks, including bankruptcy, may actually result in increases in the quality and stability of a couple's relationship. Burr and Klein (1994) observed that 33 percent of the respondents who experienced bankruptcy reported a roller-coaster pattern of adjustment that ended in an improved relationship. Sharing the event, talking about it, and feeling greater commitment to each other helped to produce a positive outcome. The unemployed also have more time for their spouses and their children. And time has become a valued commodity for many people.

NATIONAL DATA   A United States News/Bozell poll revealed that 51 percent of 1,009 adult respondents reported that they would rather have more free time even if it means having less money (Marks, 1995, 86).

Spouses don't have to actually lose their jobs to experience negative effects. Just living in an insecure job situation often results in an increase in stress and a decrease in marital adjustment, family communication, family problem-solving skills, and clarity of roles (Larson, Wilson, & Belely, 1994; Wilson & Larson, 1994). Changing jobs may also be stressful.

Drug Abuse

Spouses, parents, and children who abuse drugs contribute to the stress experienced in marriages and families. Although some individuals abuse drugs to escape from the stress of family problems, drug abuse inevitably adds to the family's problems when it results in health and medical problems, legal problems, loss of employment, financial ruin, school failure, relationship conflict, and even death. Country-and-western singer Lorrie Morgan told about her husband Keith Whitley, an alcoholic, who died of alcohol poisoning—with twenty shots of 100-proof whiskey in his system. She said of his alcoholism, "You can't love someone into sobriety. It just doesn't work" (Morgan, 1997).

Family crises involving alcohol and/or drugs are not unusual. Ten percent of 1,393 adults aged 18 to 55 reported that their spouses, partners, or children had been or were addicted to alcohol or drugs (Turner & Lloyd, 1995).

As indicated in Table 11.4, drug use is most prevalent among 18-to 25-year-olds. Drug use among teenagers under age 18 is also high. Because teenage drug use is more common than drug use among older adults, we focus on the problems parents confront when their teenagers abuse drugs.

Teenage drug abuse Various factors associated with teenage drug abuse include peers who use drugs, low self-esteem, alcoholic parents, low grades, low IQ, being male, being from a single-parent home, and lacking support from parents (Tuttle, 1995). For the drug abuser, the use of alcohol or other drugs may cause a problem in many areas of life—health, school, work, and social relationships. Parents can best prevent their children from using/abusing drugs by creating and maintaining close, healthy, supportive relationships with their children (Tuttle, 1995). Other preventive measures on the part of parents include not abusing alcohol or other drugs themselves and keeping communication channels open with their children. However, if prevention efforts fail, parents should be ready to respond. Several steps are involved in responding to a teenager's drug use.

1. *Confront Your Teenager*

   With the available evidence (your teenager is drunk or in possession of marijuana, cocaine, pills, or some other drug), make your son or daughter aware that you know of the drug use. Cutting through your teenager's de-

Table 11.4

Drug Use, by Type of Drug and Age Group			
		Ever Used	
Type of Drug	**12 to 17 years old**	**18 to 25 years old**	**26 + years old**
Marijuana	16.0%	43.4%	35.0%
Cocaine	1.3	9.6	10.9
Inhalants	6.2	8.3	4.4
Hallucinogens	4.0	11.7	8.0
Heroin	0.4	0.2	1.3
Stimulants	2.6	5.3	6.0
Sedatives	3.2	1.9	3.9
Tranquilizers	2.0	4.5	4.3
Analgesics	4.7	7.1	4.7
Alcohol	41.2	86.8	91.0
Cigarettes	33.5	68.6	76.8

Source: Adapted from *Statistical Abstract of the United States: 1997.* 117th ed. Washington D.C.: U.S. Bureau of the Census.

INSIGHT

Parents who monitor where their adolescents are and what they are doing have fewer problems with adolescent drinking and delinquency. Parental monitoring in conjunction with making it clear to the adolescents that they are loved, valued, and accepted is particularly predictive of fewer such problems (Barnes & Farrell, 1992).

Single, employed parents often have more difficulty monitoring their children; thus, children of these parents are more likely to use marijuana and participate in serious illegal activity (Haurin, 1992). Single parents might consider forming a network with other families to help monitor their children.

nial that he or she uses drugs is difficult but important for both teenager and parent.

2. *Ask for Your Teenager's Point of View*
   Be careful not to criticize or belittle your teenager but ask for an explanation of why he or she drinks, smokes, snorts, or whatever. It is not unusual for teenagers to feel very guilty about what they are doing. Once confronted by their parents, some teenagers are anxious to stop. "I drifted in over my head," said one teenager to her parents, "and I really am sick of it and want to stop."

3. *Make an Agreement*
   One alternative is to make it clear to your child that you will not tolerate drug use. One parent told her 16-year-old:

   > Your father and I know that you get tired of us butting into your life. But we feel that drugs can harm you, and we ask that you stop as long as you are living with us. To ensure that you are not using drugs, we want you to have your urine analyzed weekly. If Steve Howe, the two-million-dollar pitcher for the New York Yankees, can have his urine tested every other day to help him remain clear of drugs, this may also be helpful for you. If we discover that you are still using drugs, we will send you to an in-patient drug rehabilitation center.

Currently, technology provides parents with three alternatives to detect drug use by their children. PDT-90, available through Psychemedics, Inc. of Cambridge, Massachusetts, involves an analysis of a child's hair for trace amounts of illegal drugs that become trapped in the hair shaft. Parents send fifty or sixty strands of hair to the company, which reports whether the child has consumed any cocaine, marijuana, opiates, methamphetamines, or PCP in the past ninety days. DrugAlert, marketed by Barringer Technologies of New Providence, New Jersey, can detect trace elements of these same drugs by analyzing a premoistened "swipe" from clothing worn by the child. Finally, a urine test can be given, but it detects drug use only within the past few days. "Dr. Brown's Home Drug Testing System" has been approved by the Food and Drug Administration and is available in stores for around $30.

Considerable controversy rages over the use of such technology. Children feel that their parents don't trust them and resent intrusion into their lives. But not to detect drug use can be more devastating than the child feeling that his or her parents do not trust him/her.

Drug abuse support groups   If the substance-abuse problem is alcohol, Alcoholics Anonymous (AA) is an appropriate support group (national headquarters mailing address: AA General Service Office, P.O. Box 459, Grand Central Station, New York, NY 10017). There are over 15,000 AA chapters nationwide; the one in your community can be found through the Yellow Pages. The only requirement for membership is the desire to stop drinking.

Former drug (other than alcohol) abusers meet regularly in local chapters of Narcotics Anonymous (NA), patterned after Alcoholics Anonymous, to help each other continue to be drug free. As with AA, the premise of NA is that the best person to help someone stop abusing drugs is someone who once abused drugs. NA members of all ages, social classes, and educational levels provide a sense of support for each other to remain drug free.

Al-Anon is an organization that provides support for family members and friends of alcohol abusers. Such support is also often helpful for parents coping with a teenage drug abuser.

In an attempt to improve parent-child communication in families where parents use drugs, the Strengthening Families Program provides specific social skills training for both parents and children (Wyman, 1997). After families attend a five-hour retreat, parents and children are involved in face-to-face skills training over a four-month period. A twelve-month follow up revealed that parenting skills remained improved and reported heroin and cocaine use had declined.

Death

Even more devastating than drug abuse are family crises involving death—of one's child, of one's teenager or young adult child, of one's spouse, or of one's parent. Death is the family crisis the highest proportion of individuals experience. Almost 40 percent (38.5%) of 1,393 adults aged 18 to 55 reported that they had experienced the death of a child, spouse, or loved one (Turner & Lloyd, 1995). Indeed, death is a part of life. And we might consider viewing life as a process of growth and development that ends in death.

Gender differences in coping with the death of a loved one Although individual reactions and coping mechanisms for dealing with the death of a loved one vary, several reactions to death are common. These include shock, disbelief and denial, confusion and disorientation, grief and sadness, anger, numbness, physiological symptoms such as insomnia or lack of appetite, withdrawal from activities, immersion in activities, depression, and guilt. Eventually, surviving the death of a loved one involves the recognition that life must go on, the need to make sense out of the loss, and the establishment of a new identity. Grief and sadness often return on the anniversary of the death, on the deceased's birthday, and other special occasions.

Women and men tend to have different ways of reacting to and coping with the death of a loved one. Women are more likely than men to express and share feelings with family and friends and are also more likely to seek and accept help, such as attending support groups of other grievers. Initial responses of men are often cognitive rather than emotional. From early childhood males are taught to be in control, to be strong and courageous under adversity and to be able to take charge and fix things. Showing emotions is labeled as weak. According to Sobieski (1994), men tend to feel responsible for their grief symptoms, "almost as if mourning were an illness they need to 'get over' as soon as they can" (p. 7). In contrast, most women give themselves permission to be disoriented and to grieve.

Although men are less likely than women to seek and accept help and share their grief, they often engage in solitary or secret mourning. Sobieski (1994) notes that "men see mourning in solitude as a choice of strength because it shields others from their pain" (p. 9). But their suffering in silence is often mis-

Every one can master a grief, but he that has it.
Shakespeare

RECENT RESEARCH

Two researchers (Henderson & Scott, 1997) compared bereaved parents and spouses and found that the former experienced greater grief intensity.

Beyond the door there's peace I'm sure,
And I know there'll be no more tears in heaven.
From *Tears in Heaven* by Eric Clapton, written about the loss of his 6-year-old son

Grief fills up the room with my absent child.
Shakespeare
Playwright and poet

interpreted by others as withdrawal, defensiveness, or lack of caring about the death of their loved one.

Men often respond to the death of a loved one in behavioral, rather than emotional ways. Sometimes they immerse themselves in work or become involved in legal or physical action in response to the loss. For example, after his 20-year-old daughter lost control of her car and was killed, one father spent several weeks rebuilding a neighbor's fence that was damaged in the accident. Later, he described this activity as crucial to "getting me through those first two months" (Martin & Doka, 1996, 165). Another father expressed his grief by carving his son's memorial stone.

Abuse of alcohol and other drugs increases among grieving women and men, but more so among men. Drugs may numb painful emotions, but they also impair judgment and create additional individual and family problems.

Regarding the differences between masculine and feminine styles of grieving, Martin and Doka (1996) suggest:

> There are advantages and strengths in expressing emotions and seeking help. But there are complementary strengths in stoically continuing in the face of loss and in seeking resolution in cognitive and active approaches. . . . Different ways of coping are just that—differences, not deficiencies (p. 171).

Furthermore, Martin and Doka (1996) note that the masculine and feminine patterns of grieving are not always gender specific. Many men are comfortable in expressing and sharing emotional pain with others and many women exhibit stoicism in response to loss of a loved one. For example, after the 1963 assassination of President John F. Kennedy, Jackie Kennedy maintained emotional control at the funeral and kept busy with her regular schedule afterward. Another widow revealed how, after her husband's sudden death, she felt compelled to carry on his work, so she became involved in his business and successfully negotiated several contracts. This woman chose action and problem solving as a way of coping with her husband's death, but, "sadly, she reported that many of her friends and associates believed that she was cold, wooden, and unfeeling" (Martin & Doka, 1996, 166).

Death of one's child   The death of Ennis Cosby, Bill Cosby's son, in early 1997 reminded us of the possibility of a parent's worst fear—the death of a child. Most people expect the death of their parents but not the death of their children. Parents expect to outlive their children.

Many parents experience the loss of their child even before it is born. About 15 percent of pregnancies end in a miscarriage (Nielsen & Hahlin, 1995). Although some parents may be relieved by a miscarriage if the pregnancy was unwanted, many parents feel sadness, frustration, disappointment, and anger. Some women blame themselves for the miscarriage and believe that they are being punished for something they have done in the past.

Among infants under 1 year of age, the leading causes of death in the United States are congenital anomalies, sudden infant death syndrome (SIDS), disorders relating to short gestation and low birth weight, and respiratory distress syndrome. Table 11.5 presents the leading causes of death for United States youths in different age groups.

Table 11.5

Leading Causes of Death for U.S. Youths, by Age Group*		
Ages 1 to 4	**Ages 5 to 14**	**Ages 15 to 24**
accidents	accidents	accidents
congenital anomalies	cancer	homicide
cancer	congenital anomalies	suicide
homicide	suicide	cancer
heart disease	homicide	heart disease
pneumonia and influenza	heart disease	chronic pulmonary disease
	pneumonia and influenza	congenital anomalies

*Based on 1994 data.
Source: *Statistical Abstract of the United States: 1997*, 117th ed. Washington D.C.: U.S. Bureau of the Census. Table 130.

INSIGHT

In order to reduce the risk of sudden infant death syndrome (SIDS), the surgeon general advises parents to put their babies to sleep on their backs or sides, rather than on their stomachs.

RECENT RESEARCH

Thuen (1997) surveyed 251 parents in Norway whose infants had died and found that mothers received more emotional support from their friends than did fathers.

Death is more universal than life; everyone dies but not everyone lives.

A. Sachs

As noted in Table 11.5, accidents are the number one cause of death among youth in all age groups. Types of accidental deaths include drowning and poisoning from ingesting household products or medication. But the most common cause of accidental death among youth is motor vehicle accidents.

Sobieski (1994) suggests that spouses who experience the loss of a child also experience changes in their marital relationship. Their marriage may change in emphasis, direction, or quality. It may grow stronger or it may deteriorate. Previous differences in values and beliefs of spouses may become more apparent following the death of their child. For example, one partner may begin to attend church and seek solace through religion. The other partner may question religion and lose faith in God for allowing such a tragedy to happen. Sobieski (1994) advises couples who have experienced the death of a child to discuss the following issues (p. 16):

- Intimacy and sexual needs
- Views and feelings about having other children
- Methods of child rearing to be used for the surviving child or children

Earlier, we noted that women and men tend to grieve in different ways. When parents respond differently to the death of a child, they sometimes interpret these differences in negative ways, leading to relationship conflict and unhappiness. For example, after the death of their 17-year-old son, one wife accused her husband of not sharing in her grief. The husband explained that he couldn't allow himself to grieve until he figured out how to help his family cope with the death. The husband reacted cognitively to his son's death, while the wife responded emotionally. To deal with these differences, spouses might need to practice patience and tolerance in allowing each other to grieve in their own ways.

RECENT RESEARCH

Moss et al. (1997) investigated the reactions of 212 adult children to the
death of their last surviving parent and
found that daughters expressed more
somatic response, emotional upset,
and continuing tie to the deceased parent than sons.

*Roswell Gilbert shot his wife of
51 years twice in the head,
allegedly in order to end her
suffering.*

Death of one's spouse   The death of one's spouse is one of the most traumatic
events in a person's life. A widow, known to the authors, said of the death of
her husband:

> I felt like I had stepped through a trap door and landed on quicksand. The
> overwhelming feeling of disbelief resulted in months of physical illness—I
> couldn't stop throwing up and found it impossible to eat anything more
> than Jell-O. My mornings were spent crying in the fetal position until I
> was exhausted.

Because women tend to live longer than men, and because women are often younger than their husbands, they are more likely than men to experience
the death of their marital partners.

NATIONAL DATA   In 1996, there were over 11 million widows in the United States (11.1 percent of the female population), compared with only 2.5 million widowers (2.5 percent of the male population) (*Statistical Abstract of the United States: 1997*, Table 59).

Although the death of a spouse may be very traumatic, in some cases it may
relieve more stress than it creates. For example, caring for a spouse who is terminally ill and in great pain may be very stressful. The level of stress may actually decrease after the death of the terminally ill spouse. The level of stress involved in losing a spouse also depends on the quality of the marital relationship
before the spouse's death. Wheaton (1990) found that the distress and grief felt
after a spouse's death may be less intense if the marriage is problematic.

NATIONAL DATA   Five percent of first-year college and university students have experienced
the death of a parent (*American Council on Education and University of California, 1997*).

Death of one's parent   Spouses are likely to experience the death of their parents. Umberson (1995) studied the marital effects of the death of a parent on
forty-two individuals and compared them with a group of spouses who had not
lost a parent. She observed that marital quality declined with the death of one of
the spouse's parents:

> Results suggest that this decline may occur because the partner fails to
> provide desired emotional support, the partner cannot comprehend the
> significance and meaning of the loss, the partner is disappointed in the
> bereaved individual's inability to recover quickly, because individuals have
> new support needs following the death of a parent and these needs may
> be difficult to meet, and because some partners feel imposed upon—
> particularly by continuing distress and depression in the bereaved person
> (Umberson, 1995, 721).

Surviving the suicide of a loved one   Research suggests that those who experience the death of a friend or loved one by suicide (i.e., suicide survivors) tend
to experience different grief reactions than those whose loved ones die from accidents or natural causes (Stillion, 1996). Many family members of suicide victims experience higher levels of guilt, shame, rejection, and anger. Families of
suicide victims are often left with questions about why their loved ones killed

themselves and what could have been done to prevent the suicide (see Table 11.6 for Impending Suicide Signs). Although such questions are generally unanswerable, they often linger for years, prolonging the grieving process. In addition, survivors of suicide victims struggle with how others will view them and their family. Because some religions consider suicide a mortal sin, surviving

Table 11.6

Impending Suicide Signs	
Verbal Signs:	Direct statements about suicide, such as:
	"I am going to commit suicide."
	"I don't want to live anymore."
	Indirect statements or subtle hints indicating a wish to die, such as:
	"I wish I could go to sleep and never wake up."
	"They would be better off without me."
	"Do you think dying hurts?"
	"I don't think I can take it much longer."
	"I wish I had never been born."
	"You won't have to put up with me much longer."
Behavioral Signs:	Sadness and crying.
	Change in sleep patterns, chronic fatigue.
	Change in eating patterns, weight gain or loss.
	Drop in grades.
	Withdrawal from social interaction.
	Low self-esteem (evidenced perhaps in neglect of personal hygiene and appearance).
	Inability to concentrate or make decisions.
	Giving away personal possessions.
	Taking risks, frequent accidents.
	Prior suicide attempts.
	Abusing alcohol or other drugs.
	Angry outbursts or increased agitation.
	Obtaining weapons of destruction, such as guns, pills, ropes.
Situational Signs:	Loss of a relationship.
	Conflictual relationship with parents.
	Trouble with the law or at school.
	Unwanted pregnancy.
	Disliked and/or stigmatized by peers.
	Presence of mental or serious physical illness.

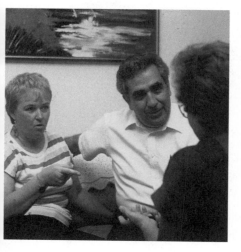

Over 70 percent of couples say they would consult a marriage therapist if they became unhappy.

family members may also struggle with what the fate of their dead loved one might be in the hereafter.

While some family members cope with the suicide of a loved one, others cope with the dilemma of keeping a family member alive on life support. This chapter's Social Policy section discusses the issue of life support/physician assisted suicide for terminally ill family members.

Marriage and Family Therapy

Couples and families experiencing a crisis may turn to marriage and family therapists for help in coping with the crisis. But there is sometimes reluctance to turn to therapy with one's problems. A sample of couples who had been married from one month to thirty-six years who had not sought therapy identified some of the reasons for not seeking therapy. These included "it won't do any good," "it's no one else's business," and "only sickies go to therapy." Wives were more willing than husbands to seek marital counseling (Bringle & Byers, 1997).

Seeing a therapist does not mean that one is mentally ill. On the contrary, we are never more mentally and emotionally healthy than when we can acknowledge that we have a problem and seek help for it.

The fact that marriage is a personal and private affair does not mean we cannot discuss our concern with a specialist. Our bodies are also personal and private, but this does not stop us from seeing a physician when we have a physical problem. Our mental health is as important to our feeling good as is our physical health. Both physicians and marriage therapists can be expected to treat the information we share with them with strict confidentiality. This is required by their code of professional ethics.

Signs to look for in your own relationship that suggest you might consider seeing a therapist include feeling distant and not wanting or being unable to communicate with your partner, avoiding each other, drinking heavily or taking other drugs, privately contemplating separation, being involved in an affair, and feeling depressed.

If you are experiencing one or more of these concerns with your partner, it may be wise not to wait until the relationship reaches a stage beyond which repair is impossible. Relationships are like boats. A small leak will not sink it. But if left unattended, the small leak may grow larger or new leaks may break through. Marriage therapy sometimes serves to mend relationship problems early by helping the partners to sort out values, make decisions, and begin new behaviors so that they can start feeling better about each other.

In spite of the potential benefits of marriage therapy, some valid reasons exist for not consulting a counselor. Not all spouses who become involved in marriage therapy regard the experience positively. Some feel that their marriage is worse as a result. Saying things the spouse can't forget, feeling hopeless at not being able to resolve a problem "even with a counselor," and feeling resentment over new demands made by the spouse in therapy are reasons some spouses cite for negative outcomes.

Life Support/Physician-Assisted Suicide for Terminally Ill Family Members

An individual may experience a significant drop in quality of life and a total loss of independence due to illness, accident, or aging. Family members are often asked their recommendations about withdrawing life support (food, water or mechanical ventilation), starting medications to end life (intravenous vasopressors), or withholding certain procedures to prolong life (cardiopulmonary resuscitation). The recommendation of family members is sometimes given considerable weight in the ultimate decision of the caregiver/attending physician. Seventy-six percent of the caregivers in one study said that "family preference" was the most important factor influencing the restriction of life-support interventions (Randolph et al., 1997). However, most family recommendations are based on the physician's recommendations to limit care or request such limitation (Luce, 1997). When the physician and family member are in conflict about what to do, the physician usually defers to the preference of the family member (Prendergast & Luce, 1997).

Though all fifty states now have living-will statutes permitting individuals who are not terminally ill (or family members) to refuse artificial nutrition (food) and hydration (water), there is no constitutional right of a family member or physician to end the life of another. *Euthansia* is defined as the deliberate taking of an individual's life at his or her request (Zalcberg & Buchanan, 1997). The issue has received nationwide attention in reference to Dr. Jack Kevorkian, who has been involved in over 30 physician-assisted suicides (PAS) at the patient's request.

The Supreme Court has ruled that state law will apply in regard to physician-assisted suicide. As of 1998, only Oregon recognizes the right of PAS with its Death with Dignity Act. Two physicians must agree that the patient is terminally ill and is expected to die within six months, the patient must ask three times for death both orally and in writing, and the patients must swallow the barbiturates themselves rather than be injected with a drug by the physician. However, the Drug Enforcement Administration has said that it will revoke the federal narcotics licenses of doctors who prescribe drugs to help terminally ill patients take their own lives.

The American Medical Association takes the position that while physicians must respect the patient's decision to forgo life-sustaining treatment, they should not participate in patient-assisted suicide. They disagree with Dr. Kevorkian. Arguments against PAS focus on who has the right to decide that a person may die. The practice is subject to abuse: for example, one spouse may encourage a physician to kill the other when he or she may thereby be relieved of a burden and inherit a lot of life insurance money.

Euthanasia remains controversial. Between 1935 and 1994, 144 people have been charged with mercy killing. One of these was Roswell Gilbert, who shot his wife of fifty-one years twice in the head, allegedly to end her suffering. He was convicted of first-degree murder but after five years in prison was paroled.

A team of researchers identified the attitudes of elderly individuals (patients) and their family members toward physician-assisted suicide (Koenig et al., 1996). While 40 percent of the elderly individuals had positive attitudes toward PAS, 59 percent of the relatives expressed favorable attitudes. Patients who opposed PAS were more likely to be women, black individuals, and those with less education.

Physician-assisted suicide has been legal in Holland for ten years. But there have been some abuses. Though physicians are required by law to report their involvement in physician-assisted suicide, 60 percent of physicians in one study did not do so. Also, more than half had suggested euthanasia to patients who were not necessarily terminally ill. One-fourth of the physicians did not have the consent of the patient even though many of the patients were competent to make such a decision (Hendin et al., 1997).

REFERENCES

Hendin, Herbert, C. Rutenfrans, and Z. Zylicz. 1997. Physician-assisted suicide and euthanasia in the Netherlands: Lessons from the Dutch. *Journal of the American Medical Association* 277: 1720–23.

Koenig, H. G., D. Wildman-Hanlon, and K. Schmader. 1996. Attitudes of elderly patients and their families toward physician-assisted suicide. *Archives of Internal Medicine* 156: 2240–48.

Luce, J. M. 1997. Withholding and withdrawal of life support: Ethical, legal, and clinical aspects. *New Horizons* 5: 30–37.

Nasser, H. E., and John Ritter. 1997. Officials: Ore. suicide law in effect. *USA Today*. 6 November, 3A.

Prendergast, T. J., and J. M. Luce. 1997. Increasing incidence of withholding and withdrawal of life support from the critically ill. *American Journal of Respiratory and Critical Care Medicine* 155: 15–20.

Randolph, A. G., M. B. Zollo, R. S. Wigton, and T. S. Yeh. 1997. Factors explaining variability among caregivers in the intent to restrict life-support interventions in a pediatric intensive care unit. *Critical Care Medicine* 25: 435–39.

Zalcberg, J. R., and J. D. Buchanan. 1997. Clinical issues in euthanasia. *Medical Journal of Australia* 166: 150–52.

INSIGHT

Spouses most likely to benefit from marriage therapy come to therapy when they experience recurrent conflicts they have been unable to communicate about and resolve. Spouses who wait until they are ready to divorce have usually waited too long. In such cases, some therapists accept that not all marriages can be salvaged.

"We need to remember that not all couples should remain married and that helping a couple to decide to divorce is a perfectly valid function of marital therapy. For many couples, therapy provides a way to break up. There is nothing wrong with this. Therapists may also be able to help a couple understand what went wrong with the marriage and assess whether it is possible to change the marriage and whether or not they want to change it" (Gottman, 1994, 426).

Therapists also may give clients an unrealistic picture of loving, cooperative, and growing relationships in which partners always treat each other with respect and understanding, share intimacy, and help each other become whoever each wants to be. In creating this idealistic image of the perfect relationship, therapists may inadvertently encourage clients to focus on the shortcomings in their relationship and to expect more of their marriage than is realistic. Couples and families in therapy must also guard against assuming that therapy will be a quick and easy fix. As noted earlier, checking the credentials of one's marriage and family therapist to ensure that she or he is certified is an important consideration.

Cost and Success of Marriage and Family Therapy

The cost of private marriage and family therapy is around $100 per hour. However, managed care has resulted in private therapists lowering their fees so as to compete with what insurance companies will pay. Many mental health centers offer marital and family therapy on a sliding-fee basis so that spouses, parents, and their children can be seen for as little as $5. The average number of sessions per couple varies. Four was the median number of sessions (range one to eighteen) for 152 couples who were seen at a university marriage and family therapy clinic (Poche et al., 1997). Couples most likely to drop out of therapy (about two-thirds) before a planned termination reported high levels of individual distress, lower levels of marital satisfaction, and lower household incomes (Poche et al., 1997).

Since some states have no laws governing the practice of marriage and family therapy, it is important to see a therapist who is a member of the American Association for Marriage and Family Therapy. Phone number is 1-202-452-0109; address is AAMFT, 1133 15th Street, N.W., Suite 300, Washington, DC 20005-2710. Clients are customers and should feel comfortable with their therapists and the progress they are making. If not, they should switch therapists.

To what degree do spouses, parents, and children benefit from marriage and family therapy? Two researchers (Doherty & Simmons, 1996) surveyed 526 clinical members of AAMFT and 492 of their clients with regard to the effectiveness of therapy. About 60 percent of both therapists and clients agreed that the relationship between the partners who sought therapy had improved.

Whether a couple in therapy remain together will depend on their motivation to do so, how long they have been in conflict, the severity of the problem, and whether one or both partners are involved in an extramarital affair. Two moderately motivated partners with numerous conflicts over several years are less likely to work out their problems than a highly motivated couple with minor conflicts of short duration.

Some couples come to therapy with the goal of separating amicably. The therapist then discusses the couple's feelings about the impending separation, the definition of the separation (temporary or permanent), the "rules" for their interaction during the period of separation (e.g., see each other, date others), and whether to begin discussions with a divorce mediator or attorneys.

One alternative for enhancing one's relationship is the Association for Couples in Marriage Enrichment. This organization provides conferences for couples to enrich their marriage. The address of ACME is P.O. Box 10596, Winston-Salem, NC 27108. Phone: 1-800-634-8325. Web address: wwwacme@aol.com.

People have different feelings about their vulnerability to crisis events. The following scale addresses the degree to which you feel you have control, or feel others have control, or feel that chance has control of what happens to you.

Internality, Powerful Others, and Chance Scales

To assess the degree to which you believe that you have control over your own life (I = Internality), the degree to which you believe that other people control events in your life (P = Powerful Others), and the degree to which you believe that chance affects your experiences or outcomes (C = Chance), read each of the following statements and select a number from minus 3 to plus 3.

−3 Strongly Disagree	−2 Disagree	−1 Slightly Disagree	+1 Slightly Agree	+2 Agree	+3 Strongly Agree

Subscale

+2 +2	I	1. Whether or not I get to be a leader depends mostly on my ability.
+3 +3	C	2. To a great extent my life is controlled by accidental happenings.
−3 −3	P	3. I feel like what happens in my life is mostly determined by powerful people.
−3 +1	I	4. Whether or not I get into a car accident depends mostly on how good a driver I am.
+3 +1	I	5. When I make plans, I am almost certain to make them work.
+3 +3	C	6. Often there is no chance of protecting my personal interests from bad luck happenings.
+3 +3	C	7. When I get what I want, it's usually because I'm lucky.
+3 +2	P	8. Although I might have good ability, I will not be given leadership responsibility without appealing to those in positions of power.
+3 −2	I	9. How many friends I have depends on how nice a person I am.
+3 +3	C	10. I have often found that what is going to happen will happen.
−3 −2	P	11. My life is chiefly controlled by powerful others.
−1 −1	C	12. Whether or not I get into a car accident is mostly a matter of luck.
−3 −1	P	13. People like myself have very little chance of protecting our personal interests when they conflict with those of strong pressure groups.
+3 +1	C	14. It's not always wise for me to plan too far ahead because many things turn out to be a matter of good or bad fortune.
+3 +1	P	15. Getting what I want requires pleasing those people above me.
−1 +1	C	16. Whether or not I get to be a leader depends on whether I'm lucky enough to be in the right place at the right time.
−3 −2	P	17. If important people were to decide they didn't like me, I probably wouldn't make many friends.

-3 -2 I 18. I can pretty much determine what will happen in my life.

+3 +2 I 19. I am usually able to protect my personal interests.

+3 -1 P 20. Whether or not I get into a car accident depends mostly on the other driver.

+3 +2 I 21. When I get what I want, it's usually because I worked hard for it.

+3 +1 P 22. In order to have my plans work, I make sure that they fit in with the desires of other people who have power over me.

+3 +2 I 23. My life is determined by my own actions.

-3 -1 C 24. It's chiefly a matter of fate whether or not I have a few friends or many friends.

I -35
C -34
P -24

I - 30
C - 36
P - 19

SCORING: Each of the subscales of Internality, Powerful Others, and Chance is scored on a six-point Likert format from minus 3 to plus 3. For example, the eight Internality items are 1, 4, 5, 9, 18, 19, 21, 23. A person who has strong agreement with all eight items would score a plus 24; strong disagreement, a minus 24. After adding and subtracting the item scores, add 24 to the total score to eliminate negative scores. Scores for Powerful Others and Chance are similarly derived.

NORMS: For the Internality subscale, means range from the low 30s to the low 40s, with 35 being the modal mean (SD values approximating 7). The Powerful Others subscale has produced means ranging from 18 through 26, with 20 being characteristic of normal college student subjects (SD = 8.5). The Chance subscale produces means between 17 and 25, with 18 being a common mean among undergraduates (SD = 8).

Source: From "Differentiating Among Internality, Powerful Others and Chance" by H. Levenson, 1981. In H. M. Lefcourt (ed.) *Research with the Locus of Control.* Vol. 1, pp. 57–59. Copyright © by Academic Press, Inc. Used by permission.

Brief Solution-Based Therapy

RECENT RESEARCH

Park (1997) emphasized (and provided case examples for) the value of "solution-focused brief therapy" by physicians in an urban teaching hospital in Seoul, Korea. He noted that the efficiency of the approach appeared remarkable in view of the fact that families visiting the medical clinic continually managed to overcome their difficulties after receiving brief therapy or counseling.

In the past, marriage and family therapy models were not constrained by the economics of managed care. As the bottom line has become the driving force determining what services insurance companies will pay for, increasingly marriage and family therapy has become more time-bound, brief, and solution-focused. Psychoanalytic and psychodynamic marital and family therapy that could take years has been replaced by brief therapy that now takes only weeks (or until the money the insurance company is willing to pay runs out). Therapists adopting these models focus on what behaviors family members want increased or decreased, initiated or terminated and try to negotiate ways to accomplish these behavioral goals. Others may focus on the cognitions or assumptions that underlie a marriage or family with the goal of ensuring that these are accurate and functional. There is considerable concern among therapists that managed-care economics is dictating therapy models.

Al-Anon An organization that provides support for family members and friends of alcohol abusers.

crisis An event that requires changes in normal patterns of behavior.

extradyadic Pertaining to emotional/sexual involvement with someone other than the person with whom one is involved in an emotional/sexual relationship.

extramarital involvement Emotional/sexual involvement of a spouse with someone other than the mate.

open marriage A marriage in which the spouses agree that each may have sex with others.

stress Reactions made by a person's body to demands made on him or her. Stress sometimes results in elevated blood pressure, irritability, and nervousness.

SUMMARY

All couples and families are confronted with crisis events that are regarded as a normal part of life. In their study of fifty successful marriages, Wallerstein and Blakeslee (1995) found that every couple had experienced at least one crisis and some had experienced several traumatic events. No spouse, marriage, or family was immune to stress or crisis.

Stress and Resilient Families

Stress and crises are a normal part of family living. Stress is the response of the body to demands made on it. A crisis is an event that requires changes in normal patterns of behavior. Stressors or sources of crises can be external (e.g., flood, downsizing) or internal (alcoholism, infidelity) and both expected (death of parents) and unexpected (suicide of teenager). Families with strengths such as commitment, communication, and flexibility are resilient in their response to stress and crisis.

Family Stress Theories

The ABC-X theoretical framework for conceptualizing family crises emphasizes that the impact of a crisis will be affected by a family's management skills and resources and by its perception of the event. A family with limited resources and a pessimistic view will experience a more difficult crisis than one with skills and resources in place as well as a positive view. Nearly 20 percent of families improve, 15 percent experience no change, and about 5 percent get worse in responding to a family crisis. About 11 percent can best be described as a mixed model.

Stress Management Strategies

Some management strategies that families employ are more helpful than others. Reevaluating basic values as a result of a crisis, being understanding with each other, and talking with others in similar situations seem to be among the most helpful strategies. Having a realistic perspective, avoiding blame, and seeking out opportunities for fun also seem to be helpful, as do exercise and sleep. Least helpful strategies include suppressing one's feelings, denying the existence of the problem, and withholding affection. Spouses also seem to profit from talking about issues of intimacy and loss, boundaries, and the need for separateness.

Five Family Crises

Physical illness is a crisis with which most individuals, spouses, and families are confronted. Illness influences the partners' intimacy with each other and requires their negotiating issues of togetherness and separateness.

Infidelity also represents a major crisis event and jeopardizes those marriages in which it occurs. Though most couples try to work through the crisis, it presents a major challenge to the spouses. Other family crises include unemployment, drug abuse, and death. Unemployment affects both spouses by causing them to adjust their standard of living and reevaluate the importance of material goods. Parents can help to prevent teenage drug abuse by maintaining a good relationship with their children and by confronting the drug issue head-on if their teenager becomes involved with drugs. One of the ultimate family crises is the death of one's child, the suicide of one's teenager, or the death of one's spouse or parents. While each has its own valence of emotional devastation, the death of one's child is the least expected.

Marriage and Family Therapy

Some couples seek marital and family therapy to help them cope with various family crises. Brief solution-based therapies are provided that focus on behavioral and/or cognitive change.

REFERENCES

ABC (American Broadcasting Company) 1995. Journey of Christopher Reeve. *20/20.* 27 September.

Amato, Paul R., and Stacy J. Rogers. 1997. A longitudinal study of marital problems and subsequent divorce. *Journal of Marriage and the Family* 59: 612–24.

American Council on Education and University of California. 1997. *The American freshman: National norms for fall, 1997.* Los Angeles: Los Angeles Higher Education Research Institute.

Barnes, G. M., and M. P. Farrell. 1992. Parental support and control as predictors of adolescent drinking, delinquency, and related problems. *Journal of Marriage and the Family* 54: 763–76.

Booth, Alan, and David R. Johnson. Declining health and marital quality. 1994. *Journal of Marriage and the Family* 56: 218–23.

Burr, Wesley R. and Shirley R. Klein and Associates. 1994. *Reexamining family stress: New Theory and research.* Thousand Oaks, Calif.: Sage.

Bringle, Robert G., and Dianne Byers. 1997. Intentions to seek marriage counseling. *Family Relations* 46: 299–304.

Capri, John. 1996. Stress: It's worse than you think. *Psychology Today.* January/February, 34–41.

Clark, W. M., and J. M. Serovich. 1997. Twenty years and still in the dark?: Content analysis of articles pertaining to gay, lesbian, and bisexual issues in marriage and family therapy journals. *Journal of Marriage and Family Therapy* 23: 239–53.

Doherty, William J., and D. S. Simmons. 1996. Clinical practice patterns of marriage and family therapists: A national survey of therapists and their clients. *Journal of Marital and Family Therapy* 22: 9–25.

Finucane, P., L. C. Giles, R. T. Withers, C. A. Silagy, A. Sedgwick, P. A. Hamdorf, J. A. Halbert, L. Cobiac, M. S. Clark, and G. R. Andrews. 1997. Exercise profile and subsequent mortality in an elderly Australian population. *Australian and New Zealand Journal of Public Health* 21: 155–58.

Frankenfield, P. K. 1996. The power of humor and play as nursing interventions for a child with cancer. *Journal of Pediatric Oncology Nursing* 13: 15–20.

Glyptis, S. 1989. *Leisure and unemployment.* Philadelphia: Open University Press.

Gottman, John Mordechai. 1994. *What predicts divorce? The relationship between marital processes and marital outcomes.* Hillsdale, N.J.: Lawrence Erlbaum Associates.

Greaves, K. M., A. M. Zvonkovic, L. S. Evans, and L. D. Hall. 1995. Economic resources, influence, and stress among married couples. *Family and Consumer Sciences Research Journal* 24: 47–70.

Haurin, R. J. 1992. Patterns of childhood residence and the relationship to young children outcomes. *Journal of Marriage and the Family* 54: 846–60.

Henderson, B. Janette, and Jean Pearson Scott. 1997. Older adult bereavement: A comparison of bereaved parents and spouses. Paper presented at the 59th Annual Conference of the National Council on Family Relations, Crystal City, Va., November. Used by permission.

Hong, Jinkuk, and M. M. Seltzer. 1995. The psychological consequences of multiple roles: The nonnormative case. *Journal of Health and Social Behavior* 36: 386–98.

Jacob, T., and S. L. Johnson. 1997. Parent-child interaction among depressed fathers and mothers: Impact on child functioning. *Journal of Family Psychology* 11: 391–409.

Jones, L. 1992. His unemployment and her reaction: The effect of husbands' unemployment on wives. *Affilia: Journal of Women and Social Work* 7: 7–20.

Kamei, T., H. Kumano, and S. Masumura. 1997. Changes of immunoregulatory cells associated with psychological stress and humor. *Perceptual and Motor Skills* 84: 1296–98.

Kate, Nancy Ten. 1994. To reduce stress, hit the hay. *American Demographics* 16: 14–16.

Kessler, R., K. McGonale, S. Zhao, C. Nelson, M. Hughes, S. Eshelman, H. Wittchen, and K. Kendler. 1994. Lifetime and twelve-month prevalence of DSM-III-R psychiatric disorders in the United States. *Archives of General Psychiatry* 51: 8–19.

Kieren, D. K., and R. Delehanty. 1995. Just doing it a little differently: A case study analysis of family problem solving after brain injury. *Family Perspective* 29: 251–68.

Knox, D., and M. Zusman. 1998. Unpublished data collected for this text.

Koos, E. L. 1946. *Families in trouble.* Morningside Heights, N.Y.: King's Crown.

Langehough, Steven O., C. Walters, D. Knox, and M. Rowley. 1997. Spirituality and religiosity as factors in adolescents' risk for antisocial behaviors and use of resilient behaviors. Paper presented at the 59th Annual Conference of the National Council on Family Relations, Crystal City, Va., November. Used by permission.

Larson, Jeffrey H., Stephan M. Wilson, and Rochelle Belely. 1994. The impact of job insecurity on marital and family relationships. *Family Relations* 43: 138–43.

Liem, J. H., and G. R. Liem. 1990. Understanding the individual and family effects of unemployment. In *Stress between work and family*, edited by J. Eckenrode and S. Gore. New York: Plenum, 175–204.

Marks, John. 1995. Time out. *U.S. News and World Report.* 11 December, 85–96.

Martin, T., and K. Doka. 1996. Masculine grief. In *Living with Grief after Sudden Loss*, edited by K. J. Doka. Bristol, Pa.: Taylor and Francis, 161–71.

Miller, W. R., and E. A. Cervantes. 1997. Gender and patterns of alcohol problems: Pretreatment responses of women and men to the comprehensive drinker profile. *Journal of Clinical Psychology* 53: 263–77.

Morgan, Lorrie. 1997. *Forever yours, faithfully: My love story.* Westminster, Md.: Ballentine Books.

Morton, Andrew. 1997. *Diana: Her true story.* New York: Pocket Star Books.

Moss, M. S., N. Resch, and S. Z. Moss. 1997. The role of gender in middle-age children's responses to parent death. *Omega—Journal of Death and Dying* 35: 43–65.

Nielsen, Sven, and Mats Hahlin. 1995. Expectant management of first-trimester spontaneous abortion. *The Lancet* 345: 84–86.

Park, E. 1997. An application of brief therapy to family medicine. *Contemporary Family Therapy* 19: 81–88.

Peterson, K. S. 1996. On-line adultery. *USA Today.* 6 February, 4d.

Poche, R. Scott, Mark B. White, and Thomas A. Smith, Jr. 1997. Correlates of therapeutic dropout. Paper presented at the Annual Conference of the National Council on Family Relations, Crystal City, Va., November.

Prather, Gayle. 1996. *I will never leave you: How couples can achieve the power of lasting love.* New York: Bantam.

Quatman, T. 1997. High functioning families: Developing a prototype. *Family Therapy* 24: 143–65.

Richardson, G. E., and S. R. Hawks. 1995. A practical approach for enhancing resiliency within families. *Family Perspective* 29: 235–51.

Rolland, John S. 1994. In sickness and in health: The impact of illness on couples' relationships. *Journal of Marital and Family Therapy* 20: 327–47.

Ross, Catherine E., and John Mirowsky. 1995. Does employment affect health? *Journal of Health and Social Behavior* 36: 230–43.

Rounds, K. A., M. Weil, and K. K. Bishop. 1994. Practice with culturally diverse families of young children with disabilities. *Families in Society* 75: 3–15.

Saad, Lydia. 1995. Children, hard work taking their toll on baby boomers. *Gallup Poll Monthly.* April, 21–24.

Sobieski, R. 1994. *Men and mourning: A father's journey through grief.* Mothers Against Drunk Driving (MADD), 511 E. John Carpenter Freeway, Suite 700. Irving, TX 75062-8187.

Statistical Abstract of the United States: 1997. 117th ed. Washington, D.C.: U.S. Bureau of the Census.

Stets, Jan E. 1995. Job autonomy and control over one's spouse: A compensatory process. *Journal of Health and Social Behavior* 36: 244–58.

Stillion, J. M. 1996. Survivors of suicide. In *Living with grief after sudden loss,* edited by K. J. Doka. Bristol, Pa.: Taylor & Francis, 41–51.

Straus, M., R. Gelles, and S. Steinmetz. 1980. *Behind closed doors.* New York: Doubleday.

Syre, T. R., J. M. Martino-McAllster, L. M. Vanada. 1997. Alcohol and other drug use at a university in the southeastern United States: Survey and implications. *College Student Journal* 31: 373–81.

Thayer, Robert E., J. Robert Newman, and Tracey M. McClain. 1994. Self-regulation of mood: Strategies for changing a bad mood, raising energy, and reducing tension. *Journal of Personality and Social Psychology* 67: 910–16.

Thoits, Peggy A. 1995. Stress, coping, and social support processes: Where are we? What next? *Journal of Health and Social Behavior.* Extra issue, 53–79.

Thomas, Linda T., and Daniel C. Ganster. 1995. Impact of family-supportive work variables on work-family conflict and strain: A control perspective. *Journal of Applied Psychology* 80: 6–16.

Thuen, F. 1997. Social support after the loss of an infant child: A long-term perspective. *Scandinavian Journal of Psychology* 38: 103–10.

Turkle, S. 1996. *Life on the screen: Identity in the age of the Internet.* New York: Simon and Schuster.

Turner, J. Blake. 1995. Economic context and the health effects of unemployment. *Journal of Health and Social Behavior* 36: 213–29.

Turner, R. J., and D. A. Lloyd. 1995. Lifetime traumas and mental health: The significance of cumulative adversity. *Journal of Health and Social Adversity* 36: 360–76.

Tuttle, Jane. 1995. Family support, adolescent individuation, and drug and alcohol involvement. *Journal of Family Nursing* 1: 303–26.

Umberson, Debra. 1995. Marriage as support or strain? Marital quality following the death of a parent. *Journal of Marriage and the Family* 57: 709–23.

Voydanoff, P. 1991. Economic distress and family relations: A review of the eighties. In *Contemporary families: Looking forward, looking back,* edited by A. Booth. Minneapolis: National Council on Family Relations, 429–45.

Wallerstein, Judith S., and Sandra Blakeslee. 1995. *The good marriage.* Boston: Houghton Mifflin.

Whisman, M. A., A. E. Dixon, and B. Johnson. 1997. Therapists' perspectives of couple problems and treatment issues in couple therapy. *Journal of Family Psychology* 11: 361–66.

Wiederman, M. W. 1997. Extramarital sex: Prevalence and correlates in a national survey. *Journal of Sex Research* 34: 167–74.

Wheaton, Blair. 1990. Life transitions, role histories, and mental health. *American Sociological Review* 55: 209–23.

Williams, P., and S. R. Lord. 1997. Effects of group exercise on cognitive functioning and mood in older women. *Australian and New Zealand Journal of Public Health* 21: 45–52.

Wilson, S. M., and J. H. Larson. 1994. The impact of job insecurity on marital and family relations. In *Families and justice: From neighborhoods to nations.* Vol. 4 of *Proceedings.* Annual Conference of the National Council on Family Relations, 34.

Wyman, J. R. 1997. Multifaceted prevention programs reach at-risk children through their families. *National Institute on Drug Abuse* 12, no. 3: 5–7.

CHAPTER 12 Divorce and Remarriage

It is less a breach of wedlock to part with wise and quiet consent betimes, than still to soil and profane that mystery of joy and union with a polluting sadness and perpetual distemper.

John Milton, Marriage

There is little agreement about the appropriateness of divorce. Divorce foes see it as a selfish scourge devastating the lives of our children. Divorce advocates see it as a necessary outlet for a doomed relationship (Whitehead, 1997). One divorced man said, "It's like you've got gangrene and you have to cut off your leg to save your body . . . sometimes you have to end your marriage to save your life" (authors' files).

Divorce is the legal ending of a valid marriage contract. Divorce, with its consequences, has been one of the most researched issues in marriage and the family. In this chapter we review societal and individual reasons for divorce, the effects on children and parents, and the involvement of divorced spouses in new relationships. Since most remarriages involve children from previous relationships, we examine the strengths of often maligned stepfamilies.

Macro Factors Contributing to Divorce

Sociologists emphasize that social contexts create outcomes. This is no better illustrated than by the fact that there was an average of only one divorce per year in Massachusetts among the Puritans from 1639 to 1760 (Morgan, 1944). The social context, reflected in strict divorce laws and strong social pressure to stay married, kept couples married. In contrast, divorce today is a frequent phenomenon. Various structural and cultural factors, also known as macro factors, help to account for our high divorce rate.

NATIONAL DATA
There are about 1 million (1,154,000) divorces each year (National Center for Health Statistics, 1998). Half of all marriages entered into in recent years in this country will end in divorce or separation if recent marital dissolution rates continue (Glenn, 1997).

Increased Economic Independence of Women

In the past, the unemployed wife was dependent on her husband for food and shelter. No matter how unhappy her marriage was, she stayed married because she was economically dependent on her husband. Her husband literally represented her lifeline.

Finding gainful employment outside the home made it possible for the wife to afford to leave her husband if she wanted to. Now that almost 70 percent of all wives are employed (and this number is increasing), fewer and fewer wives are economically trapped in an unhappy marriage relationship. Wives who earn an income may feel able to survive financially on their own; thus, employed wives are more likely to leave an unhappy marital relationship. Alternatively, the stress of employment brought home to the marriage may increase marital tension and unhappiness. Finally, unhappy husbands may be more likely to divorce if their wives are employed and able to be financially independent.

Changing Family Functions and Structure

Many of the protective, religious, educational, and recreational functions of the family have largely been taken over by outside agencies. Family members may now look to the police, the church or synagogue, the school, and commercial recreational facilities rather than to each other for fulfilling these needs. The result is that although meeting emotional needs remains a primary function of the family, fewer reasons exist to keep the family together.

In addition to the changing functions of the family brought on by the Industrial Revolution, the family structure has changed from that of larger extended families in rural communities to smaller nuclear families in urban communities. In the former, individuals could turn to a lot of people in times of stress; in the latter, more stress necessarily fell on fewer shoulders.

Liberalized Divorce Laws

All states now recognize some form of no-fault divorce. Although the legal terms are *irreconcilable differences* and *incompatibility*, the reality is that spouses can get a divorce if they want to without having to prove that one of the partners is at fault (e.g., through adultery). The effect of *no-fault divorce* laws (neither party is assigned blame for the divorce) has been to increase the divorce rate across the fifty states (Nakonezny, Shull, & Rodgers, 1995). Indeed, a backlash has occurred in response to the fact that divorce is too easy too obtain and a movement is afoot to make divorce harder to get. One divorced spouse said, "I should have stayed married when the hard times hit . . . it was just to easy to walk out" (author's files).

INTERNATIONAL DATA   The ease of obtaining a divorce has increased not only in the United States but also in other countries such as England, Wales, France, and Sweden. In contrast, Italy and Spain have very restrictive divorce laws, and Ireland has only recently allowed divorce under any conditions (Fine & Fine, 1994).

OTHER CULTURES   Although legal grounds for divorce in the United States have moved away from finding one partner at fault, divorce laws in other societies may still target a guilty party. Grounds for divorce among the Hindus in India include adultery, conversion to another religion, and having an STD in a communicable form. In prerevoluntionary China, a husband could divorce his wife (not vice versa) if she disobeyed his parents, if she displayed jealous behavior, or if she could not have children (Engel, 1982).

Fewer Moral and Religious Sanctions

Many priests and clergy recognize that divorce may be the best alternative in a particular marital relationship and attempt to minimize the guilt that members of their congregation may feel at the failure of their marriage. Increasingly, marriage is more often viewed in secular rather than religious terms. Hence, divorce has become more acceptable.

Others contend that "the real problems for the kids will begin when the antidivorce movement starts getting its way."

If the idea is to help children of divorce, then the goal should be to de-stigmatize divorce among all who interact with them—teachers, neighbors, playmates. . . . Maybe the reformers should concentrate on improving the quality of divorces (Ehrenreich, 1996, p. 80).

More Divorce Models

As the number of divorced individuals in our society increases, the probability increases that a person's friends, parents, siblings, or children will be divorced. The more divorced people a person knows, the more normal divorce will seem to that person. The less deviant the person perceives divorce to be, the greater the probability the person will divorce if that person's own marriage becomes strained. *Divorce Magazine* is a new magazine for the divorced.

NATIONAL DATA   Over one-fourth (26%) of all first-year college students in the United States have parents who are divorced (American Council on Education and University of California, 1997).

Mobility/Anonymity

That individuals are highly mobile results in fewer "roots" in a community and greater anonymity. People are less influenced by the expectations of others. Divorce thrives when social expectations are not operative.

Individualistic Cultural Goal of Happiness

Unlike familistic values in Asian cultures, individualistic values in American culture emphasize the goal of personal happiness in marriage. When spouses stop having fun (when individualistic goals are no longer met), they sometimes feel that there is no reason to stay married.

OTHER CULTURES   Asian-Americans and Mexican-Americans have lower divorce rates than whites or African-Americans because they consider the family unit to be of greater value than their individual interests. Personal unhappiness is less likely to result in movement toward divorce for these spouses (Mindel, Habenstein, & Wright, 1998). ●

Expectation That Marriage May Not Be Permanent

One consequence of personally knowing others who have divorced is to acknowledge the possibility that one's own marriage may end in divorce. Indeed, data exist which suggest that most people think that when a friend marries, divorce will follow sooner or later. Sixty-four percent of the respondents in one study said that this is the outcome they expect when one of their friends marries. Thirty-one percent expect the marriage to last forever (Boech & Ward, 1995).

Alcohol abuse is associated with subsequent divorce.

My work shows that men, women, and children have very different interests at divorce. I found that for most mothers divorce was a necessity. For example, many mothers in my random sample had to leave their marriages because of their husbands' use of drugs and alcohol and their husbands' violence.

Demie Kurz

Sociologist

Micro Factors Contributing to Divorce

Although macro factors may make divorce a viable cultural alternative to marital unhappiness, they are not sufficient to "cause" a divorce. One spouse must make a choice to divorce and initiate proceedings. Such a view is micro in that it focuses on the individual decisions and interactions within specific family units. The following subsections discuss some of the micro factors that may be operative in influencing a couple toward divorce.

Negative Behavior

People marry because they anticipate greater rewards from being married than from being single. During courtship, each partner engages in a high frequency of positive verbal and nonverbal behavior (compliments, eye contact, physical affection) toward each other. The good feelings the partners share as a result of this high frequency of positive behaviors encourage them to get married to ensure that each will be able to share the same experiences tomorrow.

Just as love feelings are based on positive behavior from the partner, hostile feelings are created when the partner engages in a high frequency of negative behavior. Amato and Rogers (1997) interviewed a national sample of spouses (1,748) and identified some of these behaviors. Negative behaviors that wives reported their husbands engaging in included getting angry easily, being moody, not talking, displaying irritating habits, not being home enough, and spending money foolishly.

Negative behaviors that husbands reported their wives engaged in included getting her feelings hurt easily, being moody, not talking, being critical, and getting angry easily. Specific behaviors associated with a couple's subsequent divorce included sexual infidelity, jealousy, drinking, spending money, moodiness, not communicating, and anger (p. 622).

When a spouse's negative behavior continues to the point of creating more costs than rewards in the relationship, either partner may begin to seek a more reinforcing situation. Divorce (being single again) or remarriage may appear to be a more attractive alternative to being married to the present spouse. This is certainly true of a marriage in which one's spouse is an alcoholic/substance abuser or physically or emotionally abuses the partner.

Lack of Conflict Resolution Skills

While every relationship experiences conflict, not every couple has the skills to effectively resolve conflict. Some partners respond to conflict by withdrawing emotionally from their relationship; others respond by attacking, blaming, and failing to listen to their partner's point of view. Without skills to resolve conflict in their relationships, partners drift into patterns of communication that may escalate rather than resolve conflict (Markman et al., 1994). Ways to negotiate differences and reduce conflict were discussed in Chapter 6, Communication and Conflict Resolution.

Radical Changes

Both spouses change throughout the marriage. "He's not the same person I married" is a frequent observation of persons contemplating divorce. People may undergo radical changes (philosophical or physical) after marriage. One minister married and decided seven years later that he did not like the confines of the marriage role. He left the ministry, earned a Ph.D. in psychology, and began to drink and have affairs. His wife, who had married him as a minister, now found herself married to a clinical psychologist who spent his evenings at bars with other women. The couple divorced.

Because people change throughout their life, the person that one selects at one point in life may not be the same partner one would select at another. Margaret Mead, the famous anthropologist, noted that her first marriage was a student marriage; her second, a professional partnership; and her third, an intellectual marriage to her soul mate, with whom she had her only child. At each of several stages in her life, she experienced a different set of needs and selected a mate who fulfilled those needs.

Spouses may also experience profound physical changes. Christopher Reeve, the movie actor who played in three Superman movies, is now a quadraplegic who requires a machine to help him breathe. His disability necessitated considerable adjustment on the part of his wife and children. In their case, the effect has been to strengthen their relationship. But such a radical physical change in a person might also result in divorce.

Satiation

Satiation, also referred to habituation, refers to the state in which a stimulus loses its value with repeated exposure. Spouses may tire of each other. Their stories are no longer new, their sex is repetitive, and their presence no longer stimulates excitement as it did in courtship. Some persons, feeling trapped by the boredom of constancy, divorce and seek what they believe to be more excitement by a return to singlehood and, potentially, new partners. A developmental task of marriage is for couples to be creative to maintain excitement in their marriage. Going new places, doing new things, and making time for intimacy in the face of rearing children and sustaining careers become increasingly important across time. Alternatively, spouses need to be realistic and not expect every evening to be a New Year's Eve.

Extramarital Relationship

Spouses who feel mistreated by their partners or bored and trapped sometimes consider the alternative of a relationship with someone who is good to them, exciting, and new and who offers an escape from the role of spouse to the role of lover.

Extramarital involvements sometimes hurry a decaying marriage toward divorce because the partner begins to contrast the new lover with the spouse. The spouse is often associated with negatives (bills, screaming children, nagging); the lover, almost exclusively with positives (clandestine candlelight dinners,

What I'm doing in this car flying down these screaming highways is getting my tail to Juarez so I can legally rid myself of the crummy son-of-a-bitch who promised me a tomorrow like a yummy fruitcake and delivered instead wilted lettuce, rotted cucumber, a garbage of a life.
Anne Richardson Roiphe
Long Division

Adultery may or may not be sinful, but it is never cheap.
Raymond Postgate
Somebody at the Door

Register-Freeman (1995) suggested that some extramarital relationships begin on the computer. She noted that her second husband was attracted not to a model but to a modem. "Housed in a gray laptop computer, it set off no warning quiver in my ever-sensitive spousal antennae" (p. 42). Nevertheless, when her husband began to spend increasing amounts of time on the machine, she asked that he go with her to see a marriage therapist, who might refer him to an electronic addictions support group. During group meetings, members might confess to "trading actual friends for the less demanding cybercronies" (p. 42). She and her husband are still together, but, she noted, "recovery is fragile" (p. 42).

Divorce is an ongoing process beginning long before physical separation and continuing long after the process is finalized.

Stephanie Coontz
Social historian

new sex, emotional closeness). The choice is stacked in favor of the lover. Although most spouses do not leave their mates for a lover, the existence of an extramarital relationship may weaken the emotional tie between the spouses so that they are less inclined to stay married.

Most Frequent Factors in Divorce

Researchers have identified the characteristics of those most likely to divorce (Amato and Rogers, 1997; Ahrons, 1995). Some of the more significant associations include the following:

1. Courtship of less than two years
2. Having little in common
3. Marrying in teens
4. Not being religiously devout
5. Differences in race, education, religion, social class, values
6. A cohabitation history
7. Previous marriage
8. No children
9. Spending little leisure time together
10. Urban residence
11. Infidelity
12. Divorced parents
13. Poor communication skills
14. Unemployment of husband
15. Employment of wife
16. Depression, alcoholism, or physical illness of spouse
17. Having seriously ill child
18. Low self-concept of spouses
19. Limited income
20. Limited education

The more of these factors that exist in a marriage, the more vulnerable a couple is to divorce.

Consequences of Divorce for Spouses

Divorce is often an emotional and financial disaster, but it may have different effects on women and men.

Divorce puts you on the edge of sanity.

Abigail Trafford
Crazy Times

Emotional and Psychological Consequences

In spite of the prevalence of divorce and the suggestion that it may be the path to greater self-actualization or fulfillment, a study of 6,000 individuals, both married and unmarried, revealed that the divorced report being the least healthy and happy and the most depressed and suicidal (Kurdek, 1991). These characteristics are particularly present during the early stage of divorce when divorcing spouses are angry, blame each other, and feel sorry for themselves. They are also experiencing a great deal of stress and are more likely to die than individuals who remain married (Hemstrom, 1996). Once they pass through the early stressful stage of divorce (a small proportion get lodged in this stage), they tend to return to being healthy and happy individuals.

Besides feeling angry and bitter, the divorced may experience feelings of despair in reference to three basic changes in their life: termination of a major source of intimacy, disruption of their daily routine, and awareness of a new status—divorced person. Going through a divorce can also lead to feelings of failure or defeat, especially if the other partner initiated the divorce. But the partner who initiates the divorce also undergoes considerable stress and guilt in making the decision to separate.

If you are married, you probably are so because you are in love and want to share your life with another. Like most people, you need to experience feelings of intimacy in a world of secondary relationships. One reason divorce hurts is that you lose one of the few people who knows you and who, at least at one time, did care about you.

Divorce also shatters your daily routine and emphasizes your aloneness. Eating alone, sleeping alone, and driving alone to a friend's house for companionship are role adaptations made necessary by the destruction of your marital patterns. Your support system may also be disrupted as your ex-spouse's family, with whom you may have been close for years, may distance themselves from you.

Although we tend to think of divorce as an intrinsically stressful event, one researcher suggested that the amount of stress involved in a divorce is determined by how stressful the marriage was (Wheaton, 1990). In marriages that are very stressful, a divorce may be beneficial in allowing unhappy spouses to escape from a chronically stressful situation. In very stressful marriages, then, going through a divorce may actually reduce more stress than it creates.

In thirty of the forty (non-European) cultures surveyed it was impossible to detect any substantial difference in the rights of men and women to terminate an unsatisfactory alliance. The stereotype of the oppressed aboriginal woman proved to be a complete myth.
George Murdock
Anthropologist

When parents divorce, they undergo a volatile and profound personal emotional journey. At the same time, they face a fundamental structural change in their family system.
Mary F. Whiteside
Family therapist

INSIGHT

How do women and men differ in their emotional and psychological adjustment to divorce? Arendell (1995) noted that women fare better emotionally after separation and divorce than men. He explained that women are more likely than men not only to have a stronger network of supportive relationships but also to profit from divorce by developing a new sense of self-esteem and confidence, since they are thrust into a more independent role. On the other hand, men are more likely to have been dependent on their wives for domestic and emotional support and to have a weaker external emotional support system. As a result, divorced men are more likely than divorced women to date more partners sooner and to become remarried more quickly. Hence, gender differences in long-term divorce adjustment are minimal.

We have been discussing the personal emotional consequences of divorce for the respective spouses, but the extended family and friends also feel the impact of a couple's divorce. While some parents are happy and relieved that their offspring are divorcing, others grieve. Either way, the relationship with their

*Debts are the sour sediment in
the lemonade of human
existence.*

Herman Sudermann

The Song of Songs

One of the ways parents avoid paying child support is to move to another state, which makes it difficult to collect the support. It is now a federal crime to cross state lines to avoid paying child support. However, indictments are few. Of over 400,000 parents who had crossed state lines only 700 cases (less than 1 percent) were sent to a U.S. attorney's office, and only 89 persons were indicted (Streeter, 1995).

*It's a terrible, terrible thing.
You know. I don't see them.*

Woody Allen

referring to not having seen his son Dylan and daughter Satchel in 2½ years

*Most noncustodial divorced
fathers are legally allowed to
see their children less than
25% of the days in a calander
year.*

grandchildren may be jeopardized. Friends often feel torn and divided in their loyalties. The courts divide the property and identify who gets the kids. But who gets the friends?

Financial Consequences

While both women and men experience a drop in income following divorce, women may suffer more. Forty percent of divorcing women lose half of their family income, whereas fewer than 17 percent of men experience this degree of loss (Arendell, 1995). African-American divorced women are especially vulnerable economically (Pollock & Stroup, 1994). Since men usually have greater financial resources, they may take all they can with them when they leave. The only money they may continue to give to the ex-wife is court-ordered in the form of child support or spousal support. The latter is rare. Some states, such as Texas, do not provide alimony at all. However, most states do provide for an equal distribution of property. The law assumes that both spouses contributed (excluding gifts or inheritance from parents) to whatever money they have at the time of divorce and that this should be divided equally. The amount may be substantial. Lorna Wendt, the wife of Gary Wendt (CEO of GE Capital Services), was awarded $20 million in their divorce settlement (Moore, 1997).

Although 56 percent of custodial mothers are awarded child support, the amount is usually inadequate, infrequent, and not dependable, so that the woman is forced to work (sometimes at more than one job) to take financial care of her children.

NATIONAL DATA The average amount of child support received by the mother is $3,543 per year (*Statistical Abstract of the United States: 1997*, Table 609).

Fathers' Separation from Children

About 5 million divorced dads wake up every morning in an apartment or home while their children are waking up in another place. These are noncustodial fathers who may be allowed to see their children only two weekends a month (Knox, 1998). The pain of being separated from their children is enormous. Douglas MacKay (1997) is a divorced father of three children who wrote:

> I hear them briefly through thin wire.
>
> Their smell is gone from the pillow cases.
>
> A visit this weekend will provide another snapshot, a touch.
>
> Will we ever share the same ground?
>
> When will I ever be allowed to tell them stories over campfires of past warriors?
>
> O great spirit, give me the strength to keep trying, for I love them so.

These words reflect the idea that noncustodial fathers feel less competent and satisfied than fathers who are still married and live with their children (Minton & Pasley, 1996). Indeed, most divorced fathers separated from their children feel a sense of loss and believe that the courts have little interest in protecting their relationship with their children (Arendell, 1995).

Dudley (1991) identified eighty-four divorced fathers (divorced for an average of six years) who reported infrequent contact with their children. Forty percent of the fathers who saw their children infrequently reported the primary reason was interference on the part of the former spouse, who reportedly refused to provide access to the children or who talked negatively to the children about their father. Two-thirds of these fathers reported that they had to return to court over visitation issues.

 INSIGHT

Custodial mothers who are angry at their ex-husbands may attempt to turn their children against the father. Noncustodial fathers are victims of this assault and feel helpless. Robert Bly (1990), who emphasized the importance of fathers in the lives of their children, particularly sons, wrote:

> A friend told me that about 35, he began to wonder who his father really was. He hadn't seen his father in about ten years. He flew out to Seattle, where his father was living, knocked on the door, and when his father opened the door, said, "I want you to understand one thing. I don't accept my mother's view of you any longer."
>
> What happened? I asked.
>
> "My father broke into tears, and said, 'Now I can die.' Fathers wait." What else can they do? (1990, 25).

Although father-child closeness usually suffers with divorce (Cooney, 1994), particularly if the father is the noncustodial parent and if the ex-wife attempts to turn the children against him, the long-term contact and closeness may not necessarily be affected. "Once reaching adulthood, children who leave the custodial parent's home have freedom to initiate and sustain contact with parents that they did not have while growing up. This may be especially likely in cases in which fathers avoided contact because of conflict with ex-wives" (Booth, 1994, 31). Fathers who want to stay connected with their children must make this a conscious goal (making it a priority in their life), try to get along with the child's mother, and hire a domestic attorney as necessary to ensure regular and frequent contact or custody (Knox, 1998).

Some divorced dads may also have negative motives for involvement with their children. These motives include wanting to annoy the ex-wife and lower child support if joint custody is awarded.

Behaviors Associated with Successful Divorce Adjustment

Ahrons (1995) emphasized the importance of having a "good divorce." While acknowledging that divorce is usually an emotional and economic disaster, she submits that it is possible to have a good divorce—that couples can part without destroying their lives or the lives of their children. In effect "the divorced parents continue to have good relationships with their children" (and vice versa) (Ahrons, 1995, xi). Almost half of Ahrons's sample of ninety-eight couples who divorced were able to do so.

INSIGHT

Ahrons asks why our culture perpetuates divorcism—the belief that divorce
must be a disaster?

Perhaps we need to
perpetuate the myth that
divorce is an unmitigated
disaster because we feel in
doing so we may preserve
marriage. Perhaps we feel
that if there's such a thing as
a good divorce, then too
many people will flee their
marriages. . . . Perhaps we
fear that nobody normal
would be left behind, no-
body who believed in duty,
home, and family (1995,
10).

*We own what we learned back
there; the experience and the
growth are grafted onto our
lives.*
Ellen Goodman
The Washington Post

RECENT RESEARCH

Harrist and Ainslie (1998) studied a
sample of 45 five-year-olds and found
that conflict between parents predicted
lower quality parent-child relationships and higher levels of child aggression.

Given that divorce is, for most adults, a difficult experience, what are some
ways to make it a good experience? The following are some of the behaviors
spouses can engage in to have a "good divorce" (Ahrons, 1995; Arendell, 1995;
Hammond, 1992; Goodman, 1992).

1. *Mediate rather than litigate the divorce.* Divorce mediators encourage a civil,
cooperative, compromising relationship while moving the couple toward
an agreement on the division of property, custody, and child support. By
contrast, attorneys make their money by encouraging hostility so that
spouses will prolong the conflict, thus running up higher legal bills. In addition, money spent on attorneys (average is more than $12,000) cannot
be divided by the couple (Hauser, 1995).

2. *Coparent with your ex-spouse.* Setting negative feelings about your ex-
spouse aside so as to cooperatively coparent not only can reduce your
personal stress but also can facilitate a successful adjustment on the part
of the children going through the divorce (Blau, 1995). Although it may be
unsettling for children to go through a transition from nuclear to binuclear
family, Ahrons (1995) observed that "it doesn't follow that you have damaged your children for life . . . that you can provide your children with
love and nurturance no matter what form your family takes. Not coparenting with your ex-spouse works against such a possibility."

3. *Take some responsibility for the divorce.* Since marriage is an interaction between the spouses, one person is seldom totally to blame for a divorce.
Rather, both spouses share in the demise of the relationship. Take some
responsibility for what went wrong.

4. *Learn from the divorce.* View the divorce as an opportunity to improve
yourself for future relationships. What did you do that you might consider
doing differently in the next relationship?

5. *Create positive thoughts.* Divorced people are susceptible to feeling as
though they are failures. They see themselves as Divorced persons with a
capital D, sometimes referred to as "hardening of the categories" disease.
Improving their self-esteem is important for divorced persons. They can
do this by systematically thinking positive thoughts about themselves.
One technique is to write down twenty-one positive statements about
yourself ("I am honest," "I am a good cook," "I am a good parent," etc.)
and transfer them to seven 3 × 5 cards, each containing three statements.
Take one of the cards with you each day and read the thoughts at three
regularly spaced intervals (e.g., ten in the morning, four in the afternoon,
ten at night). This ensures that you are thinking good things about yourself and are not allowing yourself to drift into a negative set of thoughts.

6. *Avoid alcohol and other drugs.* The stress and despair that some people feel
following a divorce make them particularly vulnerable to the use of alcohol or other drugs. Alcohol and other drugs should be avoided because
they produce an endless negative cycle. For example, stress is relieved by
alcohol; alcohol produces a hangover and negative feelings; the negative
feelings are relieved by more alcohol, producing more negative feelings,
etc.

7. *Relax without drugs.* Deep muscle relaxation can be achieved by systematically tensing and relaxing each of the major muscle groups in the body. Alternatively, yoga, transcendental meditation, and getting a massage can induce a state of relaxation in some people. Whatever the form, it is important to schedule a time each day to get relaxed.

8. *Engage in aerobic exercise.* Exercise helps one not only to counteract stress but also to avoid it. Jogging, swimming, riding an exercise bike, or other similar exercise for thirty minutes every day increases the oxygen to the brain and helps facilitate clear thinking. In addition, aerobic exercise produces endorphins in the brain, which create a sense of euphoria ("runner's high"). In addition, good health in general is associated with life satisfaction in older divorced individuals (Hammond, 1992).

9. *Engage in fun activities.* Some divorced people sit at home and brood over their "failed" relationship. This only compounds their depression. Doing what they have previously found enjoyable—swimming, horseback riding, skiing, sporting events, etc.—provides an alternative to sitting on the couch alone.

10. *Continue interpersonal connections.* Adjustment to divorce is easier when intimate interaction with friends and family is continued (Bursik, 1991). This is particularly true for individuals who divorce past the age of 45 (Goodman, 1992) and for men. When divorced women and men are compared regarding life satisfaction, men report more dissatisfaction. One researcher suggested that such lower satisfaction among men is related to their lack of skills in maintaining social supports (Hammond, 1992).

11. *Let go of the ex-partner.* Ex-spouses who stay negatively attached to their ex by harboring resentment and trying to "get back at the ex" and who do not make themselves available to new relationships limit their ability to adjust to a divorce (Tschann et al., 1989).

12. *Allow time to heal.* Since self-esteem usually drops after divorce, a person is often vulnerable to making commitments before working through feelings about the divorce. The time period most people need to adjust to divorce is between twelve and eighteen months. Although being available to others may help to repair one's self-esteem, getting remarried during this time should be considered cautiously.

> *I got my exercise acting as pallbearer to my friends who exercise.*
> Chauncey Depew

> *What a lovely surprise to finally discover how unlonely being alone can be.*
> Ellen Burstyn
> Actress

INSIGHT

Ahrons suggested that negative societal assumptions about divorce are influential in making it a drawn-out and painful experience. She identified some of the assumptions and emphasized the need to develop new assumptions (1995, 248):

Old Assumptions	New Assumptions
Divorce ends the family.	Divorce redefines the family.
Divorce ruins children.	Children in binuclear families can be as healthy as children in nuclear families.
Divorce is abnormal.	Divorce is normal.
Divorce is mysterious.	The process of divorce is predictable.

Consequences of Divorce for Children

Children are involved in approximately half of divorces.

The Personal Application (Children's Beliefs about Parental Divorce Scale) at the end of the chapter provides information about the potential impact of divorce on a child.

Effects of Divorce on Children

A team of researchers (Stewart et al., 1997) interviewed 160 children (privately) between the ages of 6 and 12 about their parents' separation within six months of its occurring and then one year later. Though 80 percent of the children were initially saddened and angry at their parents' divorce, a year later most were just as emotionally healthy as other U.S. children their age. Only about 15 percent continued to have significant emotional problems eighteen months after the separation. Many of these problems predated the separation, suggesting that the bickering of the parents had a negative impact on the children before the parents split.

However, Wolfinger (1997) analyzed data from a national sample of 8,590 interviews and found that individuals whose parents divorced were more vulnerable to divorce themselves. There was also a cumulative effect so that individuals whose parents had divorced more than once were even more likely to have their own marriages end in divorce. The researcher attributed the negative consequences of divorce to the strain of experiencing family structure transitions rather than to not having a male role model or experiencing a decline in their standard of living.

A review of the literature (Knox, 1998) on the effects of divorce on children reveals that children whose parents divorce (and particularly where the father drops out of the child's life) are more vulnerable to loss of self-esteem ("Don't my parents love me enough to stay together?"), drop in school grades (due to the stress associated with the divorce and distracted parents), and increase in drug use (supervision of children drops with divorce). Blankenhorn (1995) and Popenoe (1996) have emphasized the negative consequences of fatherlessness.

Minimizing Negative Effects of Divorce on Children

Researchers have identified the conditions under which a divorce has the least negative consequences for children (Stewart et al., 1997, Ahrons, 1995). Some of these follow.

1. *A cooperative relationship between the parents.* The most important variable in a child's positive adjustment to divorce is that the child's parents continue to maintain a cooperative relationship throughout the separation, divorce, and post-divorce period. In contrast, bitter parental conflict places the children in conflict. One daughter of divorced parents said:

What about the love we made? She's upstairs in her room crying over you and me.

Shelby Lynn

Singer

RECENT RESEARCH

Holroyd and Sheppard (1997) studied the effects of parental separation on 28 children of separated parents. None of the children in the study welcomed the divorce of their parents and all expressed the wish that their parents be reunited.

RECENT RESEARCH

Billingham and Abrahams (1998) compared 75 women from intact families with 34 women from divorced families and found that women from divorced families were significantly more likely to rate themselves as less physically attractive than the women from intact families rated themselves. In addition, women from divorced families had higher body dissatisfaction scores than did women from intact families.

Children whose parents separate typically suffer disadvantages compared to children whose parents live together.

Judith Seltzer

Sociologist

There are no hard and fast links between family structure, parental behaviors, and children's outcomes. . . . Indeed the worst problems for children stem from parental conflict, before, during, and after divorce—or within marriage.

Stephanie Coontz

Social historian

My father told me, "If you love me, you would come visit me," and my mom told me, "If you love me, you won't visit him."

Couples who are able to get along after the divorce and coparent their children are most likely to have grown apart or to have had stable long-term issues such as a spouse's laziness or career conflicts. Couples who are conflictual and unable to coparent had high anger in leaving the marriage and the anger continued. Issues of the divorce were more likely to be infidelity and drug abuse (Christensen et al., 1996).

2. *Parents' attention to the children.* Both the custodial and the noncustodial parent continue to spend time with the children and to communicate to them that they love them and are interested in them.

3. *Encouragement to see noncustodial parent.* Children who usually live with custodial mothers following divorce are encouraged by the mother to maintain a regular and stable visitation schedule with their father.

4. *Attention from the noncustodial parent.* Noncustodial parents, usually the fathers, establish frequent and consistent times to be with the children. Noncustodial parents who do not show up at regular intervals exacerbate their children's emotional insecurity by teaching them, once again, that parents cannot be depended on. Parents who show up often and consistently teach their children to feel loved and secure.

5. *Assertion of parental authority.* Both parents continue to assert their parental authority with their children and continue to support the discipline practices of each other to their children.

6. *A temperament on the part of the child that allows the child to adjust to change easily.* Kalter (1989) found that the reaction of a child to divorce is influenced by the child's temperament. Some children are not easily frustrated and readily adapt to change; others have difficulty with even minor changes.

7. *Regular and consistent child support payments.* Support payments (usually from the father to the mother) are associated with enhanced well-being of the children of divorced parents (King, 1994).

> The economic effect of child support might act through a variety of mechanisms. It could directly increase the resources available to the child. More money could be spent on educational activities and materials. The extra money might allow the child to receive better health care or to live in a better neighborhood (91).

8. *Stability.* The parents don't move the children to a new location. Moving them causes them to be cut off from their friends, neighbors, and teachers. It is important to keep their life as stable as possible.

Because the greatest damage to children from a divorce is a continuing hostile and bitter relationship between their parents, some states require divorce mediation as a mechanism to encourage civility in working out differences and to clear the court calendar from protracted court battles. The social policy of this chapter focuses on divorce mediation.

Divorce Mediation?

DEFINITION OF DIVORCE MEDIATION

Mediation is a process in which spouses who have decided to separate or divorce meet with a neutral third party (mediator) to negotiate the issues of (1) child custody and visitation, (2) child support, (3) property settlement, and (4) spousal support. Mediation is not for everyone. It does not work where there is a history of spouse abuse, where the parties do not disclose their financial information, where one party is controlled by someone else (e.g., a parent), where there is the desire for revenge, or where the mediator is biased. The latter situation can be mitigated by selecting a professional who has specific training and experience in divorce mediation (Hauser, 1995).

BENEFITS OF MEDIATION

1. *Better relationship.* Spouses who choose to mediate their divorce have a better chance for a more civil relationship because they cooperate in specifying the conditions of their separation or divorce. Mediation emphasizes negotiation and cooperation between the divorcing partners. Such cooperation is particularly important if the couple has children in that it provides a positive basis for discussing issues in reference to the children across time.
2. *Economic benefits.* Mediation is less expensive than litigation. The cost of hiring an attorney and going to court over issues of custody and division of property is around $12,000. A mediated divorce costs about $1,000 (Neumann, 1989). What the couple spend in legal fees they cannot keep as assets to later divide.
3. *Less time-consuming process.* A mediated divorce takes two to three months versus two to three years if the case is litigated.
4. *Avoidance of public exposure.* Some spouses do not want to discuss their private lives and finances in open court. Mediation occurs in a private and confidential setting.
5. *Greater overall satisfaction.* Mediation results in an agreement developed by the spouses, not one imposed by a judge or the court system. A comparison of couples who chose mediation with couples who chose litigation found that those who

mediated their own settlement were much more satisfied with the conditions of their agreement. In addition, children of mediated divorces adjust better to their parents' divorce than children of litigated divorces (Marlow & Sauber, 1990).

BASIC MEDIATION GUIDELINES

1. *Children.* What is best for a couple's children is a primary concern of the mediator. Children of divorced parents adjust best under three conditions: (a) the noncustodial parent is allowed regular and frequent access, (b) the children see the parents relating in a polite and positive way, and (c) each parent talks positively about the other parent and neither parent talks negatively about the other to the children.
2. *Fairness.* It is important that the agreement is fair, with neither party being exploited or punished. It is "fair" for both parents to contribute financially to the children. It is "fair" for the noncustodial parent to have regular access to his or her children.
3. *Open disclosure.* The spouses will be asked to disclose all facts, records, and documents to ensure an informed and fair agreement regarding property, assets, and debts.
4. *Other professionals.* During mediation the spouses may be asked to consult an accountant regarding tax laws. In addition, spouses are encouraged to consult an attorney throughout the mediation and to have the attorney review the written agreements that result from the mediation. However, during the mediation sessions, all forms of legal action against each other (the spouses) should be stopped.
5. *Confidentiality.* The mediator will not divulge anything the spouses say during the mediation sessions without the permission of the spouses. The spouses are asked to sign a document stating that should they not complete mediation, they agree not to empower any attorney to subpoena the mediator or any records resulting from the mediation for use in any legal action. Such an agreement is necessary for the spouses to feel free to talk about all aspects of their relationship without fear of legal action against them for such disclosures.

SOCIAL POLICY (continued)

Divorce mediation is not for every couple getting divorced. Divorcing couples in which at least one of the spouses is abusive, is hiding assets, or wants to "punish" the partner will not benefit from mediation. Mediation requires that the partners not feel afraid of each other, be open/honest with each other about finances, and be conciliatory in demeanor toward each other.

James, Paula. 1997. *The divorce mediation handbook.* San Francisco: Jossey-Bass.

Marlow, L., and S. R. Sauber. 1990. *The handbook of divorce mediation.* New York: Plenum.

Neumann, D. 1989. *Divorce mediation: How to cut the cost and stress of divorce.* New York: Holt.

REFERENCES

Hauser, Joyce. 1995. *Good divorces, bad divorces: A case for divorce mediation.* Lanham, Md.: University Press of America.

What they (those who predict doom for divorced children) really mean to say is not that children in divorced families have more problems but that more children of divorced parents have problems . . . the large majority of children of divorce do not experience severe or long term problems: Most do not drop out of school, get arrested, abuse drugs, or suffer long-term emotional distress.

Stephanie Coontz

Social historian

INSIGHT

Remarriage rates are lower among Hispanics and African-Americans. Only about a third of Hispanics and 20 percent of African-Americans will eventually remarry. Remarriage rates among Hispanics may be depressed because Hispanics are predominantly Catholic (the Catholic Church opposes remarriage). Lower remarriage rates among African-Americans are "also consistent with the lesser place of marriage in the African-American family"(Cherlin, 1996, 384).

Remarriage

NATIONAL DATA About two-thirds of divorced women and three-fourths of divorced men will remarry (U.S. Bureau of the Census, 1992). Close to half a million remarriages occur annually (National Center for Health Statistics, 1998).

Remarriage for the Divorced

Ninety percent of remarriages consist of persons who are divorced rather than widowed. The majority of the divorced get remarried for many of the same reasons as those getting married for the first time—love, companionship, emotional security, and a regular sex partner. Other reasons are unique to remarriage and include financial security (particularly for the wife), help in rearing one's children, the desire to provide a "social" mother or father for one's children, avoidance of the stigma associated with the label "divorced person," and legal threats regarding the custody of children. With regard to the latter, a parent seeking custody of a child is viewed more favorably by the courts if he or she is married. Regardless of the reason, half of the remarriages occur within five years of one's divorce, with men marrying sooner than women. Older divorced women (over 40) are less likely than younger women to remarry. Not only are there fewer available men, but the mating gradient whereby men tend to marry women younger than themselves is operative. In addition, some women are economically independent, enjoy the freedom of singlehood, and want to avoid the restrictions (and constrictions) of marriage.

Divorced persons getting remarried are usually about ten years older than those marrying for the first time.

NATIONAL DATA The median age for divorced men who remarry is 37; for women, 34 *(Statistical Abstract of the United States: 1997,* Table 148).

A substantial number of
divorces are in the best
interest of children.

Paul Amato
Sociologist

If parents who divorce could
manage to maintain their
child's standard of living and
avoid a residential move, their
child would have about the
same risk of dropping out of
high school and becoming a
teen mother as an average
child with similar characteristics whose parents do not
divorce.

Sara McLanahan
Sociologist

There is no help for misfortune
but to marry again.

Thompson
Body, Boots and Britches

Nothing's wrong that can't be
cured with a new love.

Patty Loveless
Country-and-western singer

Persons in their mid-30s who are considering remarriage have usually finished school and are established in a job or career. Courtship is usually short and must take into account the individuals' respective work schedules and career commitments. Because each partner may have children, much of the couple's time together includes their children. Going out on an expensive dinner date during courtship before the first marriage is replaced with eating pizza at home and renting a PG-rated movie to watch with the kids.

The principle of homogamy is illustrated in the selection of remarriage partners. Sixty percent of divorced people marry other divorced people, 35 percent marry single men and women, and 4 percent marry widowed individuals (Ganong & Coleman, 1994).

INSIGHT

Most divorced individuals are open to the possibility of becoming involved in another love relationship. Sometimes, rather than try to meet someone new, they try to locate a previous love and rekindle the relationship. Warren Bennis, twice divorced, sent a note to Grace Gabe, to whom he had been engaged thirty years earlier. The note read, "Any chance you can have dinner with me on October 13? Yours, Warren." Grace was also single again and open to the possibility of seeing her old flame. The couple met for dinner and eventually married.

The following are some of the guidelines for rekindling an old flame suggested by Dr. Gabe (1993):

1. *Anticipate stages of reentry.* Rather than plunge headlong into the old love affair, most couples have "periods of moving very close that alternate with periods of needing some emotional distance. . . . Your old sweetheart may wait longer than you want between communications or might cancel a date because the reconnecting may be too much too soon" (p. 36).

2. *Be aware that the original problems will reactivate.* After the euphoria of the reunion has passed, the old issues will resurface. This is healthy, and the partners have the life experience of years to help work through these old issues together.

3. *Review the original breakup and reach consensus.* "Reviewing the reasons for the breakup of the old relationship may be painful or at least difficult. But it must take place between the sweethearts. If you can't do this, don't start anything with an idea of getting back together" (p. 38). Consensus implies that both partners develop a shared view of why the relationship failed and why it would not have worked back then.

4. *Respect the partner's previous love history.* During the years apart, each partner likely had other important love relationships. Validate these rather than minimize them. Recognize that you do not "have a monopoly on the partner's attachments" (p. 63).

OTHER CULTURES Although getting remarried is regarded as an option by both women and men in the United States, such is not the case in all societies. In a study of 200 divorced or separated men and women in a major urban area in central India, the majority of the women had no interest in getting remarried. One reason is that divorced women were allowed to remarry only widowers, whereas divorced men could marry any single woman of their choice (Pothern, 1989). ●

Preparation for Remarriage

It is not uncommon for persons who are divorced to live together with a new partner before remarriage (Ganong & Coleman, 1994). Like other cohabitants, they drift into living together by gradually spending more time together. Aside from living together, they (like most couples in courtship, whether first or second marriage) do little else to prepare for their new marriage. In a study of couples who remarried, fewer than 25 percent sought remarriage counseling, attended support groups, or discussed their impending marriage with friends. Even though children are a critical topic for remarrieds, half of the couples did not discuss the issue (Ganong & Coleman, 1988).

Issues Involved in Remarriage

People who remarry must confront several issues (Goetting, 1982; Ganong & Coleman, 1994).

Boundary maintenance Movement from divorce to remarriage is not a static event that happens in a brief ceremony and is over. Rather, ghosts of the first marriage in terms of the ex-spouse and, possibly, the children must be dealt with. A parent must decide how to relate to his or her ex-spouse in order to maintain a good parenting relationship for the biological children while keeping an emotional distance to prevent problems from developing with the new partner. Some spouses continue to be emotionally attached to the ex-spouse and have difficulty breaking away. However, boundary ambiguity does not appear to be a major problem for remarried spouses (Pasley & Ihinger-Tallman, 1989).

Emotional remarriage Remarriage involves beginning to trust and love another person in a new relationship. Such feelings may come slowly as a result of negative experiences in the first marriage.

Psychic remarriage Divorced individuals considering remarriage may find it difficult to give up the freedom and autonomy of being single again and to develop a mental set conducive to pairing. This transition may be particularly difficult for people who sought a divorce as a means to personal growth and autonomy. These individuals may fear that getting remarried will put unwanted constraints on them.

Community remarriage This stage involves a change in focus from single friends to a new mate and other couples with whom the new pair will interact. The bonds of friendship established during the divorce period may be particularly valuable because they have lent support at a time of personal crisis. Care should be taken not to drop these friendships.

Parental remarriage Because most remarriages involve children, people must usually work out the nuances of living with someone else's children. Since mothers are usually awarded primary physical custody, this translates into the new stepfather's adjusting to the mother's children. If a person has children from a previous marriage who do not live primarily with him or her, the new spouse must adjust to these children on weekends, holidays, vacations, or other visitation times.

There is only one way to have a happy marriage and as soon as I learn what it is, I'll get married again.
Clint Eastwood
Actor

Well, being divorced is like being hit by a Mack truck. If you live through it, you start looking very carefully to the right and to the left.
Jean Kerr
Mary, Mary
Act I

A jury consists of 12 persons chosen to decide who has the better lawyer.
Robert Frost
Poet

Economic and legal remarriage The second marriage may begin with economic responsibilities to the first marriage. Alimony and child support often threaten the harmony and sometimes even the economic survival of second marriages. One wife said that her paycheck was endorsed and mailed to her husband's first wife to cover his alimony and child support payments. "It irritates me beyond description to be working for a woman who lived with my husband for seven years," she added. In another case, a remarried woman who was receiving inadequate child support from her ex-spouse felt too embarrassed to ask her new husband to pay for her son's braces.

Persons who remarry may also sign a prenuptial or premarital agreement that specifies what assets will go to whom in the event of a divorce. Though such an agreement is not romantic, neither is divorce. Wills specifying who gets what at the death of the respective parties are drawn up after the marriage, since the terms *husband* and *wife* are used in the document.

Remarriage for the Widowed

Only 10 percent of remarriages consist of widows or widowers. Nevertheless, our population has a number of widowed people.

NATIONAL DATA There are 11 million widowed women and 2.5 million widowed men. The median age of the widow who remarries is 54; of the widower, 63 (*Statistical Abstract of the United States: 1997*, Tables 59 and 148).

Remarriage for the widowed is usually very different from remarriage for the divorced. Aside from the fact that the widowed are less likely to remarry, when they do remarry they are usually much older and their children are grown.

A widow or widower may marry someone of similar age or someone who is considerably older or younger. Marriages in which one spouse is considerably older than the other are referred to as May-December marriages. Here, we will discuss only "December marriages," in which both spouses are elderly.

A study of twenty-four elderly couples found that the need to escape loneliness or the need for companionship was the primary motivation for remarriage (Vinick, 1978). The men reported a greater need to remarry than the women. Most of the spouses (75 percent) met through a mutual friend or relative and married less than a year after their partner's death (63 percent).

The children of the couples had mixed reactions to their parent's remarriage. Most of the children were happy that their parent was happy and felt relieved that the companionship needs of their elderly parent would now be met by someone on a more regular basis. But some children also disapproved of the marriage out of concern for their inheritance rights. "If that woman marries Dad," said a woman with two children, "she'll get everything when he dies. I love him and hope he lives forever, but when he's gone, I want the farm." While children may be less than approving of the remarriage of their widowed parent, adult friends of the couple are usually very approving, including the kin of the deceased spouses (Ganong & Coleman, 1994).

Stability of Remarriages

Recent national data suggest that remarriages are more likely to end in divorce in the early years of marriage, but as time passes, the rates converge (Clarke & Wilson, 1994). Within the first two to five years (Rutter, 1994), remarried spouses are more likely to divorce than first-married spouses. But as the spouses age, their chances of staying together remain the same as for those in first marriages. Indeed, after fifteen years, husbands and wives who have both been married before have a lower divorce rate than first-married spouses (25%–29%) (Clarke & Wilson, 1994). Two researchers (Booth & Edwards, 1992) noted that quick divorce in the early years of remarriage reflects less tolerance on the part of those who have been married before for staying in an unhappy relationship. As a result, some in second marriages who remain married may be those who have chosen to stay together for emotional rather than practical reasons.

Blended Families

The myth of instant love and its antithesis of never-ending, unsolvable problems need to be dispelled.

Step Family Association of America

Blended families, also known as binuclear, step, remarried, reconstituted, reorganized, recycled, combined, merged, and second-time-around families, represent the fastest-growing type of family in the United States (Ahrons, 1995). A blended family is one in which the spouses in a new marriage relationship are blended with the children of at least one of the spouses from a previous marriage. (Another popular term is *binuclear,* which refers to a family that spans two households—when a married couple with children divorce, their family unit

Remarried spouses often begin their marriage with children.

spreads into two households.) We will use the terms *blended family* and *step-family* interchangeably. This section examines how blended families differ from nuclear families; how they are experienced from the viewpoints of women, men, and children; and the developmental tasks that must be accomplished to make a successful blended family.

Definition and Types of Blended Families

Although a blended family can be created when a never-married or a widowed parent with children marries a person with or without children, most blended families today are composed of spouses who were once divorced. This is different from blended families characteristic of the early twentieth century, which more often were composed of spouses who had been widowed.

NATIONAL DATA About 20 percent of all married couples today with children are a blended family. Over a third of children born in the United States will be in a stepfamily before reaching age 18 (Ganong & Coleman, 1994).

There are several types of stepfamilies:

1. Stepfamilies in which a child lives with his or her married parent and stepparent.
2. Stepfamilies in which the children from a previous marriage visit with their remarried parent and stepparent.
3. An unmarried couple living together in which at least one of the partners has children from a previous relationship who live with or visit them.
4. A remarried couple in which each of the spouses brings children into the new marriage from the previous marriage.
5. A couple who not only bring children from a previous marriage but also have a child or children of their own.
6. A remarried couple, both of whom have children from a previous marriage who may live in another state and have very little contact with the remarried couple.

Unique Aspects of Stepfamilies

Stepfamilies differ from nuclear families in a number of ways. To begin with, the children in a nuclear family are biologically related to both parents, whereas the children in a stepfamily are biologically related to only one parent. Also, in a nuclear family, both biological parents live with their children, while only one biological parent in a stepfamily lives with the children. In some cases, the children alternate living with each parent.

While nuclear families are not immune to loss, everyone in a stepfamily has experienced the loss of a love partner, which results in grief. About 70 percent of children are living without their biological father (whom some children desperately hope will reappear and reunite the family). The respective spouses may also have experienced emotional disengagement and physical separation from a once loved partner. Stepfamily members may also be experiencing

INSIGHT

Preserving original relationships is helpful in reducing grief. It is sometimes helpful for the biological parent and child to take time to nurture their relationship apart from stepfamily activities. This will reduce the child's sense of loss and any feelings of jealousy toward new stepsiblings.

Particularly for children, stepfamilies extend the potential for warm, nurturing relationships with more people.

Chester Winton
Sociologist

INSIGHT

A parent's emotional bond with children (particularly if the children are young and dependent) from a previous marriage may weaken a remarriage from the start. As one parent says, "Nothing and nobody is going to come between me and my kids." However, new spouses may view such bonding differently. One spouse said that such concern for one's own children was a sign of a caring and nurturing person. "I wouldn't want to live with anyone who didn't care about his kids." But another said, "I feel left out and that she cares more about her kids than me. I don't like the feeling of being an outsider."

losses because of having moved away from the house in which they lived, their familiar neighborhood, and their circle of friends.

Children in nuclear families have also been exposed to a relatively consistent set of beliefs, values, and behavior patterns. When children enter a stepfamily, they "inherit" a new parent, who may bring a new set of values and beliefs and a new way of living into the family unit.

Likewise, the new parent now lives with children who may have been reared differently from the way in which the stepparent would have reared them if he or she had been their parent all along. One stepfather explained:

> It's been a difficult adjustment for me living with Molly's kids. I was reared to say "Yes sir" and "Yes ma'am" to adults and taught my own kids to do that. But Molly's kids just say "yes" or "no." It rankles me to hear them say that, but I know they mean no wrong with "yes" and "no" as long as it is said politely, and that it is just something that I am going to have to live with.

Another uniqueness in stepfamilies is that the relationship between the biological parent and the children has existed longer than the relationship between the adults in the remarriage. Jane and her twin children have a nine-year relationship and are emotionally bonded to each other. But Jane has known her new partner only a year, and although her children like their new stepfather, they hardly know him.

In addition, the relationship between the biological parent and his or her children is longer than that of the stepparent and stepchildren. The short history of the relationship between the child and the stepparent is one factor that may contribute to increased conflict between these two during the child's adolescence. Children may also become confused and wonder if they are disloyal to their biological parent if they become friends with the stepparent.

Another unique feature of stepfamilies is that unlike children in the nuclear family, who have one home they regard as theirs, children in stepfamilies have two homes they regard as theirs. In some cases of joint custody, children spend part of each week with one parent and part with the other; they live with two sets of adult parents in two separate homes.

Money, or lack of it, from the ex-spouse (usually the husband) may be a source of conflict. In some stepfamilies, the ex-spouse (usually the father) is expected to send child support payments to the parent who has custody of the children. Less than one-half of these fathers send any money; those who do may be irregular in their payments. Fathers who pay regular child support tend to have higher incomes, be remarried, live close to their children, and visit them regularly (Teachman, 1991). They are also more likely to have legal shared or joint custody, which helps to ensure that they will have access to their children (Dudley, 1991).

Fathers who do not voluntarily pay child support and are delinquent by more than one month may have their wages garnished by the state. Some fathers change jobs frequently and move around to make it difficult for the government to keep up with them. Such dodging of the law is frustrating to custodial mothers who need the child support money. Added to the frustration is the fact that fathers are legally entitled to see their children even though they do not pay

*That's why we are called
STEPMOTHERS. I have
footprints all over my body.*
A stepmother

INSIGHT

One of the reasons stepfamilies are stigmatized is that they are thought to be wrought with conflict. But researchers have compared the conflict in families with and without stepchildren and found just the opposite. On the basis of nationally representative data on 2,655 African-American and white married couples with children, MacDonald and DeMaris (1995) concluded that "marital conflict occurs less frequently in double remarriages than in first marriages. Thus, contrary to hypothesis, remarriage does not increase the frequency of open disagreements between spouses. In fact, if both spouses are remarried, disagreements are likely to be fewer than if both spouses are in their first marriage" (MacDonald & Maris, 1995, 394). The researchers reasoned that persons who have been married before are more likely to be realistic about expectations in their new marriage.

court-ordered child support. This angers the mother who must give up her child on weekends and holidays to a man who is not supporting his child financially. Such distress on the part of the mother is probably conveyed to the child.

New relationships in stepfamilies experience almost constant flux. Each member of a new stepfamily has many adjustments to make. Issues that must be dealt with include how the mate feels about the partner's children from a former marriage, how the children feel about the new stepparent, and how the newly married spouse feels about the spouse's sending alimony and child support payments to an ex-spouse. In general, it takes at least two years for newly remarried spouses to feel comfortable together and five to seven years for the whole family to feel comfortable. Some marriages and families feel comfortable much more quickly; some never do.

Stepfamilies are also stigmatized. *Stepism* is the assumption that stepfamilies are inferior to biological families. Stepism, like racism, involves prejudice and discrimination (Darden & Zimmerman, 1992). We are all familiar with the wicked stepmother in *Cinderella*. The fairy tale certainly gives us the impression that to be in a stepfamily with stepparents is a disaster. Even textbooks in marriage and the family emphasize a deficit model when discussing stepfamilies. After reviewing twenty-six such introductory texts, researchers concluded that the books "contain more sources of stress than potential strengths/benefits, and positive outcome variables are ignored in favor of a focus on problems" (Coleman, Ganong, & Goodwin, 1994, 289).

Stepparents also have no childfree period. Unlike the newly married couple in the nuclear family, who typically have their first child about two and one-half years after their wedding, the remarried couple begin their marriage with children in the house.

Profound legal differences exist between nuclear and blended families. While biological parents in nuclear families are required in all states to support their children, only five states require stepparents to provide financial support for their stepchildren. However, when there is a divorce, this and other discretionary types of economic support usually stop (Ganong, Coleman, & Mistina, 1995). Other legal matters with regard to nuclear families versus stepfamilies involve inheritance rights and child custody.

Finally, extended family networks in nuclear families are smooth and comfortable, whereas those in stepfamilies often become complex and strained. Table 12.1 summarizes the differences between nuclear families and stepfamilies.

Strengths of Stepfamilies

Stepfamilies have both strengths and weaknesses. Strengths include children's exposure to a variety of behavior patterns, their observation of a happy remarriage, adaptation to stepsibling relationships inside the family unit, and greater objectivity on the part of the stepparent.

Exposure to a variety of behavior patterns Children in stepfamilies experience a variety of behaviors, values, and lifestyles. They have had the advantage of living on the inside of two families. One 12-year-old said:

Table 12.1

Differences Between Nuclear Families and Stepfamilies	
Nuclear Families	**Stepfamilies**
1. Children are (usually) biologically related to both parents.	1. Children are biologically related to only one parent.
2. Both biological parents live together with children.	2. As a result of divorce or death, one biological parent does not live with children. In the case of joint physical custody, the children may live with both parents, alternating between them.
3. Beliefs and values of members tend to be similar.	3. Beliefs and values of members are more likely to be different because of different backgrounds.
4. Relationship between adults has existed longer than relationship between children and parents.	4. Relationship between children and parents has existed longer than relationship between adults.
5. Children have one home they regard as theirs.	5. Children may have two homes they regard as theirs.
6. The family's economic resources come from within the family unit.	6. Some economic resources may come from ex-spouse.
7. All money generated stays in the family.	7. Some money generated may leave the family in the form of alimony or child support.
8. Relationships are relatively stable.	8. Relationships are in flux: new adults adjusting to each other; children adjusting to stepparent; stepparent adjusting to stepchildren; stepchildren adjusting to each other.
9. No stigma is attached to nuclear family.	9. Stepfamilies are stigmatized.
10. Spouses had childfree period.	10. Spouses had no childfree period.
11. Inheritance rights are automatic.	11. Stepchildren do not automatically inherit from stepparents.
12. Rights to custody of children are assumed if divorce occurs.	12. Rights to custody of stepchildren are usually not considered.
13. Extended family networks are smooth and comfortable.	13. Extended family networks become complex and strained.
14. May not have experienced loss.	14. Experienced with loss.

> *Only about one in five children has a bad relationship with a stepparent. My research and other studies show a lot of people flourish and do better because they've had good stepparents.*
>
> James Bray
> Psychologist

My real mom didn't like sports and rarely took me anywhere. My stepmother is different. She likes to take me fishing and roller skating. She recently bought me a tent and is going to take me camping this summer.

Happier parents Single parenting can be a demanding and exhausting experience. Remarriage can ease the stress of solo parenting and provide a happier context for the parent. Research by Kurdek and Fine (1991) suggests that wives are happier in stepfamilies than husbands and are more optimistic about stepfamily living. One daughter said:

Looking back on my parents' divorce, I wish they had done it long ago. While I miss my dad and am sorry that I don't see him more often, I was always upset listening to my parents argue. They would yell and scream, and it would end with my mom crying. It was a lot more peaceful (and I know my mom was a lot more happy) after they got divorced. Besides, I like my stepdaddy. Although he isn't my real dad, I know he cares about me.

Opportunity for new relationship with stepsiblings While some children reject their new stepsiblings, others are enriched by the opportunity to live with a new person to whom they are now "related." One 14-year-old remarked, "I have never had an older brother to do things with. We both like to do the same things and I couldn't be happier about the new situation." Some stepsibling relationships are maintained throughout adulthood. Two researchers (White & Riedmann, 1992) analyzed national data and assessed the degree to which full and step/halfsiblings keep in touch as adults. They found that while step/halfsiblings see each other less often, "adults report substantial contact with their step/halfsiblings and only 0.5% of stepsiblings were so estranged that they did not at least know where their step/halfsiblings lived" (p. 206). Characteristics of those more likely to maintain contact included being female, African-American, younger, and geographically closer.

RECENT RESEARCH

Ganong et al. (1998) found that stepfamily adoption of a stepchild was related to an uninvolved biological parent, the desire to be a "regular" family, and desires to sever relationships with the nonresidential parent.

More objective stepparents Because of the emotional tie between a parent and a child, some parents have difficulty discussing certain issues or topics. A stepparent often has the advantage of being less emotionally involved and can relate to the child at a different level. One 13-year-old said of the relationship with her father's new wife:

She went through her own parents' divorce and knows what it's like for me to be going through my dad's divorce. She is the only one I can really talk to about this issue. Both my dad and mom are too emotional about the subject to be able to talk about it.

According to Papernow (1988), a stepparent may be the ideal person for stepchildren to talk with about sex, their feelings about their parents' divorce, career choices, drugs, and other potentially charged subjects.

The foregoing represent only a few of the many strengths of stepfamilies that are often given limited visibility. Other stepfamily strengths include the fact that the adults are often more mature (and therefore better role models) than they were in their first marriage, the children are more adaptable (good for their self-esteem), and the spouses adhere less rigidly to stereotyped gender roles (also good modeling) (Coleman et al., 1994). Still other strengths include that the circle of resource adults for children widens and that children observe a variety of parenting styles.

Developmental Tasks for Blended Families

A developmental task is a skill that, if mastered, allows the family to grow as a cohesive unit. Developmental tasks that are not mastered will bring the family closer to the point of disintegration. Some of the more important developmental tasks for stepfamilies are discussed in this section.

Acknowledge losses and changes   As noted earlier, each member of a blended family is experiencing a loss of a spouse or of a biological parent in the home. These losses are sometimes compounded by home, school, neighborhood, and job changes. Feelings about these losses and changes should be acknowledged as important and consequential. In addition, children should not be required to love their new stepparent or stepsiblings (and vice versa). Such feelings will develop only as a consequence of positive interaction over an extended period of time.

Nurture the new marriage relationship   Critical to the healthy functioning of a new stepfamily is the importance of the new spouses' nurturing each other and forming a strong unit. From this base, the couple can communicate, cooperate, and compromise with regard to the various issues in their new blended family. Too often spouses become child-focused and neglect the relationship on which the rest of the family depends. Such nurturing translates into spending time alone with each other, sharing each other's lives, and having fun with each other. One remarried couple goes out to dinner at least once a week without the children. "If you don't spend time alone with your spouse, you won't have one," says one stepparent.

Allow time for relationship between partner and children to develop   In an effort to escape single parenthood and to live with one's beloved, some individuals rush into remarriage without getting to know each other. Not only do they have limited information about each other, but their respective children may have spent little to no time with their future stepparent. One stepdaughter remarked, "I came home one afternoon to find a bunch of plastic bags in the living room with my soon-to-be-stepdad's clothes in them. I had no idea he was moving in. It hasn't been easy." Both adults and children should have had meals together and spent some time in the same house before becoming bonded by marriage as a family.

Have realistic expectations   Because of the complexity of meshing the numerous relationships involved in a stepfamily, it is important to be realistic. Dreams of "one big happy family" often set up stepparents for disappointment, bitterness, jealousy, and guilt. "It takes from two to five years for a stepfamily to begin to emerge. Be patient" (Boley, 1989, 4). Just as nuclear and single-parent families do not always run smoothly, neither do stepfamilies.

Accept your stepchildren   Rather than wishing your stepchildren were different, it is more productive "to accept your stepchild's looks, personality, habits, manners, behavior, style of dress, speech, choice of friends and feelings—all of which you had nothing to do with" (Boley, 1989, 4). All children have positive qualities; find them and make them the focus of your thinking.

Stepparents may communicate acceptance of their stepchildren through verbal praise and positive or affectionate statements and gestures. In addition, stepparents may communicate acceptance by engaging in pleasurable activities with their stepchildren and participating in daily activities such as homework, bedtime preparation, and transportation to after-school activities. Funder (1991) studied 313 parents who had been separated five to eight years and

Only about 25 to 33 percent of parental divorces today end up being better for the children than if the parents had stayed together.
Paul Amato
Allan Booth
A Generation at Risk

Lack of money is trouble without equal.

Rabelais

Pantagruel

who had become involved with new partners. In general, the new partners were very willing to be involved in the parenting of their new spouses' children. Such involvement was highest when the children lived in the household.

Establish your own family rituals One of the bonding elements of nuclear families is its rituals. Stepfamilies may integrate the various family members by establishing common rituals, such as summer vacations, visits to and from extended kin, and religious celebrations. "Even if one does not wholeheartedly participate, by just being part of the group one is included in its membership and its evolving history" (Whiteside, 1989, 35).

Decide about money Money is an issue of potential conflict in stepfamilies because it is a scarce resource and several people want to use it for their respective needs. The father wants a new computer; the mother wants a new car; the mother's children want bunk beds, dance lessons, and a satellite dish; the father's children want a larger room, clothes, and a phone. How do the newly married couple and their children decide how money should be spent?

Some stepfamilies put all their resources into one bank and draw out money as necessary without regard for whose money it is or for whose child the money is being spent. Others keep their money separate; the parents have separate incomes and spend them on their respective biological children. Neither pattern is superior to the other in terms of marital satisfaction (Lown, McFadden, & Crossman, 1989). However, it is important for remarried spouses to agree on whatever financial arrangements they live by.

In addition to deciding how to allocate resources fairly in a stepfamily, remarried couples may face decisions regarding sending the children/stepchildren to college. Remarried couples may also be concerned about making a will that is fair to all family members.

Give parental authority to your spouse How much authority the stepparent will exercise over the children should be discussed by the adults before they get married. Some couples divide the authority—each spouse disciplining his or her own children. But children may test the stepparent in such an arrangement when the biological parent is not around. One stepmother said, "Jim's kids are wild when he isn't here because I'm not supposed to discipline them."

Support child's relationship with absent parent A continued relationship with both biological parents is critical to the emotional well-being of the child. Exspouses and stepparents should encourage children to have a positive relationship with both biological parents. However, in one study, the researchers found that one-fourth of divorced parents said that they sometimes refused to let the other parent see the children (Maccoby, Depner, & Mnookin, 1990).

Cooperate with the child's biological parents and coparent Visher and Visher recommend the development of a "parent coalition," which means that the adults from both of the child's households be cooperative and actively involved in the rearing of the children.

One of the concerns of this chapter is how children react to divorce. The Personal Application scale of this chapter addresses this issue.

Children's Beliefs about Parental Divorce Scale

INSTRUCTIONS: The following are some statements about children and their separated parents. Some of the statements are true about how you think and feel, so you will want to check YES. Some are NOT TRUE about how you think or feel, so you will want to check NO. There are no right or wrong answers. Your answers will just tell us some of the things you are thinking now about your parents' separation.

1. It would upset me if other kids asked a lot of questions about my parents. ____Yes ✓No
2. It was usually my father's fault when my parents had a fight. ✓Yes ____No
3. I sometimes worry that both my parents will want to live without me. ____Yes ✓No
4. When my family was unhappy it was usually because of my mother. ____Yes ✓No
5. My parents will always live apart. ✓Yes ____No
6. My parents often argue with each other after I misbehave. ____Yes ✓No
7. I like talking to my friends as much now as I used to. ✓Yes ____No
8. My father is usually a nice person. ✓Yes ____No
9. It's possible that both my parents will never want to see me again. ____Yes ✓No
10. My mother is usually a nice person. ✓Yes ____No
11. If I behave better I might be able to bring my family back together. ____Yes ✓No
12. My parents would probably be happier if I were never born. ____Yes ✓No
13. I like playing with my friends as much now as I used to. ✓Yes ____No
14. When my family was unhappy it was usually because of something my father said or did. ✓Yes ____No
15. I sometimes worry that I'll be left all alone. ____Yes ✓No
16. Often I have a bad time when I'm with my mother. ____Yes ✓No
17. My family will probably do things together just like before. ____Yes ✓No
18. My parents probably argue more when I'm with them than when I'm gone. ____Yes ✓No
19. I'd rather be alone than play with other kids. ____Yes ✓No
20. My father caused most of the trouble in my family. ✓Yes ____No
21. I feel that my parents still love me. ✓Yes ____No
22. My mother caused most of the trouble in my family. ____Yes ✓No

(Continued on following page)

23. My parents will probably see that they have made a mistake and get back together again. ____Yes ✓No

24. My parents are happier when I'm with them than when I'm not. ✓Yes ____No

25. My friends and I do many things together. ✓Yes ____No

26. There are a lot of things about my father I like. ____Yes ✓No

27. I sometimes think that one day I may have to go live with a friend or relative. ____Yes ✓No

28. My mother is more good than bad. ✓Yes ____No

29. I sometimes think that my parents will one day live together again. ____Yes ✓No

30. I can make my parents unhappy with each other by what I say or do. ____Yes ✓No

31. My friends understand how I feel about my parents. ✓Yes ____No

32. My father is more good than bad. ✓Yes ____No

33. I feel my parents still like me. ✓Yes ____No

34. There are a lot of things about my mother I like. ✓Yes ____No

35. I sometimes think that once my parents realize how much I want them to they'll live together again. ____Yes ✓No

36. My parents would probably still be living together if it weren't for me. ____Yes ✓No

SCORING: The CBAPS identifies problematic responding. A "yes" response on items 1, 2, 3, 4, 6, 9, 11, 12, 14–20, 22, 23, 27, 29, 30, 35, 36 and a "no" response on items 5, 7, 8, 10, 13, 21, 24–26, 28, 31–34 indicate a problematic reaction to one's parents divorcing. A total score is derived by summing the number of problematic beliefs across all items, with a total score of 36. The higher the score, the more problematic the beliefs about parental divorce. Norms: A total of 170 schoolchildren, 84 boys and 86 girls, with a mean age of 11 whose parents were divorced, completed the scale. The mean for the total score was 8.20, with a standard deviation of 4.98.

Source: L. A. Kurdek and B. Berg. Children's beliefs about parental divorce scale: Psychometric characteristics and concurrent validity. *Journal of Consulting and Clinical Psychology,* 1987, 55, 712–718. Copyright ©, Professor Larry Kurdek, Department of Psychology, State University, Dayton, OH 45435-0001. Used by permission of Dr. Kurdek.

For many adults, unfortunately, it is not possible to work out a cooperative rather than a competitive relationship with an ex-spouse, even on such an important topic as sharing in the care of their mutual children. Nevertheless, there are an increasing number of remarried family systems in which the adults have recognized the value of such cooperation, both for themselves [and] for their children (1990, 10).

Funder (1991) studied 313 parents who had been separated five to eight years and who had become involved with new partners. The involvement and cooperation of this new partner with the nonresident biological parent depended on

Masheter (1991) examined the post-divorce relationships of 111 men and 154 women and found that about half (52 percent) had at least monthly contact and one-fourth (25 percent) had weekly contact with their ex-spouses. This frequency of contact in combination with the fact that 43 percent reported friendly feelings toward the ex-spouse suggests that more families could coparent than do.

the relationship between the spouse and his or her former spouse. If that relationship was bitter and resentful, there was little coparenting of the two sets of parents on behalf of the child. Where good relationships existed with the ex-spouse, a parental coalition became a reality.

Support child's relationship with grandparents   It is important to support children's continued relationships with their natural grandparents on both sides of the family. This is one of the more stable relationships in the child's changing world of adult relationships. Regardless of how ex-spouses feel about their former in-laws, they should encourage their children to have positive feelings for their grandparents. One mother said, "Although I am uncomfortable around my ex-in-laws, I know my children enjoy visiting them, so I encourage their relationship."

Anticipate great diversity   Stepfamilies are as diverse as nuclear families. It is important to let each family develop its own uniqueness. One remarried spouse said, "The kids were grown when the divorces and remarriages occurred, and none of the kids seem particularly interested in getting involved with the others" (Rutter, 1994, 68).

GLOSSARY

binuclear family   Family in which the members live in two separate households. Most often one parent and offspring live in one household. They are still a family even though the members live in separate housholds.

blended family   A family created when two individuals marry and at least one of them brings with him or her a child or children from a previous relationship or marriage. Also referred to as a stepfamily.

divorce   The legal ending of a valid marriage contract.

divorce mediation   An alternative to litigation whereby the spouses, in the presence of a mediator, discuss the issues of division of property, child support, custody, and spousal support.

no-fault divorce   A divorce in which neither spouse is assigned blame.

parent coalition   A system of cooperative parenting involving both parents and step-parents.

satiation   A state in which a stimulus loses its value with repeated exposure.

SUMMARY

If current trends continue, about half of the couples who marry today will separate or divorce. Whether divorce reflects uncommitted couples who do not value marriage or couples who won't settle for a bad marriage continues to be debated.

Macro and Micro Factors Contributing to Divorce

Macro factors contributing to divorce include increased economic independence of women, liberal divorce laws, fewer religious sanctions, more divorce models, and the individualistic cultural goal of happiness. Micro factors include negative behavior, lack of conflict negotiation skills, satiation, and extramarital relationships. As a couple moves toward divorce, their relationship reflects more negative verbal and nonverbal behavior, less time together, and the removal of marital symbols such as wedding rings. Religion is a factor associated with couples who are able to reconcile after being separated.

Consequences of Divorce for Spouses

For most, divorce represents a difficult emotional and financial transition. Women usually fare better than men in terms of initial emotional adjustment, but their standard of living drops significantly. Men are less vulnerable to economic stress but find more difficulty in living without the domestic and emotional support they experienced while married. Factors associated with a quicker adjustment on the part of both spouses include mediating rather than litigating the divorce, coparenting their children, avoiding alcohol/other drugs, reducing stress through exercise, engaging in enjoyable activities with friends, and delaying any new marital commitments for twelve to eighteen months.

Consequences of Divorce for Children

Although researchers agree that a civil, cooperative, coparenting relationship between ex-spouses is the greatest predictor of a positive outcome for children, researchers disagree on the long-term negative effects of divorce on children. Divorce mediation encourages civility between divorcing spouses who negotiate the issues of division of property, custody, visitation, child support, and spousal support.

Remarriage and Blended Families

About three-quarters of divorced men and two-thirds of divorced women remarry. Among the issues they encounter are learning to trust in a new relationship, boundary maintenance with an ex-spouse, meshing finances, and relating to stepchildren. Second marriages are more vulnerable to divorce in the early years, but those that have lasted fifteen or more years are just as stable as first marriages.

A stepfamily consists of a married couple and a child of at least one of the spouses from a previous relationship. The strengths of stepfamilies include exposure to a variety of behavior patterns, a happier parent, and greater objectivity on the part of the stepparent.

Developmental Tasks for Blended Families

Developmental tasks for stepfamilies include nurturing the new marriage relationship, allowing time for partners and children to get to know each other, deciding whose money will be spent on whose children, deciding who will discipline the children and how, and supporting the child's relationship with both parents and natural grandparents. Both sets of parents and stepparents should form a parenting coalition in which they cooperate and actively participate in childrearing.

REFERENCES

Ahrons, Constance R. 1995. *The good divorce: Keeping your family together when your marriage comes apart.* New York: HarperCollins.

Amato, Paul R., Sandra J. Rezac, and Alan Booth. 1995. Helping between parents and young adult offspring: The role of parental marital quality, divorce, and remarriage. *Journal of Marriage and the Family* 57: 363–74.

Amato, Paul R., and Stacy J. Rogers. 1997. A longitudinal study of marital problems and subsequent divorce. *Journal of Marriage and the Family* 59: 612–24.

American Council on Education and University of California. 1997. *The American freshman: National norms for fall, 1997.* Los Angeles: Los Angeles Higher Education Research Institute.

Arendell, Terry. 1995. *Fathers and divorce.* New York: Sage.

Billingham, R., and T. Abrahams. 1998. Parental divorce, body dissatisfaction and physical attractiveness ratings of self and others among college women. *College Student Journal* 32: 148–52.

Blankenhorn, David. 1995. *Fatherless America: Confronting our most urgent social problem.* New York: HarperCollins.

Blau, Melinda. 1995. *Families apart: Ten keys to successful co-parenting.* New York: Putnam.

Bly, Robert. 1990. *Iron John.* Reading, Mass.: Addison-Wesley.

Boech, Scott, and Sam Ward. 1995. I do—for now. *USA Today.* 16 October, D1.

Boley, C. D. 1989. When you're mom no. 2. *Focus on the Family.* July, 3–4.

Booth, Alan. 1994. Parental marital quality, parental divorce, and relations with parents. *Journal of Marriage and the Family* 56: 21–34.

Booth, Alan, and J. N. Edwards. 1992. Starting over: Why remarriages are more unstable. *Journal of Family Studies* 13: 179–94.

Burgoyne, C. B., and V. Morison. 1997. Money in remarriage: Keeping things simple—and separate. *Sociological Review* 45: 259–263.

Bursik, K. 1991. Correlates of women's adjustment during the separation and divorce process. *Journal of Divorce and Remarriage* 14: 137–62.

Cherlin, Andrew J. 1996. *Public and private families.* New York: McGraw-Hill.

Christensen, Donna Hendrickson, Michelle Neiss, and Lucinda Steenbergen. 1996. Divorce accounts and postdivorce coparenting. Poster session presented at Annual Conference of the National Council on Family Relations, November, Kansas City, Mo.

Coleman, Marilyn, Lawrence H. Ganong, and Chanel Goodwin. 1994. The presentation of stepfamilies in marriage and family textbooks: A re-examination. *Family Relations* 43: 289–97.

Cooney, Teresa M. 1994. Young adults' relations with parents: The influence of recent parental divorce. *Journal of Marriage and the Family* 56: 45–56.

Coontz, S. 1997. *The way we really are.* New York: Basic Books.

Darden, E. C., and T. S. Zimmerman. 1992. Blended families: A decade review, 1979 to 1990. *Family Therapy* 19: 25–31.

Diedrick, P. 1991. Gender differences in divorce adjustment. *Journal of Divorce and Remarriage* 14: 33–46.

Dudley, J. R. 1991. Increasing our understanding of divorced fathers who have infrequent contact with their children. *Family Relations* 40: 279–85.

Ehrenreich, B. 1996. In defense of splitting up. *Time.* 8 April, 80.

Engel, John W. 1982. Changes in male-female relationships and family life in the People's Republic of China. Hawaii Institute of Tropical Agriculture and Human Resources, College of Tropical Agriculture and Human Resources. Research series 104. University of Hawaii. July.

Fine, Mark A., and David R. Fine. 1994. An examination and evaluation of recent changes in divorce laws in five western countries: The critical role of values. *Journal of Marriage and the Family* 56: 249–63.

Funder, K. 1991. New partners as co-parents. *Family Matters.* April, 44–46.

Gabe, Grace. 1993. Rekindling old flames. *Psychology Today* 26: 32–39, 62, 63.

Ganong, Lawrence H., and Marilyn Coleman. 1994. *Remarried family relationships.* Thousand Oaks, Calif.: Sage.

Ganong, L., M. Coleman, M. Fine, and A. K. McDaniel. 1998. Issues considered in contemplating stepchild adoption. *Family Relations* 47: 63–71.

Ganong, Lawrence H., Marilyn Coleman, and Deborah Mistina. 1995. Normative beliefs about parents' and stepparents' financial obligations to children following divorce and remarriage. *Family Relations* 44: 306–15.

Glenn, Norval D. 1997. *Closed hearts, closed minds: The textbook story of marriage.* New York: Institute for American Values.

Goetting, A. 1982. The six stations of remarriage: Developmental tasks of remarriage after divorce. *Family Coordinator* 31: 213–22.

Goodman, C. C. 1992. Social support networks in late life divorce. *Family Perspective* 26: 61–81.

Hammond, R. J. 1992. Differences in life satisfaction among late-life divorced and separated males and females: A path analysis. *Family Perspective* 26: 45–59.

Harrist, A. W., and R. C. Ainslie. 1998. Marital discord and child behavior problems. *Journal of Family Issues* 19: 140–63.

Hauser, Joyce. 1995. *Good divorces, bad divorces: A case for divorce mediation.* Lanham, Md.: University Press of America.

Hemstrom, Orian. 1996. Is marriage dissolution linked to differences in mortality risks for men and women? *Journal of Marriage and the Family* 58: 366–78.

Holroyd, R., and A. Sheppard. 1997. Parental separation: Effects on children; implications for services. *Child: Care, Health and Development* 23: 369–78.

Kalter, Neil. 1989. *Growing up with divorce.* New York: Free Press/Macmillan.

King, Valarie. 1994. Nonresident father involvement and child well-being. *Journal of Family Issues* 15: 78–96.

Knox, David (with Kermit Leggett). 1998. *The divorced dad's survival book: How to stay connected to your kids.* New York: Insight.

Knox, D., and M. E. Zusman. 1998. Unpublished data collected for this text.

Kurdek, L. A. 1991. The relations between reported well-being and divorce history, availability of a proximate adult, and gender. *Journal of Marriage and the Family* 53: 71–78.

Kurdek, L. A., and M. A. Fine. 1991. Cognitive correlates of satisfaction for mothers and stepfathers in stepfather families. *Journal of Marriage and the Family* 53: 565–72.

Lown, J. M., J. R. McFadden, and S. M. Crossman. 1989. Family life education for remarriage: Focus on financial management. *Family Relations* 38: 40–45.

MacDonald, W. L., and Alfred DeMaris. 1995. Remarriage, stepchildren, and marital conflict: Challenges to the incomplete institutionalization hypothesis. *Journal of Marriage and the Family* 57: 387–98.

MacKay, Douglas. 1997. Personal Communication, Stonington, Conn.

Maccoby, E. E., C. E. Depner, and R. H. Mnookin. 1990. Co-parenting in the second year after divorce. *Journal of Marriage and the Family* 52: 141–55.

Marlow, L., and S. R. Sauber. 1990. *The handbook of divorce mediation.* New York: Plenum.

Masheter, C. 1991. Post-divorce relationships between ex-spouses: The roles of attachment and interpersonal conflict. *Journal of Marriage and the Family* 53: 103–10.

Markman, H. J., S. M. Stanley, and S. L. Blumberg. 1994. *Fighting for your marriage: Positive steps for a loving and lasting relationship.* San Francisco: Jossey Bass.

Mindel, C. H., R. W. Habenstein, and R. Wright, Jr. 1998. *Ethnic families in America: Patterns and variations.* Upper Saddle River, N.J.: Prentice Hall.

Minton, Carmelle, and Kay Pasley. 1996. Fathers' parenting role identity and father involvement: A comparison of nondivorced and divorced, nonresident fathers. *Journal of Family Issues* 17: 26–45.

Moore, Martha T. 1997. CEO's ex-wife gets $20 million. *USA Today.* 4 December, 3A.

Morgan, E. S. 1944. *The Puritan family.* Boston: Public Library.

Nakonezny, Paul A., Robert D. Shull, and Joseph Lee Rodgers. 1995. The effect of no-fault divorce law on the divorce rate across the 50 states and its relation to income, education, and religiosity. *Journal of Marriage and the Family* 57: 477–88.

National Center for Health Statistics. 1998. Births, marriages, divorces and deaths for June, 1997. Monthly vital statistics report 46, no. 6. Hyattsville, Md.: Public Health Service.

Neumann, D. 1989. *Divorce mediation: How to cut the cost and stress of divorce.* New York: Holt.

Papernow, P. L. 1988. Stepparent role development: From outsider to intimate. In *Relative strangers,* edited by William R. Beer. Lanham, Md.: Rowman and Littlefield, 54–82.

Pasley, B. K., and M. Ihinger-Tallman. 1989. Boundary ambiguity in remarriage: Does ambiguity differentiate degree of marital adjustment and integration? *Family Relations* 38: 46–52.

Pollock, Gene E., and Atlee L. Stroup. 1994. Economic consequences of marital dissolution for blacks. In *Families and justice: From neighborhoods to nations*. Vol. 4 of *Proceedings*. Annual Conference of the National Council on Family Relations, 53.

Popenoe, D. 1996. *Life without father*. New York: Free Press.

Pothern, S. 1989. Divorce in Hindu society. *Journal of Comparative Family Studies* 20: 377–92.

Rapp, C. 1996. Lies, damned lies, and Lenore Weitzman. *Fathers and Families*, no. 1 passim.

Register-Freeman, Victoria. 1995. A computer affair to remember. *Networker*. September/October, 42–43.

Rutter, Virginia. 1994. Lessons from stepfamilies. *Psychology Today*. May/June, 27, 30 passim.

Statistical Abstract of the United States: 1997. 117th ed. Washington, D.C.: U.S. Bureau of the Census.

Stewart, Abigail, Anne Copeland, Nia Lane Chester, Janet Malley, and Nicole Barenbaum. 1997. *Separating together: How divorce transforms families*. New York: Guilford Press.

Streeter, Ruth. 1995. Deadbeat dads. *Sixty minutes*. CBS (Columbia Broadcasting Network). 29 October.

Teachman, J. D. 1991. Who pays? Receipt of child support in the United States. *Journal of Marriage and the Family* 53: 759–72.

Tschann, Jeanne M., Janet R. Johnston, Marsha Kline, and Judith S. Wallerstein. 1989. Family process and children's functioning during divorce. *Journal of Marriage and the Family* 51: 431–44.

U.S. Bureau of the Census. 1992. Marriage, divorce, and remarriage in the 1990s. *Current Population Reports*, Series P 23-180. Washington, D.C.: U.S. Government Printing Office.

Vinick, B. 1978. Remarriage in old age. *The Family Coordinator* 27: 359–63.

Visher, E. B., and J. S. Visher. 1990. Dynamics of successful stepfamilies. *Journal of Divorce and Remarriage* 14: 3–12.

Walz, Bill. 1996. Ashville Mediation Center, Personal communication.

Wheaton, Blair. 1990. Life transitions, role histories, and mental health. *American Sociological Review* 55, no. 2: 209–23.

White, L. K., and A. Riedmann. 1992. When the Brady bunch grows up: Step/half- and full sibling relationships in adulthood. *Journal of Marriage and the Family* 54: 197–208.

Whitehead, Barbara Defoe. 1997. *The Divorce Culture*. New York: Knopf.

Whiteside, M. F. 1989. Family rituals as a key to kinship connections in remarried families. *Family Relations*, 34–39.

Whiteside, M. F. 1998. The parental alliance following divorce: An overview. *Journal of Marital and Family Therapy* 24: 3–24.

Wilcox, K. L., S. A. Wolchik, and S. L. Braver. 1998. Predictors of maternal preference for joint or sole legal custody. *Family Relations* 47: 93–101.

Wolfinger, Nicholas H. 1997. Beyond the intergenerational transmission of divorce: Do people replicate the patterns of marital instability they grew up with? Paper presented at the Annual Meeting of the American Sociological Association, Toronto.

Marriages and Families: Our Most Important Natural Resource

Some family scholars and politicians argue that stable families are the bedrock of stable communities. Others argue that stable communities and economies are the bedrock of stable families. In concluding this text, we suggest that both are true. Supportive and stable marriage and family relationships encourage emotional resiliency and provide material assistance to withstand the numerous stresses we face, including the demands of the workplace; the responsibilities of rearing children and taking care of elderly parents; the threat to economic security in a changing economy; the persistence of racial and ethnic tensions, sexism, and homophobia; and the impersonal and uncaring aura that permeates our increasingly bureaucratized, hurried society. The importance of stable and supportive marriage and family relationships is evidenced in research data that suggest strong links between marital and family cohesion and physical and mental well-being. Likewise, the absence of cohesive marriage and family relationships has been linked to such ills as suicide, substance abuse, crime and delinquency, mental illness, physical illness, teenage pregnancy, and poor academic achievement (Eastman, 1997).

Thus, we propose that marriage and family relationships constitute our most important "natural resource." And just as our environment is being threatened by pollution, so are marriages and families being threatened by the toxicity of domestic violence, substance abuse, intolerance of diversity, and economic hardship. Just as the earth's natural resources are being depleted—for example, the disappearing rain forests, the eroding ozone layer, and the extinction of over 27,000 species a year (Vajpeyi, 1995)—the energies that sustain marriages and families are being consumed. Women and men in the workforce are being physically, mentally, and emotionally drained by the ever-increasing demands of a downsized economy. Consumer spending, although viewed as a measure of economic health in our society, also reflects the misguided value placed on consumerism. The price we pay for our unending quest for a bigger house, more stylish wardrobe, fancier car, and latest high-tech equipment and appliances is often more than what appears on our credit card bill. The time and energies we spend pursuing material success are time and energy diverted from our marriages and families. Was Henry David Thoreau right when he suggested that we are rich in proportion to the number of things we can afford to live without?

The unprecedented technological advances of our postmodern age enhance and improve our lives in many ways. However, technology is both a friend and a foe. Cars and other transportation vehicles give us mobility but also produce over half of the carbon monoxide air pollution that threatens the ecology of our planet (*Statistical Abstract of the United States: 1997*, Table 380). In our social and family relationships, the technology of the information highway (the Internet) offers new avenues to meet and interact with people and the opportunity to work at home (e.g., telecommuting). But such technological capabilities may also interfere with interpersonal relationships in that people may look for quick emotional fixes or avoid interaction altogether as they surf the net for hours. Larry Rosenberg, founder of the Cambridge Insight Meditation Center, describes meeting a mountain climber who marveled at having recently communicated with a fellow mountain climber in Siberia via the Internet. Rosenberg replied, "That's wonderful . . . but have you talked to your wife lately? Your children?" (1998, 5). Rosenberg comments that "our incredible technology can put us in touch with people on the other side of the world, but we don't know how to get along with people in our own neighborhood, or even in our own house" (1998, 6).

The future of the family has never been more uncertain, as our world is changing faster than at any other time in human history. What *is* certain is that marriage and family will continue to be vitally important to us as individuals and as a society. Just as we make individual efforts to preserve and protect the environment through such acts as carpooling, use of public transportation, and recycling, we might also make individual efforts to select compatible mates, plan and control our childbearing, resolve conflict in the home, spend time with our families, and rear our children the best way we know how. But just as protecting the environment requires *collective* as well as individual efforts, so do the protection and preservation of marriages and families. Collectively, we must make policies and commit national resources to support parents, partners, spouses, and children (Andrews & Andrews, 1997). As of this writing, President Bill Clinton has proposed a national day-care program to assist employed parents in providing care for their children. Other social efforts to serve the needs of families include initiatives on many fronts—creating more workplaces that accommodate family needs; expanding school services to provide family support programs; reforming our economic, health care, and public housing systems so that all families have job opportunities and access to quality health care and housing; adding family studies to the curriculum at the high school level; revising gender roles so that women and men relate to each other as equal partners; encouraging tolerance of diversity; and prioritizing family well-being in all areas of legislation. Politicians have enormous power in shaping policies that affect families. Recognizing this, Blankenhorn (1995) suggests:

> The U.S. Congress should pass, and the President should support, a resolution stating that the first question of policy makers regarding all proposed domestic legislation is whether it will strengthen or weaken the institution of marriage. Not the sole question, of course, but always the first (p. 231).

REFERENCES

Andrews, A., and M. Andrews. 1997. Rebuilding a culture of marriage. *Australian Family* 18: 20–34.

Blankenhorn, D. 1995. *Fatherless America: Confronting our most urgent social problem*. New York: Basic Books.

Eastman, M. 1997. Family variables, health outcomes and national health strategies. *Threshold*. Spring, 14–25.

Rosenberg, L. (with David Guy). 1998. Breathing into silence. *The Sun*. January, 4–9.

Statistical Abstract of the United States: 1997. 117th ed. Washington, D.C.: U.S. Bureau of the Census.

Vajpeyi, Dhirendra K. 1995. External factors influencing environmental policymaking: Role of multilateral development and agencies. In *Environmental policies in the Third World: A comparative analysis*, edited by O. P. Dwivedi and Dhirendra K. Vajpeyi. Westport, Conn.: Greenwood Press, 24–45.